Interest Rate Models

Interest Rate Models

An Introduction

Andrew J. G. Cairns

Princeton University Press

Princeton and Oxford

Published by Princeton University Press,
41 William Street, Princeton, New Jersey 08540

In the United Kingdom: Princeton University Press,
3 Market Place, Woodstock, Oxfordshire OX20 1SY

Library of Congress Cataloguing-in-Publication Data

Cairns, Andrew (Andrew J. G.)
  Interest rate models: an introduction / Andrew J. G. Cairns.
    p.cm.
  Includes bibliographical references and index.
  ISBN 0-691-11893-0 (cl.: alk. paper) — ISBN 0-691-11894-9 (pbk.: alk. paper)
  1. Interest rates—Mathematical models. 2. Bonds—Mathematical models.
  3. Securities—Mathematical models. 4. Derivative securities—Prices—Mathematical models.
  I. Title.

HG1621.C25 2004
332.8′01′51—dc22   2003062309

British Library Cataloguing-in-Publication Data

A catalogue record for this book is available from the British Library

This book has been composed in Times and typeset by T&T Productions Ltd, London
Printed on acid-free paper ∞
www.pupress.princeton.edu

Printed in the United States of America

10 9 8 7 6 5 4 3

ISBN-13: 978-0-691-11894-9 (pbk.)

ISBN-10: 0-691-11894-9 (pbk.)

Contents

Preface

The past thirty years or so have seen considerable development in the field of financial mathematics: first, in the field of equity derivatives following on from the work of Black, Scholes and Merton; and then in the theory of bond pricing and derivatives following on, for example, from Vasicek's work. If we wish to model a stock market in which prices evolve in a way that is free of arbitrage, the move from equities to bonds adds a whole new level of complexity and interest for the modeller. This is because we have a large number of tradable assets whose price dynamics typically depend upon the same (small number of) random factors. As a result, we must ensure with any model that the prices of these assets all evolve in a way that avoids arbitrage.

In recent years a considerable number of textbooks have been written that cover this now broad field. These range widely in their level of comprehensiveness and technical difficulty.

The origin of this book lies within a graduate-level lecture course on bond pricing given to students on the MSc in Financial Mathematics at Heriot-Watt University and Edinburgh University. While there exist textbooks that cover this topic, it was felt that none were entirely appropriate for the present course.

The book is aimed at people who are just starting to learn the subject of interest rate modelling. Thus, the primary readership is intended to include students on advanced taught courses, doctoral students and financial market practitioners learning about bond pricing and bond-derivative pricing for the first time. Other readers who are familiar with the basics of interest rate modelling will hopefully also find much of interest in the second half of the book, where I move on to more advanced and more recent topics. Finally, there are other practitioners in areas such as insurance who are not involved with the day-to-day running of a bond portfolio or a derivatives operation but who nevertheless need good interest rate models. I am primarily thinking of my original field of actuarial science, where having a good model for interest rates is becoming increasingly important. These practitioners should also find the book useful.

The level at which I have written the book is intended to make it accessible and helpful to masters-level students in financial mathematics. Students typically will have a good first degree in the mathematical sciences and will already be comfortable with probability, stochastic processes, stochastic differential equations and arbitrage-free pricing of equity derivatives (including the Fundamental Theorem of Asset Pricing). As a reminder, however, the main results in probability theory are summarized in the appendixes. The level of detail I have provided is such that students should finish up with the skills necessary to follow the majority of published research in the field of bond pricing and to carry out their own research and development in this field.

The book stops short of a completely rigorous treatment of the subject as experience suggests that this will make it inaccessible to the typical masters student. However, the mathematics will be presented at a sufficiently high level to allow readers to apply the skills learned in a useful way upon completion of the book. In other words, I have aimed to strike the right balance between too much mathematics, so that students are not able to make headway, and too little, in which case students may finish up with a broad knowledge of the subject but little ability to develop and apply what they have learned. Time will tell if I have succeeded in this aim!

In my development of this book there were five other books that I referred to most frequently. First, there are the textbooks by Hull (2000, and its earlier editions) and Baxter and Rennie (1996), which I find complement each other very nicely. These provided me with the basics of derivative pricing in my formative years in financial mathematics and I still recommend them to students learning about equity derivatives. I might highlight here Baxter and Rennie's 'three steps to replication'. In this book the three steps have turned into five, but I hope to have retained the clarity of the approach they used to developing pricing and hedging strategies. Second, there are the more advanced textbooks by James and Webber (2000) (a good, wide-ranging reference book), Rebonato (1998) (the practitioner's point of view) and Bielecki and Rutkowski (2002) (a specialized book on credit risk).

The chapters in the book proceed as follows. Chapter 1 sets the scene by introducing the government bond market and the various traditional types of interest rates that we see in the bonds market. It finishes with a description of the various theories of interest rate dynamics.

Chapter 2 continues in a more mathematical vein. It introduces the Fundamental Theorem of Asset Pricing for interest rate derivatives which lies at the core of arbitrage-free pricing theory. The chapter then discusses some model-free results in bond pricing.

Chapter 3 is where the modelling starts. Here we focus on discrete-time binomial models. The aim here is to familiarize readers with one of the main themes: the switch from the real-world measure P to the risk-neutral measure Q. The advantage of using the binomial model is that it provides a straightforward proof of why the risk-neutral approach, as a computational tool, produces the unique no-arbitrage price for a bond or a derivative. We also introduce some simple derivative contracts at this stage and illustrate the difference between the no-arbitrage group of models and the time-homogeneous models for the risk-free rate of interest.

Chapter 4 is perhaps the heart of the book. We move now into continuous time and focus on the celebrated papers of Vasicek (1977) and Cox, Ingersoll and Ross (1985) (CIR). Particular attention is paid here to the alternative proofs of the Fundamental Theorem of Asset Pricing using the original partial differential equation (PDE) approach developed by Vasicek and the martingale approach more heavily favoured today. To illustrate some of the many issues in modelling, there is a detailed comparison of the characteristics of the Vasicek and CIR models. Some of the more detailed proofs relating specifically to the Vasicek and CIR models are placed in an appendix to avoid interrupting the flow too much. Technically, the CIR model is much tougher to develop and in most other textbooks pricing formulae are just stated or left as an exercise for the reader. One such 'exercise' takes up about 10 pages in Appendix B and this is not for the faint-hearted! However, I was convinced that a detailed, textbook account of the CIR model was long overdue.

Chapter 5 brings in the no-arbitrage models much favoured by market practitioners. These models take the reasonable point of view that today's observed prices must match today's theoretical prices. We look first at some Markov models which can be regarded as satisfactory for pricing and hedging of derivative contracts that are both simple in structure and short-term. We then move on to describe the Heath, Jarrow and Morton (1992) (HJM) approach, which provides practitioners with a general framework within which a variety of no-arbitrage models can be developed.

Chapter 6 on multifactor models is the last of the theoretical chapters that rely on the risk-neutral approach to pricing. Here we introduce some of the multifactor models that have been published over the years. It provides an important bridge between the perhaps unrealistic world of single-factor models such as Vasicek and CIR and real-world applications where often at least two or three factors are essential.

Chapter 7 introduces yet another probability measure. Not satisfied with having just two measures we look at how the use of other measures equivalent to P and Q can significantly assist with derivative-pricing calculations. In this chapter we use the zero-coupon bond maturing at the same time as the derivative contract as the numeraire. However, the bigger idea is the concept that other pricing measures might be helpful. This idea provides us with the key to Chapters 8 and 9.

Chapter 8 describes a relatively new and general framework under which it is reasonably straightforward to develop models that guarantee that interest rates stay positive. This issue had long been a problem, with the CIR model standing out in the earlier literature as being the one tractable, positive interest model.

Chapter 9 describes the market-model approach, which was developed around the same time as the positive interest framework. These models—most famously, perhaps, the model of Brace, Gatarek and Musiela (1997) (BGM)—focus directly on the interest rates, such as LIBOR and swap rates, which dictate derivative pay-offs. This is in contrast to earlier approaches which typically focus on the 'wrong' rates, such as the instantaneous risk-free rate of interest. This chapter shows how modelling the 'right' quantity directly can significantly simplify pricing calculations.

Chapter 10 looks at numerical methods for pricing bonds and derivatives on the assumption that no closed-form solutions exist. The first part of the chapter focuses on the popular lattice methods for solving PDEs, with ample illustration of how the different methods relate to one another. The second part of the chapter focuses on Monte Carlo methods and variance-reduction techniques. In particular, there is an introduction to the topic of quasi-Monte Carlo methods, which are becoming popular for dealing with complex, multifactor pricing problems.

Chapter 11 gives an introduction to the important subject of credit risk. The two main approaches are described: structural and reduced-form models. The latter class of models—in particular, the approach advocated by Jarrow, Lando and Turnbull (1997) (JLT)—requires quite a different mathematical toolkit to tackle pricing. Thus, we are introduced to the world of multiple-state models and counting processes.

Finally, Chapter 12 looks at model calibration. Here we discuss how to fit a smooth yield curve to observed interest rates and we provide a detailed comparison of the different parametric and non-parametric approaches. Within the book, the object of this task is to provide input for a derivative-pricing problem. Along with many other techniques in the book, however, the methods can be applied in a variety of fields including the development of monetary policy by central banks.

Acknowledgements

There are many people I would like to acknowledge who have contributed in the development of this book: some in a substantial way, others through the making of casual remarks which have influenced my thinking. First, there are many colleagues in finance and insurance: David Wilkie, David Forfar, Samuel Garcia, Gary Parker, Philippe Artzner, Phelim Boyle, Mary Hardy, Ken Seng Tan, Keith Feldman, Geoff Chaplin, Mark Deacon, Andrew Smith, Mark Davis, Nick Webber, Riccardo Rebonato, Michael Dempster, Martin Baxter, David Heath, Freddy Delbaen, Stuart Hodges, Lane Hughston, John Hibbert and colleagues at Barrie & Hibbert, and Yakoub Yakoubov and colleagues at AON. Among these, particular thanks go to Mary and Phelim for acting as wonderful hosts during the traumatic times in September 2001. As part of this same group I wish to include seminar and conference audiences in Cairns, Stockholm, Cambridge, Waterloo, Nuremberg, Oslo, Tokyo, Barcelona, Manchester, Lisbon, where I have been talking on various interest-rate-related topics. Next come students who have attended my MSc lectures on interest rate modelling since 1998, particularly those who moved on to work with me on interest-rate-related projects: Gavin Falk, Paul Wilson, David Lonie, Nelius du Plessis, Andriani Kampisopoulou, Jiangchun Bi, Eleni Galatsanou, Aidan McGovern and Phil Wood. Last, but by no means least, come my colleagues in the Department of Actuarial Mathematics and Statistics at Heriot-Watt University: Howard Waters (for being a very supportive head of department), Terence Chan, Angus Macdonald, Mark Owen, Delme Pritchard, Anke Wiese, Julia Wirch.

I am very grateful to David Ireland and Richard Baggaley at Princeton University Press for nurturing me as a first-time author and never showing irritation at the consistent delays in producing this work and also Jonathan Wainwright for doing an excellent job with the typesetting.

Posthumous thanks must go to my late father, Jim, from whom I inherited both the mathematical and actuarial genes. Finally, I would like to thank my wife Susan for her patience and support during the writing of this book, and also 'Coire' for helping to ease out the stresses of the human world.

1

Introduction to Bond Markets

1.1 Bonds

A *bond* is a securitized form of loan.

The buyer of a bond lends the issuer an initial price P in return for a predetermined sequence of payments. These payments may be fixed in nominal terms (a fixed-interest bond) or the payments may be linked to some index (an index-linked bond): for example, the consumer (or retail) prices index.

In the UK, government bonds are called *gilt-edged securities* or *gilts* for short. In other countries they have other names such as *treasury bills* or *treasury notes* (both USA). Since government bonds are securitized, they can be traded freely in the stock market. Additionally, the main government bond markets are very liquid because of the large amount of stock in issue and the relatively small number of stocks in issue. For example, in the UK the number of gilts is less than 100 and the total value of the gilts market is about £300 billion.

Bonds are also issued by institutions other than national governments, such as regional governments, banks and companies (the latter giving rise to the name *corporate bonds*). Bonds that have identical characteristics but are sold by different issuers may not have the same price. For example, consider two bonds that have a term of 20 years and pay a coupon of 6% per annum payable twice yearly in arrears. One is issued by the government and the other by a company. The bond issued by the company will probably trade at a lower price than the government bond because the market makers will take into account the possibility of default on the coupon payments or on the redemption proceeds. In countries such as the USA and the UK it is generally assumed that government bonds are default free, whereas corporate bonds are subject to varying degrees of default risk depending upon the financial health of the issuing company.

We can see something of this in Figure 1.1. Germany, France and Italy all operate under the umbrella of the euro currency.[1] With standardized bond terms and in

---

[1] The group of European countries that use the euro currency are officially referred to as the *Eurozone*. However, market practitioners and others often refer to *Euroland*.

the absence of credit risk the three yield curves should coincide. Some differences (especially between Germany and France) may be down to differences in taxation and the terms of the contracts. Italy lies significantly above Germany and France, however, and this suggests that the international bond market has reduced prices to take account either of a perceived risk of default by the Italian government or that Italy will withdraw from the euro.

1.2 Fixed-Interest Bonds

1.2.1 Introduction

We will concentrate in this book on fixed-interest government bonds that have no probability of default.

The structure of a default-free, fixed-interest bond market can generally be characterized as follows. We pay a price P for a bond in return for a stream of payments $c_1, c_2, \ldots, c_n$ at times $t_1, t_2, \ldots, t_n$ from now respectively. The amounts of the payments are fixed at the time of issue.

For this bond, with a nominal value of 100, normally:

g = coupon rate per 100 nominal;

n = number of coupon payments;

Δt = fixed time between payments

(equal to 0.5 years for UK and USA government bonds);

t_1 = time of first payment ($t_1 \leqslant \Delta t$);

$t_j = t_{j-1} + \Delta t$ for $j = 2, \ldots, n$;

t_n = time to redemption;

$$c_1 = \begin{cases} g\Delta t & \text{normally the first coupon or interest payment,} \\ 0 & \text{if the security has gone ex-dividend;} \end{cases}$$

$c_j = g\Delta t$ for $j = 2, \ldots, n - 1$ (i.e. subsequent interest payments);

$c_n = 100 + g\Delta t$ (final interest payment plus return of nominal capital).

Some markets also have irredeemable bonds (that is, $n = \infty$), but these bonds tend to trade relatively infrequently making quoted prices less reliable. Furthermore, they often have option characteristics which permit early redemption at the option of the government.

Details of government bond characteristics (Δt and ex-dividend rules) in many countries are given in Brown (1998).

1.2.2 Clean and Dirty Prices

Bond prices are often quoted in two different forms.

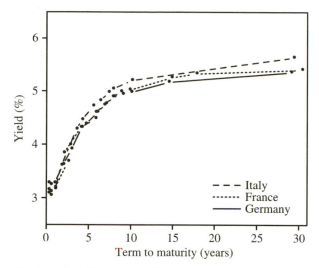

Figure 1.1. Benchmark yield curves for Germany, France and Italy on 3 January 2002.

The *dirty* price is the actual amount paid in return for the right to the full amount of each future coupon payment and the redemption proceeds. If the bond has gone ex-dividend, then the dirty price will give the buyer the right to the full coupon payable in just over six months (assuming twice-yearly coupon payments) but not the coupon due in a few days. As a consequence, the dirty price of a bond will drop by an amount approximately equal to a coupon payment at the time it goes ex-dividend. In addition, the dirty price of a bond will (everything else in the market being stable) rise steadily in between ex-dividend dates.

The *clean* price is an artificial price which is, however, the most-often-quoted price in the marketplace. It is equal to the dirty price minus the *accrued interest*. The accrued interest is equal to the amount of the next coupon payment multiplied by the proportion of the current inter-coupon period so far elapsed (according to certain conventions regarding the number of days in an inter-coupon period).[2] The popularity of the clean price relies on the fact that it does not jump at the time a bond goes ex-dividend, nor does it vary significantly (everything else in the market being stable) in between ex-dividend dates.

The evolution of clean and dirty prices relative to one another can be seen in Figure 1.2. In (a) market interest rates are left constant, demonstrating the sawtooth effect on dirty prices and the relative stability of clean prices. Randomness in interest rates creates volatility in both clean and dirty prices, although the same relationship between the two sets of prices is still quite clear.

---

[2]Further details for individual countries can be found in Brown (1998).

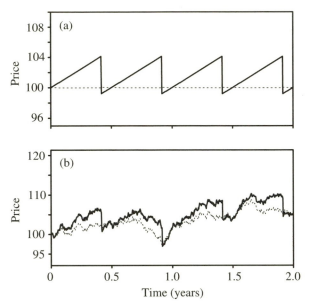

Figure 1.2. Evolution of clean (dotted line) and dirty (solid line) prices for a coupon bond over time. Coupon rate 10%. Maturity date $T = 10$. Coupons payable half-yearly. Ex-dividend period 30 days. (a) Constant market interest rates (10%). (b) Stochastic market interest rates.

1.2.3   Zero-Coupon Bonds

This type of bond has a coupon rate of zero and a nominal value of 1.

We will denote the price at time t of a zero-coupon bond that matures at time T by $P(t, T)$. In places we will call it the T-zero-coupon bond or T-bond for short.

Note that the value of 1 due immediately is, of course, 1, i.e. $P(t, t) = 1$ for all t.

Arbitrage considerations also indicate that $P(t, T) \leqslant 1$ for all T.

1.2.4   Spot Rates

The spot rate at time t for maturity at time T is defined as the yield to maturity of the T-bond:

$$R(t, T) = -\frac{\log P(t, T)}{T - t};$$

that is,

$$P(t, T) = \exp[-(T - t)R(t, T)].$$

The spot rate, $R(t, T)$, is interpreted in the following way. If we invest £1 at time t in the T-bond for $T - t$ years, then this will accumulate at an average rate of $R(t, T)$ over the whole period.

1.2.5 Forward Rates

The forward rate at time t (continuously compounding) which applies between times T and S ($t \leqslant T < S$) is defined as

$$F(t, T, S) = \frac{1}{S - T} \log \frac{P(t, T)}{P(t, S)}. \tag{1.1}$$

The forward rate arises within the terms of a forward contract. Under such a contract we agree at time t that we will invest £1 at time T in return for $e^{(S-T)F(t,T,S)}$ at time S. In other words we are fixing the rate of interest between times T and S in advance at time t.

The following simple no-arbitrage argument shows that the value of this contract must be zero. The forward contract imposes cashflows of -1 at time T and $+e^{(S-T)F(t,T,S)}$ at time S. By definition, this contract must have a value of zero at the time the contract is struck, time t, provided $F(t, T, S)$ is the fair forward rate.

We will now argue that the fair forward rate must be as defined in equation (1.1). Suppose that this is not true and that

$$F(t, T, S) > (S - T)^{-1} \log[P(t, T)/P(t, S)].$$

Then we could set up the following portfolio at time t: one forward contract (value zero at t by definition); $+1$ units of the T-bond; $-P(t, T)/P(t, S)$ units of the S-bond. The total cost of this portfolio at t is zero. This portfolio—if held to maturity of the respective contracts—will produce a net cashflow of zero at time T and $e^{(S-T)F(t,T,S)} - P(t, T)/P(t, S) > 0$ at time S. This is an example of an arbitrage: we have started with a portfolio worth zero at t and have a sure profit at S. Throughout this book we will assume that arbitrage opportunities like this do not exist. It follows that we cannot have $F(t, T, S) > (S-T)^{-1} \log[P(t, T)/P(t, S)]$. Equally, we cannot have $F(t, T, S) < (S - T)^{-1} \log[P(t, T)/P(t, S)]$ by constructing the reverse portfolio.

In summary, the forward rate $F(t, T, S)$ must satisfy equation (1.1) if we assume no arbitrage. This is an important example of a case where a price or relationship can be determined independently of the interest rate model being employed.

If the exercise date T for the forward contract is in fact equal to t, then the forward and spot rates must be equal, that is, $F(t, t, S) = R(t, S)$.

The *instantaneous* forward-rate curve (or just forward-rate curve) at time t is, for $T > t$,

$$f(t, T) = \lim_{S \to T} F(t, T, S) = -\frac{\partial}{\partial T} \log P(t, T) = -\frac{\partial P(t, T)/\partial T}{P(t, T)}$$

$$\Rightarrow \quad P(t, T) = \exp\left[-\int_t^T f(t, u) \, du\right].$$

In other words, we can make a contract at time t to earn a rate of interest of $f(t, T)$ per time unit between times T and $T + dt$ (where dt is very small). This is, of course, a rather artificial concept. However, it is introduced for convenience as bond-price modelling is carried out much more easily with the instantaneous forward-rate curve $f(t, T)$ than the more cumbersome $F(t, S, T)$.

Arbitrage considerations indicate that $f(t, T)$ must be positive for all $T \geqslant t$. Hence $P(t, T)$ must be a decreasing function of T.

1.2.6 Risk-Free Rates of Interest and the Short Rate

$R(t, T)$ can be regarded as a risk-free rate of interest over the fixed period from t to T. When we talk about *the* risk-free rate of interest we mean the *instantaneous* risk-free rate:

$$r(t) = \lim_{T \to t} R(t, T) = R(t, t) = f(t, t).$$

The easiest way to think of $r(t)$ is to regard it as the rate of interest on a bank account: this can be changed on a daily basis by the bank with no control on the part of the investor or bank account holder. $r(t)$ is sometimes referred to as the *short rate*.

1.2.7 Par Yields

The par-yield curve $\rho(t, T)$ specifies the coupon rates, $100\rho(t, T)$, at which new bonds (issued at time t and maturing at time T) should be priced if they are to be issued at par. That is, they will have a price of 100 per 100 nominal.

The par yield for maturity at time T (with coupons payable annually, $\Delta t = 1$) can be calculated as follows (for $T = t + 1, t + 2, \dots$):

$$100 = 100\rho(t, T) \sum_{s=t+1}^{T} P(t, s) + 100P(t, T).$$

That is, with a coupon rate of $\rho(t, T)$ and a maturity date of T, the price at t for the coupon bond would be exactly 100. This implies that

$$\rho(t, T) = \frac{1 - P(t, T)}{\sum_{s=t+1}^{T} P(t, s)}.$$

1.2.8 Yield-to-Maturity or Gross Redemption Yield for a Coupon Bond

This term normally applies to coupon bonds. Consider the coupon bond described in Section 1.2.1 with coupon rate g, maturity date t_n and current price P. Let δ be a solution to the equation

$$P = \sum_{j=1}^{n} c_j e^{-\delta t_j}.$$

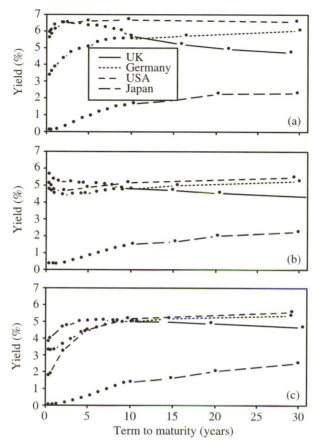

Figure 1.3. Benchmark yield curves (yields to maturity on benchmark coupon bonds) for the UK, Germany, the USA and Japan on (a) 26 January 2000, (b) 6 February 2001 and (c) 3 January 2002.

Since $P > 0$ and each of the c_j are positive there is exactly one solution to this equation in δ. This solution is the (continuously compounding) yield-to-maturity or gross redemption yield.

Typically, yields are quoted on an annual or semi-annual basis. The *Financial Times* (FT), for example, quotes half-yearly gross redemption yields. This reflects the fact that coupons on gilts are payable half yearly (that is, $\Delta t = 0.5$). Thus, the quoted rate is

$$y = 2[e^{\delta/2} - 1].$$

Thus, if $t_1 = 0.5$ and $P = 100$, y will be equal to the par yield.

Sometimes the expression *yield curve* is used, but this means different things to different people and should be avoided or described explicitly.

Gross redemption yields on benchmark bonds on 26 January 2000, 6 February 2001 and 3 January 2002 and are given for four different countries in Figure 1.3. These graphs illustrate the fact that yield curves in different currency zones can be quite different, reflecting the different states of each economy. They also show how the term structure of interest rates can vary over time. Finally, correlations between different countries can be seen, an example being a worldwide reduction in short-term interest rates following the terrorist attacks in the United States of America on 11 September 2001.

1.2.9 Relationships

For a given t, each of the curves $P(t, T)$, $f(t, T)$, $R(t, T)$, $\rho(t, T)$ (with coupons payable continuously) uniquely determines the other three. For example,

$$P(t, T) = \exp[-R(t, T)(T - t)] = \exp\left[-\int_t^T f(t, s)\,ds\right].$$

In Figure 1.4 we give examples of the forward-rate, spot-rate and par-yield curves for the UK gilts market on three dates (1 September 1992, 1 September 1993 and 1 March 1996). These curves have been derived from the prices of coupon bonds (see Cairns (1998) for further details). The par-yield curve most closely matches the information available in the coupon-bond market (that is, gross redemption yields), whereas spot rates and forward rates are implied by the way in which the par-yield curve varies with term to maturity. The relationship between the shapes of the three types of curve is determined by the fact that the spot rate is the arithmetic average of the forward rate and the par yield is a weighted average of the spot rates.

1.2.10 Example

Suppose that the (continuously compounding) forward rates for the next five one-year periods are as follows:

T	1	2	3	4	5
$F(0, T - 1, T)$	0.0420	0.0500	0.0550	0.0560	0.0530

Now the prices of zero-coupon bonds and spot rates at time 0 are given by

$$P(0, T) = \exp\left[-\sum_{t=1}^T F(0, t - 1, t)\right] \quad \text{and} \quad R(0, T) = -\frac{\log P(0, T)}{T}.$$

Hence,

T	1	2	3	4	5
$P(0, T)$	0.958 87	0.912 11	0.863 29	0.816 28	0.774 14
$R(0, T)$	0.042 0	0.046 0	0.049 0	0.050 75	0.051 2

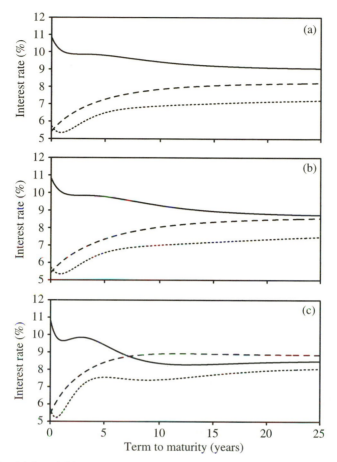

Figure 1.4. (a) Par-yield, (b) spot-rate and (c) forward-rate curves for the UK gilts market on 1 September 1992 (solid curves), 1 September 1993 (dotted curves) and 1 March 1996 (dashed curves). All curves are estimated from coupon-bond prices.

Finally, par yields (with coupons payable annually, $\Delta t = 1$) can be calculated according to the formula

$$\rho(0, T) = \frac{1 - P(0, T)}{\sum_{s=1}^{T} P(0, s)}.$$

Thus we have

T	1	2	3	4	5
$\rho(0, T)$	0.0429	0.0470	0.0500	0.0517	0.0522

(or 4.29 per 100 nominal, etc.).

1.3 STRIPS

STRIPS (Separate Trading of Registered Interest and Principal of Securities) are zero-coupon bonds that have been created out of coupon bonds by market makers rather than by the government. For example, in the UK, gilts maturing on 7 June or 7 December have been 'strippable' since November 1997. This means that a market exists in zero-coupon bonds which mature on the 7 June and the 7 December each year up to June 2032. More recently, strippable bonds have been added that have allowed the creation of zero-coupon bonds maturing on the 7 March and the 7 September each year up to March 2025. Where the date coincides with the maturity date of a strippable coupon bond there will be two types of zero-coupon bond available, depending upon whether or not it is made up of a coupon payment or a redemption payment, but they should have the same price because they are taxed on the same basis. In fact, stripped coupon interest and stripped principal do have slightly different prices, but their buying and selling spreads overlap, making arbitrage impossible.

1.4 Bonds with Built-in Options

In many countries the government bond market is complicated by the inclusion of a number of bonds which have option characteristics. Two examples common in the UK are as follows.

- Double-dated (or callable) bonds: the government has the right to redeem the bond *at par* at any time between two specified dates with three months notice. Thus, they will redeem if the price goes above 100 between the two redemption dates. This is similar to an American option. (An example is the UK gilt Treasury $7\frac{3}{4}$% 2012–15.)

- Convertible bonds: at the conversion date the holder has the right but not the obligation to convert the bond into a specified quantity of another bond.

Such bonds must be priced using the more sophisticated derivative-pricing techniques described later in this book.

1.5 Index-Linked Bonds

A number of countries including the UK and USA issue index-linked bonds. Let $CPI(t)$ be the value of the consumer prices index (CPI) at time t. (In the UK this is called the Retail Prices Index or RPI.)

Suppose that a bond issued at time 0 has a nominal redemption value of 100 payable at time T and a nominal coupon rate of g% per annum, payable twice yearly. The payment on this bond at time t will be

$$\frac{\text{CPI}(t - L)}{\text{CPI}(-L)} \times g \, \Delta t \qquad \text{for } t = \Delta t, 2\Delta t, \dots, T - \Delta t,$$

$$\frac{\text{CPI}(T - L)}{\text{CPI}(-L)} \times (100 + g \, \Delta t) \quad \text{for } t = T.$$

L is called the *time lag* and is typically about two months in most countries (including the USA), but sometimes eight months (in the UK for example).[3] The time lag of two months ensures that the relevant index value is known by the time a payment is due. The time lag of eight months ensures that the absolute amount of the next coupon payment is known immediately after the time of payment of the immediately preceding coupon. This makes the calculation of accrued interest precise (that is, the difference between the clean and dirty prices) but reduces the effectiveness of the security as a hedge against inflation.

1.6 General Theories of Interest Rates

In this section we will introduce four theories which attempt to explain the term structure of interest rates. The first three are based upon general economic reasoning, each containing useful ideas. The fourth theory, arbitrage-free pricing, introduces us to the approach that we will take in the rest of this book.

1.6.1 Expectations Theory

There are a number of variations on how this theory can be defined but the most popular form seems to be that

$$e^{F(0,S,S+1)} = E[e^{R(S,S+1)} \mid \mathcal{F}_0], \tag{1.2}$$

where $\mathcal{F}_t$ represents the information available at time t. Thus, the annualized one-year forward rate of interest for delivery over the period S to $S + 1$ is conjectured to be equal to the expected value of the actual one-year rate of interest at time S.

Assume the conjecture to be true.

- Since e^x is a convex function, Jensen's inequality implies that $F(0, S, S+1) > E[R(S, S + 1) \mid \mathcal{F}_0]$.

- Since $2F(0, S, S+2) = F(0, S, S+1) + F(0, S+1, S+2)$, it also follows from equation (1.2) that

$$e^{2F(0,S,S+2)} = E[e^{R(S,S+1)}]E[e^{R(S+1,S+2)}].$$

---

[3] Note that the Debt-Management Office in the UK is currently considering reducing the time lag of eight months to two months.

The theory also suggests that $e^{F(0,S,S+2)} = E[e^{R(S,S+2)}]$, which then implies that $e^{R(S,S+1)}$ and $e^{R(S+1,S+2)}$ must be uncorrelated. This is very unlikely to be true.

An alternative version of the theory is based upon continuously compounding rates of interest, that is, for any $T < S$,

$$F(0, T, S) = E[R(T, S)].$$

This version of the theory does allow for correlation between $R(T, U)$ and $R(U, S)$, for any $T < U < S$.

The problem with this theory, on its own, is that the forward-rate curve is, more often than not, upward sloping. If the theory was true, then the curve would spend just as much time sloping downwards. However, we might conjecture that, for some reason, a forward rate is a biased expectation of future rates of interest. This is encapsulated by the next theory.

1.6.2 Liquidity Preference Theory

The background to this theory is that investors usually prefer short-term investments to long-term investments—they do not like to tie their capital up for too long. In particular, a small investor may incur a penalty on early redemption of a longer-term investment. In practice, bigger investors drive market prices. Furthermore, there is a very liquid market in bonds of all terms to maturity.

The theory has a better explanation, although this is not related to its name. The prices of longer-term bonds tend to be more volatile than short-term bonds. Investors will only invest in more volatile securities if they have a higher expected return, often referred to as the *risk premium*, to offset the higher risk. This leads to generally rising spot-rate and forward-rate curves.

We can see, therefore, that a combination of the expectations theory and the liquidity preference theory might explain what we see in the market.

1.6.3 Market Segmentation Theory

Each investor has in mind an appropriate set of bonds and maturity dates that are suitable for their purpose. For example, life insurance companies require long-term bonds to match their long-term liabilities. In contrast, banks are likely to prefer short-term bonds to reflect the needs of their customers.

Different groups of investors can act in different ways. The basic form of market segmentation theory says that there is no reason why there should be any interaction between different groups. This means that prices in different maturity bands will change in unrelated ways. More realistically, investors who prefer certain maturities may shift their investments if they think that bonds in a different maturity band are particularly cheap. This possibility therefore draws upon the risk–return aspect of liquidity-preference theory.

1.6.4 Arbitrage-Free Pricing Theory

The remainder of this book considers the pricing of bonds in a market which is free of arbitrage. The theory (which is very extensive) pulls together the expectation, liquidity-preference and market-segmentation theories in a mathematically precise way. Under this approach we can usually decompose forward rates into three components:

- the expected future risk-free rate of interest, $r(t)$;
- an adjustment for the market price of risk;[4]
- a convexity adjustment to reflect the fact that $E(e^X) \geqslant e^{E(X)}$ for any random variable X.

For example, consider the Vasicek model (see Section 4.5): given $r(0)$ we have

$$E[r(t)] = \mu + (r(0) - \mu)e^{-\alpha t},$$

whereas the forward-rate curve at time 0 can be written as the sum of three components corresponding to those noted above. That is,

$$f(0, T) = \mu + (r(0) - \mu)e^{-\alpha T} - \lambda \sigma (1 - e^{-\alpha T})/\alpha - \tfrac{1}{2}\sigma^2 [(1 - e^{-\alpha T})/\alpha]^2,$$

where μ, α and σ are parameters in the model and λ is the market price of risk. For reasons which will be explained later, λ is normally negative.

The form of the two adjustments is not obvious. This is why we need arbitrage-free pricing theory to derive prices.

For a single-factor model—one with a single source of randomness, such as the Vasicek model—there is no place for market-segmentation theory. However, many models for the term structure of interest rates have more than one random factor (so-called multifactor models). These allow us to incorporate market-segmentation theory to some extent.

1.7 Exercises

Exercise 1.1. Prove that the gross redemption yield is uniquely defined for a fixed-interest coupon bond.

Exercise 1.2. One consequence of an arbitrage-free bond market is that the instantaneous risk-free rate, $r(T)$, must be non-negative for all T.

---

[4]We use the name *market price of risk* in the following sense. When we simulate prices under the real-world probability measure P, we will see in a later chapter that the excess expected return on a risky asset over the risk-free rate of interest is equal to the market price of risk multiplied by the volatility of the risky asset. Thus, the market price of risk is the excess expected return per unit of volatility.

(a) Why must the forward-rate curve, $f(t, t+s)$, also be non-negative?

(b) What are the consequences for $r(T)$, with $T > t$, if $f(t, t+s) = 0$ for some $s > 0$?

(c) What are the consequences for the form of $P(t, T)$?

Exercise 1.3. Show that the term structure is not necessarily arbitrage-free even if the spot-rate curve $R(t, t+s) \geqslant 0$ for all $s > 0$.

Exercise 1.4. Suppose that the UK government issues two bonds, Treasury 8% 2010–14 and Treasury 8% 2010, with earliest redemption date of the former coinciding with the fixed redemption date of the latter. Explain which bond will have the higher price?

Exercise 1.5. Suppose the UK government issues two bonds, Treasury 8% 2010 and Convertible 8% 2010. The bonds are redeemable on the same date. On 1 January 2006 (not a coupon-payment date) holders of Convertible 8% 2010 will be able to convert their stock into Treasury $8\frac{3}{4}$% 2017 on a one-for-one basis. Show that Convertible 8% 2010 will have a higher price than Treasury 8% 2010.

Exercise 1.6. Suppose that the spot rates (continuously compounding) for terms 1, 2 and 3 to maturity are

T	1	2	3
$R(0, T)$	6%	6.5%	7%

(a) Find the values of $F(0, 1, 2)$, $F(0, 1, 3)$ and $F(0, 2, 3)$.

(b) Assuming that coupons are payable annually in arrears, find the par yields for terms 1, 2 and 3 years.

Exercise 1.7. In a certain bond market coupons are payable annually. At time t, par yields $\rho(t, T)$ are given for maturities $T = t + 1, t + 2, \ldots$. Derive recursive formulae for calculating the prices of zero-coupon bonds maturing at times $T = t + 1, t + 2, \ldots$.

Exercise 1.8.

(a) In a certain bond market coupons are payable continuously. At time t, par yields $\rho(t, T)$ are given for all maturities, $T \in R$ with $t < T < t + s$. Show that the zero-coupon bond prices can be found by solving the ordinary differential equation

$$\frac{\partial P}{\partial T}(t, T) + P(t, T)\left(\rho(t, T) - \frac{1}{\rho(t, T)}\frac{\partial \rho}{\partial T}(t, T)\right) = -\frac{1}{\rho(t, T)}\frac{\partial \rho}{\partial T}(t, T).$$

(b) Hence find $P(0, T)$ given $\rho(0, T) = (1 - \frac{1}{2}T)^{-1}$ for $0 < T < 1$.

(c) Explain why, in part (b), T is limited above by 1.

2
Arbitrage-Free Pricing

In this chapter we will introduce some of the basic ideas of arbitrage and arbitrage-free pricing. The formal definition of a discrete-time arbitrage is given in equation (2.2) below. More informally, an arbitrage happens when

(a) we are able to construct at time 0 some portfolio which has net value zero (thus a non-trivial portfolio will have a mixture of positive and negative holdings which cost zero in total);

(b) at some fixed time T in the future this portfolio will give us a sure profit.

This definition has the alternative name of a *free lunch*.

The definition of an arbitrage is most clear when we consider static portfolios (or 'buy-and-hold' strategies). Suppose then that we can invest in n assets. Asset i has price $P_i(t)$ at time t per unit with no dividends or coupons payable. Suppose we have x_i units of asset i in our portfolio which therefore has value $V(t) = \sum_{i=1}^{n} x_i P_i(t)$ at time t. The definition of arbitrage above then implies that

$$V(0) = \sum_{i=1}^{n} x_i P_i(0) = 0,$$
$$\Pr(V(T) \geqslant 0) = 1,$$
$$\Pr(V(T) > 0) > 0.$$

The *Principle of No Arbitrage* states simply that we assume that such arbitrage opportunities do not exist (otherwise smart investors could make infinite amounts of money). (A more detailed look at the nature and the definition of no arbitrage is beyond the scope of this book.)

Besides the definition given above, the principle of no arbitrage has the following equivalent forms.

1. We cannot construct a riskless portfolio which returns more than the risk-free rate of return.

2. If two portfolios A and B give rise to identical (but possibly random) future cashflows with certainty, then A and B must have the same value at the present time (the *law of one price*).

2.1   Example of Arbitrage: Parallel Yield Curve Shifts

It is easy to construct a model which admits arbitrage. Suppose that

$$P(0, T) = \exp\left[-\int_0^T f(0, u) \, du\right]$$

for some initial forward-rate curve $f(0, T)$. Our model dictates that at time 1 the forward-rate curve will be

$$f(1, T) = f(0, T) + \epsilon \quad \text{for } T > 1,$$

where ϵ is some random variable distributed on the real line. Thus, the forward-rate curve has been subjected to a *parallel shift* up or down.

Suppose we have available for investment zero-coupon bonds which mature at times T_1, T_2 and T_3, with $1 < T_1 < T_2 < T_3$. At time $t = 1$ we will have

$$P(1, T) = \exp\left[-\int_1^T f(1, u) \, du\right] = \exp\left[-\int_1^T (f(0, u) + \epsilon) \, du\right]$$

$$= \frac{P(0, T)}{P(0, 1)} e^{-\epsilon(T-1)}. \tag{2.1}$$

Let x_i be the number of units held at time 0 of the bond maturing at time T_i. For an arbitrage we require

$$\sum_{i=1}^{3} x_i P(0, T_i) = 0 \quad \text{(initially, the portfolio has value zero)}, \tag{2.2}$$

$$\sum_{i=1}^{3} x_i P(1, T_i) \geqslant 0 \quad \text{with probability 1}$$

(no loss in any future scenario),

$$\sum_{i=1}^{3} x_i P(1, T_i) > 0 \quad \text{with probability greater than 0}$$

(a profit in some scenarios with positive probability).

The value of the portfolio at time 1 is

$$V_1(\epsilon) = \sum_{i=1}^{3} x_i P(1, T_i)$$

$$= \frac{e^{-\epsilon(T_2-1)}}{P(0, 1)} g(\epsilon) \quad \text{(by equation (2.1))},$$

where

$$g(\epsilon) = \sum_{i=1}^{3} x_i P(0, T_i) e^{-\epsilon(T_i - T_2)}.$$

Firstly, note that $\sum_{i=1}^{3} x_i P(0, T_i) = 0$ implies that $g(0) = 0$. Secondly, note that $V_1(\epsilon) < 0$ if and only if $g(\epsilon) < 0$.

Since $g(\epsilon)$ is continuous and twice differentiable, the requirement that $V_1(\epsilon) > 0$ for all $\epsilon \neq 0$ and the fact that $g(0) = 0$ means that we must also have

$$g'(0) = 0,$$

$$\Rightarrow \quad \sum_{i=1}^{3} x_i (T_2 - T_i) P(0, T_i) = 0,$$

$$\Rightarrow \quad \sum_{i=1}^{3} x_i T_i P(0, T_i) = 0 \quad \text{since} \quad \sum_{i=1}^{3} x_i P(0, T_i) = 0.$$

$$(2.3)$$

Thirdly, it is sufficient that $g''(\epsilon) > 0$ for all ϵ to ensure that $g(\epsilon) > 0$ for all $\epsilon \neq 0$. We have

$$g''(\epsilon) = \sum_{i=1}^{3} x_i (T_2 - T_i)^2 P(0, T_i) e^{-\epsilon(T_i - T_2)}. \tag{2.4}$$

This is greater than 0 for all ϵ if and only if x_1 and x_3 are both greater than or equal to 0 (and one of these is strictly greater than 0).

Now try $x_2 = -1$. Equations (2.2) and $x_2 = -1$ imply that at least one of x_1 and x_3 must be greater than zero. Equation (2.3) implies that x_1 and x_3 must both be positive or both negative. Hence both x_1 and x_3 must be greater than 0. (This is true for all $x_2 < 0$. If $x_2 > 0$, then x_1 and x_3 are both negative.) It follows that $g''(\epsilon) > 0$ for all ϵ. Hence, $g(\epsilon) > 0$ for all $\epsilon \neq 0$.

Example 2.1. Suppose that $P(0, t) = e^{-0.08t}$ for all $t > 0$, and that, for all $t > 0$,

$$P(1, t+1) = \begin{cases} e^{-0.1t} & \text{if } I = 1, \\ e^{-0.06t} & \text{if } I = 0, \end{cases}$$

where $I = 0$ or 1 is a random variable. In other words, the spot- and forward-rate curves will both have a shift up or down of 2%.

Suppose that we hold x_1, x_2 and x_3 units of the bonds maturing at times 1, 2 and 3 respectively, such that

$$x_2 P(0, 2) = -1,$$
$$x_1 P(0, 1) + x_2 P(0, 2) + x_3 P(0, 3) = 0,$$
$$x_1 P(0, 1) + 2x_2 P(0, 2) + 3x_3 P(0, 3) = 0,$$

$$x_2 = \frac{-1}{P(0, 2)} = -1.173\,511,$$

$$\Rightarrow \quad x_3 = \frac{1}{2P(0, 3)} = 0.635\,624,$$

$$x_1 = \frac{1}{2P(0, 1)} = 0.541\,644.$$

At time 1 the value of this portfolio is $0.000\,21$ if $I = 1$ or $0.000\,22$ if $I = 0$. So the model is not arbitrage free. Since there are only two outcomes there are other combinations of values for (x_1, x_2, x_3) (beyond simply rescaling) which also give rise to arbitrage.

From this example we can conclude that parallel shifts in the yield curve cannot occur at any time in the future.

2.2 Fundamental Theorem of Asset Pricing

Suppose that the risk-free rate $r(t)$ is stochastic. Randomness in $r(t)$ is underpinned by the probability triple $(\Omega, \mathcal{F}, P)$, where P is the *real-world* (or natural) probability measure. Let the cash account (or money-market account) be

$$B(t) = B(0) \exp\left(\int_0^t r(s)\,ds\right).$$

Note that $dB(t) = r(t)B(t)\,dt$, that is, there is no Brownian $dW(t)$ term (zero quadratic variation). This explains why the cash-account process is described as risk-free (even though $r(t)$ is stochastic).

We will now state the Fundamental Theorem of Asset Pricing, the result that is central to everything in this book. Proofs of part (i) will be given in Chapters 3 and 4.

Theorem 2.2.

(i) *Bond prices evolve in a way that is arbitrage free if and only if there exists a measure Q, equivalent to P, under which, for each T, the discounted price process $P(t, T)/B(t)$ is a martingale for all $t : 0 < t < T$.*

(ii) *If (i) holds, then the market is complete if and only if Q is the unique measure under which the $P(t, T)/B(t)$ are martingales.*

The measure Q is often referred to, consequently, as the *equivalent martingale measure* (with the cash account $B(t)$ as the numeraire). Other names for Q are also in common use: the *risk-neutral measure* and the *risk-adjusted measure*. The three names all usually mean the same thing.

Corollary 2.3. *Hence*

$$P(t, T) = E_Q\left[\exp\left(-\int_t^T r(s)\,ds\right)\,\Big|\,\mathcal{F}_t\right],$$

where $\mathcal{F}_t$ is the sigma-algebra generated by price histories up to time t, and E_Q implies expectation with respect to the equivalent martingale measure, Q.

Remark 2.4. It also follows that if X is some $\mathcal{F}_T$-measurable derivative payment payable at T, and if $V(t)$ is the fair value at time t of this derivative contract, then the discounted price process $V(t)/B(t)$ is also a martingale under Q. Hence

$$V(t) = E_Q\left[\exp\left(-\int_t^T r(u)\,du\right)X\,\Big|\,\mathcal{F}_t\right].$$

Example 2.5 (forward pricing). A forward contract has been arranged in which a price K will be paid at time T in return for a repayment of 1 at time S ($T < S$). Equivalently, K is paid at T in return for delivery at the same time T of the S-bond which has a value at that time of $P(T, S)$. How much is this contract worth at time $t < T$?

As an interest rate derivative contract, this has value $X = P(T, S) - K$ at time T. Remark 2.4 indicates that the price of this contract at time t is

$$\begin{aligned}
V(t) &= E_Q\left[\exp\left(-\int_t^T r(u)\,du\right)(P(T, S) - K)\,\Big|\,\mathcal{F}_t\right] \\
&= E_Q\left[\exp\left(-\int_t^T r(u)\,du\right)E_Q\left\{\exp\left(-\int_T^S r(u)\,du\right)\,\Big|\,\mathcal{F}_T\right\}\,\Big|\,\mathcal{F}_t\right] \\
&\qquad\qquad\qquad\qquad - KE_Q\left[\exp\left(-\int_t^T r(u)\,du\right)\,\Big|\,\mathcal{F}_t\right] \\
&= E_Q\left[\exp\left(-\int_t^S r(u)\,du\right)\,\Big|\,\mathcal{F}_t\right] - KE_Q\left[\exp\left(-\int_t^T r(u)\,du\right)\,\Big|\,\mathcal{F}_t\right]
\end{aligned}$$

$$\text{(by the Tower Property)}$$

$$= P(t, S) - KP(t, T).$$

If we choose K to ensure that $V(t) = 0$, then $K = P(t, S)/P(t, T)$. (In fact, this is the basis of the definition of forward rates $F(t, T, S)$ given in Section 1.2.5.) This contract can be hedged at no cost at time t by buying one unit of $P(t, S)$ and selling $P(t, S)/P(t, T)$ units of $P(t, T)$.

2.3 The Long-Term Spot Rate

Let $l(t) = \lim_{T\to\infty} R(t, T)$ be the long-term spot rate (if this limit exists).

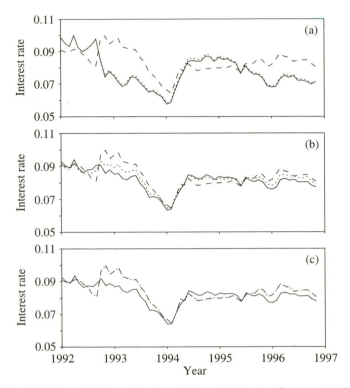

Figure 2.1. Evolution of UK market rates of interest (estimated from coupon bond data) over the period 1992 to 1996 (see Cairns 1998). Par yields (solid curves), spot rates (dotted curves) and forward rates (dashed curves) for (a) 10 and (b) 20 years to maturity and (c) with an infinite term to maturity. In (c) the spot and forward rates coincide.

Usually $l(t)$ also equals $\lim_{T \to \infty} f(t, T)$, but this is not true in general. For example, $f(t, T)$ could have undamped oscillations as $T \to \infty$ while $R(t, T)$ tends to some constant limit.

There is no standard, tradable security which allows $l(t)$ to be observed exactly, since the longest-dated zero-coupon bonds typically have a term to maturity of only 30 years. Instead, a theoretical value of $l(t)$ must be estimated from other quantities.

Empirical research into the behaviour of long-term rates of interest (see, for example, Cairns 1998) suggests that $l(t)$ fluctuates substantially over long periods of time. This is illustrated in Figure 2.1, where we have plotted the evolution of UK par yields, spot rates and forward rates over the period 1992 to 1996. In (c) the spot rates and forward rates are both equal to $l(t)$ given the assumption of an infinite term to maturity. In particular, we can identify a drop in the fitted value $l(t)$ of 4% between the end of 1992 and early 1994. We then ask ourselves if this behaviour is consistent with an arbitrage-free framework.

None of the models we will examine later in this book allow $l(t)$ to decrease over time. Indeed almost all of the arbitrage-free models result in a constant value for $l(t)$ over time. This suggests that a fluctuating $l(t)$ is not consistent with no arbitrage. However, we should be careful how we interpret Figure 2.1. The values plotted represent the results of a curve-fitting exercise which took price data on each date for coupon bonds and estimated the underlying forward-rate curve assuming that it had the parametric form

$$f(t, t + s) = b_0(t) + b_1(t)e^{-c_1 s} + \cdots + b_4(t)e^{-c_4 s}$$

(see Chapter 12 for a fuller discussion). In particular, the parameters $b_0(t)$ to $b_4(t)$ were estimated for each date without reference to other dates, so that the limiting forward and spot rate $l(t)$ is equal to $b_0(t)$. Furthermore, the possibility that $b_0(t)$ might be constant was not investigated by Cairns (1998) so it is not immediately clear from the graph if the variation in $b_0(t)$ over time is statistically significant. It is important to note that any estimate of $b_0(t)$ from coupon-bond price data will be subject to relatively high standard errors given $b_0(t)$ is far removed from the more immediately observable par yields for maturities up to 30 years. Note also that the infinite maturity par yield can vary over time even though $l(t)$ remains constant.

In summary, Figure 2.1 does not prove that $l(t)$ does vary over time even though the graph suggests that it might.

The next result proves that, under certain assumptions, $l(t)$ cannot decrease over time in an arbitrage-free world.

Theorem 2.6 (Dybvig–Ingersoll–Ross Theorem (Dybvig *et al.* 1996)). *Suppose that the dynamics of the term structure are arbitrage free. Then $l(t)$ is non-decreasing almost surely.*

Proof. We assume that we have available to us zero-coupon bonds maturing at times $1, 2, 3, \ldots$ and that we invest an amount $1/[T(T + 1)]$ in the bond maturing at time T. The total investment at time 0 is thus $V(0) = 1$, since $\sum_{T=1}^{\infty} 1/[T(T + 1)] = 1$.

Let us consider the value of our portfolio at time 1. In particular, suppose that $l(1) < l(0)$. Let $\epsilon = (l(0) - l(1))/3 > 0$. Then there exists some $T_0 < \infty$ such that $P(0, T) < \exp[-(l(0) - \epsilon)T]$ and $P(1, T) > \exp[-(l(1) + \epsilon)T]$ for all $T \geqslant T_0$:

$$V(1) = \sum_{T=1}^{\infty} \frac{P(1, T)}{T(T + 1)P(0, T)}$$

$$> \sum_{T=1}^{T_0-1} \frac{P(1, T)}{T(T + 1)P(0, T)} + \sum_{T=T_0}^{\infty} \frac{1}{T(T + 1)} \frac{e^{-(l(1)+\epsilon)T}}{e^{-(l(0)-\epsilon)T}}$$

$$= \sum_{T=1}^{T_0-1} \frac{P(1, T)}{T(T + 1)P(0, T)} + \sum_{T=T_0}^{\infty} \frac{1}{T(T + 1)}e^{\epsilon T}.$$

But

$$\sum_{T=T_0}^{T_1} \frac{1}{T(T+1)} e^{\epsilon T} \to \infty \quad \text{as } T_1 \to \infty \quad \Rightarrow \quad V(1) = \infty.$$

Since dynamics are assumed to be arbitrage free there exists an equivalent martingale measure, Q. Thus, $E_Q[V(1)/B(1) \mid \mathcal{F}_0] = V(0)$, where $B(1)$ is the cash account, $\exp[\int_0^1 r(s)\,ds]$. It follows that, under Q, the probability must be zero that (a) $1/B(1)$ is greater than zero or (b) $V(1)$ is infinite. Excluding the possibility that $B(1)$ is infinite, this means that the probability under Q that $V(1) = \infty$ is zero. Therefore under the equivalent martingale measure the probability that $l(1)$ is less than $l(0)$ must be zero. Equivalent measures must satisfy the constraint that events with probability zero under one measure must also have probability zero under the equivalent measure. Therefore, the probability that $l(1) < l(0)$ under the real-world probability measure P must also be 0.   □

A more rigorous proof of this result can be found in Hubalek *et al.* (2002).

Remarks. This is an ideal situation. In reality we could not buy arbitrarily small amounts of each security, nor are we able to buy bonds beyond about 30 years to maturity (although this means that $l(t)$ is not observable).

In contrast, most models give us a complete, theoretical picture of the term structure of interest rates, including $l(t)$.

What the Dybvig–Ingersoll–Ross (DIR) Theorem does tell us is that we will not be able to construct an arbitrage-free model for the term structure that allows the long-term rate $l(t)$ to go down. In most models $l(t)$ is either constant or infinite. The DIR theorem gives us a benchmark against which we can test any new model.

Example 2.7. Suppose under the equivalent martingale measure that

$$r(t) = \begin{cases} 0.05 & \text{for } 0 \leqslant t < 1, \\ 0.04 & \text{for } t \geqslant 1 \text{ with probability } 0.5, \\ 0.06 & \text{for } t \geqslant 1 \text{ with probability } 0.5. \end{cases}$$

Then, for $T \geqslant 1$,

$$P(0, T) = e^{-0.05} \times \{\tfrac{1}{2} e^{-0.04(T-1)} + \tfrac{1}{2} e^{-0.06(T-1)}\}$$
$$= \tfrac{1}{2} e^{-0.01-0.04T} [1 + e^{0.02-0.02T}].$$

Hence

$$l(0) = \lim_{T \to \infty} -\frac{1}{T} \log P(0, T)$$
$$= \lim_{T \to \infty} \left(0.04 - \frac{1}{T} \log \tfrac{1}{2} e^{-0.01} - \frac{1}{T} \log[1 + e^{0.02-0.02T}] \right)$$
$$= 0.04.$$

At time 1, $P(1, T)$ is equal to $\exp(-0.04(T - 1))$ or $\exp(-0.06(T - 1))$ with equal probability with $l(1)$ equal to 0.04 or 0.06 respectively. So $l(t)$ is constant or increasing, as indicated by the DIR Theorem.

In this example, the model for $r(t)$ is time inhomogeneous. More importantly, the two states 0.04 and 0.06 are absorbing. Clearly, this is not a realistic model. However, it is included here to demonstrate that we can construct models under which $l(t)$ may increase over time. In practice, many models we consider have a recurrent stochastic structure which ensures that $l(t)$ is constant. In other models $l(t)$ is infinite for all $t > 0$.

2.4 Factors

A *one-factor* model is one under which there is a single, one-dimensional source of randomness affecting bond prices (for example, one-dimensional Brownian motion). Under such a model all price changes are perfectly (but non-linearly) correlated. In other words, if we know the change in one quantity (for example, the risk-free rate $r(t)$), then we know the change in the prices of all assets.

A *multifactor* model is one under which there is more than one source of randomness. Then we find that price changes are not perfectly correlated. The drivers under a two-factor model might be two-dimensional Brownian motion and these may be used, for example, to drive the short-term rate, $r(t)$, and its volatility, $\sigma(r(t))$, or the short rate and the rate of interest on irredeemable bonds. If there are m factors, then the changes in the prices of m bonds will be sufficient for us to know the changes in the prices of all other bonds.

2.5 A Bond Is a Derivative

The obvious derivative securities in the bond market are options. However, individual bonds are themselves derivatives. For example, in a one-factor model the price of any bond is derived from our knowledge of the short (risk-free) rate, $r(t)$, which takes on the role of the underlying.

2.6 Put-Call Parity

We saw in Example 2.5 the fair price for a forward contract could be determined using a model-free argument. Another important, model-free result is the strict relationship between European put and call option prices. Thus, consider European call and put options with the same exercise date T, a strike price K, and the S-bond, price $P(t, S)$, as the underlying with $S > T$. Let $c(t)$ and $p(t)$ be the prices at t of the call and put options respectively. Consider now two portfolios.

   A. One call option plus K units of the T-bond, $P(t, T)$.

   B. One put option plus one unit of the S-bond, $P(t, S)$.

The value of A at time T is $\max\{P(T, S) - K, 0\} + K = \max\{P(T, S), K\}$. The value of B at time T is $\max\{K - P(T, S), 0\} + P(T, S) = \max\{P(T, S), K\}$.

So A and B have identical payoffs at time T. By the law of one price, the values of portfolios A and B at any earlier time must also be equal. Hence

$$c(t) + K P(t, T) = p(t) + P(t, S).$$

This model-free result is called *put-call parity*. It does not tell us what the prices $c(t)$ and $p(t)$ are individually but it does define a strict relationship between the two. Any proposed model for derivative pricing in a liquid market must satisfy this result. If it does not, then the model will contain arbitrage opportunities.

2.7 Types of Model

There are two main types of model.

Equilibrium and Short-Rate Models

Equilibrium models are built on assumptions about how the economy works. They take account of the varying risk preferences of different investors and aim to achieve a balance between the supply of bonds and other securities and the demand for these by investors. In the present context we are particularly interested in assessing how the economy affects the term structure of interest rates. In a one-factor model this normally means constructing a simple stochastic model for the evolution of the risk-free rate. This is done in a way which captures the essential characteristics of the wider economy as far as they impact on interest rates. We then invoke the Fundamental Theorem of Asset Pricing to derive a theoretical set of bond prices. Under such a model the theoretical prices evolve in a way which is free from arbitrage.

It may be that the initial theoretical set of prices is different from observed market prices, giving rise to possible arbitrage opportunities. (This is a form of cheap/dear analysis.)

Often models for the short rate are regarded as equilibrium models, although this does not always have to be true. In fact, it is generally very difficult to prove that a short-rate model has an equilibrium derivation.

No-Arbitrage Models

These models use the observed term structure at the current time as the starting point. Future prices evolve in a way which is consistent with this initial price structure and which is arbitrage free. Such models are used for the pricing of short-term derivatives.

The possible actual-minus-theoretical errors in prices in an equilibrium model tend to get magnified when we are pricing derivatives; for example, a 1% error in the theoretical price of an underlying bond may lead to a 10% error in the price of an option on the same underlying.

On the other hand, in the longer term, no-arbitrage models can imply peculiar dynamics for such quantities as $r(t)$ which are hard to justify.

2.8 Exercises

Exercise 2.1. Suppose that $f(0, s) = 0.08$ for all $s > 0$. We have available for investment three zero-coupon bonds maturing at times 5, 10 and 15. At time 1 the forward-rate curve will be $f(1, s) = f(0, s) + \epsilon$, where $\epsilon = +0.02$ with probability 0.5 and $\epsilon = -0.02$ with probability 0.5. Construct an arbitrage which will take advantage of this parallel forward-rate curve shift.

Exercise 2.2. In a particular 1-period bond-pricing model, four bonds are available that mature at times 1, 2, 3 and 4. Their prices at time 0 are 0.9, 0.81, 0.729 and 0.684 respectively. At time 1 there will be one of three outcomes ω_1, ω_2 and ω_3. The prices of the outstanding bonds for each outcome are given in the following table:

	ω_1	ω_2	ω_3
$P(1, 2)$	0.88	0.9	0.92
$P(1, 3)$	0.77	0.805	0.86
$P(1, 4)$	0.7	0.75	x

No trading is possible between times 0 and 1.

(a) Find the value of x which will make this model arbitrage free.

(b) Is this market complete?

(c) If instead $x = 0.81$, show how to create an arbitrage opportunity.

Exercise 2.3. In a particular 1-period bond-pricing model, two bonds are available that mature at times 1 and 2. Their prices at time 0 are 0.9 and 0.81 respectively. At time 1 there will be one of three outcomes ω_1, ω_2 and ω_3. The prices of the outstanding bond for each outcome are given in the following table:

	ω_1	ω_2	ω_3
$P(1, 2)$	0.88	0.9	0.92

(a) Is this market complete? Give theoretical reasons for your answer.

(b) Give an example of a derivative which illustrates your answer to part (a).

Exercise 2.4. Suppose that $P(0, T) = e^{-0.08T}$ for all T. Furthermore, $r(t) = 0.08$ for $0 \leqslant t < 1$. No trading is possible between times 0 and 1.

At time 1 the spot-rate curve will be either

$$R(1, s) = 0.08 + u(s) \qquad \text{for all } s$$

or

$$R(1, s) = 0.08 - d(s) \qquad \text{for all } s$$

for some curves $u(s)$ and $d(s)$.

(a) Suppose that $u(s)$ for $s \geqslant 0$ and $d(1)$ are given. The prices of all zero-coupon bonds maturing after time 1 evolve in an arbitrage-free way. Thus determine the form of $d(s)$ for all $s > 0$ in terms of $d(1)$ and $u(s)$. What do you notice about $d(s)$ as $s \to \infty$?

(b) Suppose instead that $d(s) = 0.01$ for all $s > 0$ and $u(1) = 0.01$. Show that it is not possible to derive values for $u(s)$ for all s which keep the model arbitrage free. Which theorem indicates that this would be the case?

Exercise 2.5. Suppose that the risk-free rate of interest $r(t)$ is governed by the stochastic differential equation

$$dr(t) = \mu(t, r(t)) \, dt + \sigma(t, r(t)) \, dW(t),$$

where $\mu(t, r)$ and $\sigma(t, r)$ satisfy the usual conditions for the existence and uniqueness of $r(t)$.

Let the value of the cash account at t be denoted by $B(t)$ with $dB(t) = r(t)B(t) \, dt$. Show that

$$B(t) = B(0) \exp\left[\int_0^t r(s) \, ds \right].$$

Exercise 2.6. Suppose that

$$r(t) = \begin{cases} r_0 & \text{for } 0 \leqslant t < 1, \\ r_0 + \epsilon & \text{for } 1 \leqslant t, \end{cases}$$

where ϵ is a positive-valued random variable with probability density function $f(\epsilon) > 0$ for all $\epsilon > 0$. Prove that

$$l(0) = \lim_{T \to \infty} R(0, T) = r_0.$$

Exercise 2.7. Suppose that for some positive-valued stopping time U (not known until time U),

$$r(t) = \begin{cases} r_0 & \text{for } 0 \leqslant t < U, \\ r_1 & \text{for } U \leqslant t, \end{cases}$$

where $r_1 > r_0$. Discuss, in general terms, whether $l(0) = r_0$, r_1 or some other value.

Exercise 2.8.

(a) Suppose that U has an exponential distribution under Q with mean $1/\lambda$ and that U is not observable until time U. The risk-free rate is

$$r(t) = \begin{cases} r_0 & \text{for } 0 \leqslant t < U, \\ r_1 & \text{for } U \leqslant t, \end{cases}$$

where r_1 is equal to $r_0 + \epsilon$ with probability p, or r_0 with probability $1 - p$, and ϵ is some known positive constant. What is $l(0)$? Discuss, using general reasoning, the shape of $f(0, T)$ for $0 \leqslant T < \infty$.

(b) Suppose instead that for the same λ, ϵ and p,

$$\Pr(U \geqslant \tau) = (1 - p) + pe^{-\lambda\tau}$$

and that

$$r(t) = \begin{cases} r_0 & \text{for } 0 \leqslant t < U, \\ r_0 + \epsilon & \text{for } U \leqslant t. \end{cases}$$

Is $f(0, T)$ the same here as in part (a) of this question?

(c) Why are the forward-rate curves, $f(t, T)$, in parts (a) and (b) not equal in general for $t > 0$?

3

Discrete-Time Binomial Models

3.1 A Simple No-Arbitrage Model

Let us now consider a simple binomial model for the dynamics of bond prices. Let $P(t, T)$ be the price at time t of a zero-coupon bond which matures at time T for $t = 1, 2, 3, \ldots$ and $T = t, t+1, \ldots$.

We know that $P(t, t) = 1$ for all t. We also specify at time 0 a set of observed prices $P(0, T)$ for $T = 1, 2, \ldots, t_1$ (where t_1 might be infinite).

For any integer t, the risk-free rate of interest between t and $t + 1$ is defined as

$$r(t + s) = -\log P(t, t + 1) \quad \text{for } 0 \leqslant s < 1;$$

that is, the one-year bond is the risk-free asset. Note that $r(t + s)$ for $0 \leqslant s < 1$ is $\mathcal{F}_t$-measurable.

Let us also at this stage introduce the cash account, $B(t)$. This evolves according to the equations

$$B(0) = 1,$$

$$B(t + 1) = \frac{B(t)}{P(t, t + 1)} = \exp\left[\int_0^{t+1} r(s)\, ds\right] = \exp\left[\sum_{s=0}^{t} r(s)\right].$$

That is, $B(t)$ grows in line with the return on the one-year zero-coupon bond $P(t, t + 1)$. Since $P(t, t + 1)$ is known at time t, $B(t + 1)$ is known at time t.

We now ask the question: *is it possible for us to develop a stochastic model for the dynamics of these bond prices which is arbitrage free?*

This is trivial to demonstrate if interest rates are deterministic. We define first the forward rates $F(0, T, T + 1) = \log[P(0, T)/P(0, T + 1)]$ for $T = 0, 1, 2, \ldots$. We then define the instantaneous risk-free rate, $r(t)$, to be equal to $F(0, T, T + 1)$ for $T \leqslant t < T + 1$. We next define

$$P(t, T) = \exp\left[-\sum_{s=t}^{T-1} F(0, s, s + 1)\right] = \frac{P(0, T)}{P(0, t)}.$$

With this structure, the prices of all bonds grow at the risk-free rate, and the model is arbitrage free.

3.2   The Ho and Lee No-Arbitrage Model

Suppose that at time 1 either all prices go up or they all go down relative to the
risk-free return on cash; that is, for all $T \geqslant 1$,

$$P(1, T) = \begin{cases} u(0, T) \dfrac{P(0, T)}{P(0, 1)} & \text{if 'up'}, \\[2ex] d(0, T) \dfrac{P(0, T)}{P(0, 1)} & \text{if 'down'}. \end{cases}$$

(The 'T' in the functions $u(0, T)$ and $d(0, T)$ refers to the outstanding term to
maturity of the zero-coupon bond at the start of the time period.) Note that $P(0, T)/$
$P(0, 1)$ is the forward price at time 0 for delivery at time 1 of the zero-coupon bond
which matures at time T. If $P(1, T)$ were in fact equal to its forward price, then it
would have delivered the same return as the risk-free, one-year bond.

Note that if $u(0, s) = d(0, s) = 1$ for all s, then prices at time 1 are deterministic.
We can repeat this step at all future times. Thus, given $P(t, \tau)$ for $\tau > t$,

$$P(t + 1, T) = \begin{cases} u(t, T - t) \dfrac{P(t, T)}{P(t, t + 1)} & \text{if 'up'}, \\[2ex] d(t, T - t) \dfrac{P(t, T)}{P(t, t + 1)} & \text{if 'down'}. \end{cases}$$

It is necessary that the $u(t, s)$ and $d(t, s)$ are known at time t. By convention, we
assume that $u(t, s) \geqslant d(t, s)$ for all t, s. It is possible that the $u(t, s)$ and $d(t, s)$ can
depend upon the history of the process up to time t; in particular, they may depend
upon on the current set of prices. However, we will assume here for convenience
that there is no dependence upon prices or upon t. Thus, we write $u(s)$ and $d(s)$ for
all $s \geqslant 1$. We require $u(1) = d(1) = 1$ to ensure that $P(t, t) = 1$ for all t.

The following theorem includes the Fundamental Theorem of Asset Pricing.

Theorem 3.1. *Consider all price changes between times 0 and 1.*

  (i) *Suppose that the model is arbitrage free. Then*

$$u(T) > 1 > d(T) > 0 \quad \text{for all } T \geqslant 2.$$

 (ii) *Suppose that the model is arbitrage free. Define*

$$q(T) = \frac{1 - d(T)}{u(T) - d(T)} \quad \text{for all } T \geqslant 2.$$

*Then there exists q, $0 < q < 1$, such that $q(T) = q$ for all $T \geqslant 2$. q defines
the equivalent martingale measure Q; that is, $\Pr_Q(\text{'up'}) = q$, $\Pr_Q(\text{'down'}) =
1 - q$.*

(iii) *Suppose there exists an equivalent martingale measure, Q; that is, a q such that $0 < q < 1$ and $E_Q[P(1, T)/B(1)] = P(0, T)/B(0)$ for all T. Then there is no arbitrage between times 0 and 1 in the binomial model.*

Proof. (i) $u(T) > d(T)$, by definition, for all $T \geqslant 2$, $d(T) > 0$, since prices are positive for all t, T. Suppose that $u(T) > d(T) > 1$. Then

$$\frac{P(1, T)}{P(0, T)} \geqslant \frac{d(T)P(0, T)}{P(0, 1)} \times \frac{1}{P(0, T)} \geqslant \frac{1}{P(0, 1)} = B(1).$$

This tells us that the T-bond is guaranteed to pay out more than the risk-free cash account at time 1. This is an arbitrage opportunity, so $u(T) > d(T) > 1$ is impossible. Similarly, we cannot have $1 > u(T) > d(T)$, otherwise the bond always pays out less than cash at the year end.

(ii) Consider the measure Q_T under which

$$\mathrm{Pr}_{Q_T}(\text{'up'}) = \frac{1 - d(T)}{u(T) - d(T)} = q(T).$$

Then

$$\begin{aligned}
E_{Q_T}[P(1, T) \mid \mathcal{F}_0] &= q(T)\frac{P(0, T)u(T)}{P(0, 1)} + (1 - q(T))\frac{P(0, T)d(T)}{P(0, 1)} \\
&= \frac{P(0, T)}{P(0, 1)}\left[\frac{1 - d(T)}{u(T) - d(T)}u(T) + \frac{u(T) - 1}{u(T) - d(T)}d(T)\right] \\
&= \frac{P(0, T)}{P(0, 1)}.
\end{aligned}$$

But

$$P(0, 1) = \frac{B(0)}{B(1)} \quad \Rightarrow \quad \frac{P(0, T)}{B(0)} = E_{Q_T}\left[\frac{P(1, T)}{B(1)} \,\middle|\, \mathcal{F}_0\right].$$

Under Q_T, $P(t, T)/B(t)$ is a martingale. Thus, $q(T)$ can be thought of as the risk-neutral probability that the actual price of the bond is higher than its forward price $P(0, T)/P(0, 1)$.

Let us replicate $P(1, 2)$ by using the T-bond (for $T \geqslant 3$) and cash; that is, at time 0 we hold x units of cash and y units of $P(0, T)$. At time 1 the value of this portfolio is $xB(1) + yP(1, T)$. This should be equal to $P(1, 2)$ regardless of whether prices go up or down:

$$\begin{aligned}
\text{up} \quad &\Rightarrow \quad xB(1) + yu(T)P(0, T)B(1) = u(2)P(0, 2)B(1), \\
\text{down} \quad &\Rightarrow \quad xB(1) + yd(T)P(0, T)B(1) = d(2)P(0, 2)B(1).
\end{aligned}$$

Hence,

$$y = \frac{(u(2) - d(2))P(0, 2)}{(u(T) - d(T))P(0, T)} \quad \text{and} \quad x = \frac{(u(T)d(2) - d(T)u(2))P(0, 2)}{u(T) - d(T)}.$$

The initial value of the portfolio is

$$x + yP(0, T) = \frac{(u(T)d(2) - d(T)u(2))P(0, 2) + (u(2) - d(2))P(0, 2)}{u(T) - d(T)}$$

$$= \left[u(2)\frac{1 - d(T)}{u(T) - d(T)} + d(2)\frac{u(T) - 1}{u(T) - d(T)} \right] P(0, 2).$$

This must equal $P(0, 2)$ to avoid arbitrage (*the law of one price*). Thus,

$$u(2)\frac{1 - d(T)}{u(T) - d(T)} + d(2)\frac{u(T) - 1}{u(T) - d(T)} = 1,$$

$$\Rightarrow \quad u(2)q(T) + d(2)(1 - q(T)) = 1,$$

$$\Rightarrow \quad q(T) = \frac{1 - d(2)}{u(2) - d(2)}.$$

That is,

$$q(T) = q(2) = q \quad \text{for all } T.$$

Finally, note that $0 < q < 1$, since $u(T) > 1 > d(T)$.

This quantity q defines an equivalent martingale measure Q such that

$$E_Q\left(\frac{P(1, T)}{B(1)} \right) = \frac{P(0, T)}{B(0)} \quad \text{for all } T \geqslant 1.$$

(ii) Take any portfolio $\{x_T\}_{T=1}^N$ with net value 0; that is, $\sum_{T=1}^N x_T P(0, T) = 0$.
Then

$$E_Q\left(\sum_{T=1}^N x_T P(1, T) \right) = \sum_{T=1}^N x_T \frac{1}{P(0, 1)} E_Q\left(\frac{P(1, T)}{B(1)} \right)$$

$$= \frac{1}{P(0, 1)} \sum_{T=1}^N x_T \frac{P(0, T)}{B(0)}$$

$$= 0.$$

Hence, if we consider the random variable $\sum_{T=1}^N x_T P(1, T)$, either both outcomes are 0 or one outcome is positive and one negative. So no arbitrage is possible between times 0 and 1.   □

Remark 3.2. The requirement that the $q(T) = q$ for all T and for some $0 < q < 1$ imposes the relationship $u(T) = q^{-1}[1 - (1 - q)d(T)]$ for all T.

3.3   Recombining Binomial Model

As in Section 3.2 we will assume that $u(T)$ does not depend upon the current time t or upon the history up to time t, $\mathcal{F}_t$; that is, $u(t, T, \mathcal{F}_t) = u(T)$ for all t, $\mathcal{F}_t$.

Furthermore, we would like the prices to be path independent. For example, suppose that, up to time t, there have been i down-steps and $t - i$ up-steps. Clearly, the price should depend upon the number of up-steps. In contrast, it is our desire that the price should not depend upon the order of the up- and down-steps. For example, at time 2 prices following the sequence up–down should be the same as prices following the sequence down–up. This allows us to build up a binomial *lattice* rather than a *tree* for prices. Note that this is a desirable characteristic for computational efficiency rather than a necessary one.

We define $P(t, T, i) \equiv P(t, T)$ given that there have been i down-steps and $t - i$ up-steps between 0 and t (for $i = 0, 1, \ldots, t$). What constraints are required on the $u(T)$ to ensure that the order of the up- and down-steps is not relevant?

Let us consider the two-year period $t = 0$ to $t = 2$. We require that all prices after the up–down sequence are equal to the prices after the down–up sequence. We have, for $T \geqslant 2$,

$$\left. \begin{aligned} P(1, T, 0) &= u(T)\frac{P(0, T, 0)}{P(0, 1, 0)}, \\ P(1, T, 1) &= d(T)\frac{P(0, T, 0)}{P(0, 1, 0)}, \end{aligned} \right\} \quad \text{for } t = 1,$$

$$\left. \begin{aligned} P(2, T, 1) &= d(T - 1)\frac{u(T)P(0, T, 0)/P(0, 1, 0)}{P(1, 2, 0)} \quad \text{up–down,} \\ &= u(T - 1)\frac{d(T)P(0, T, 0)/P(0, 1, 0)}{P(1, 2, 1)} \quad \text{down–up,} \end{aligned} \right\} \quad \text{for } t = 2,$$

$$\Rightarrow \qquad \frac{d(T - 1)u(T)}{P(1, 2, 0)} = \frac{u(T - 1)d(T)}{P(1, 2, 1)},$$

$$\Rightarrow \qquad \frac{d(T)}{u(T)} = k\frac{d(T - 1)}{u(T - 1)},$$

where

$$k = \frac{P(1, 2, 1)}{P(1, 2, 0)} = \frac{d(2)}{u(2)},$$

implying that $0 < k < 1$.

Since $u(1) = d(1) = 1$, we can deduce that $d(T)/u(T) = k^{T-1}$.

From Theorem 3.1 and Remark 3.2 we also know that $qu(T) + (1 - q)d(T) = 1$ for all T and some $0 < q < 1$. Hence we have

$$u(T) = \frac{1}{(1 - q)k^{T-1} + q} \quad \text{and} \quad d(T) = \frac{k^{T-1}}{(1 - q)k^{T-1} + q}. \tag{3.1}$$

This model can be shown (see, for example, Exercise 3.3) to be path independent over any interval 0 to t.

Example 3.3. Suppose $P(0, T) = 0.94, 0.90, 0.87, 0.84$ for $T = 1, 2, 3, 4$ respectively. Furthermore, it is known that $P(1, 2) = 0.94$ or 0.965. It follows that

$$u(2) = P(1, 2, 0)P(0, 1)/P(0, 2) = 1.007\,889,$$
$$d(2) = P(1, 2, 1)P(0, 1)/P(0, 2) = 0.981\,778,$$
$$q = (1 - d(2))/(u(2) - d(2)) = 0.697\,872,$$
$$k = d(2)/u(2) = 0.974\,093.$$

The remainder of the $u(T)$ and $d(T)$ are determined by equation (3.1):

T	1	2	3	4
$u(T)$	1.000\,000	1.007\,889	1.015\,694	1.023\,414
$d(T)$	1.000\,000	0.981\,778	0.963\,749	0.945\,917

We can then compute values for the $P(t, T, x)$ using

$$P(t, T, x) = \begin{cases} u(T - t + 1)\dfrac{P(t - 1, T, x)}{P(t - 1, t, x)} & \text{for } x = 0, \ldots, t - 1, \\[2mm] d(T - t + 1)\dfrac{P(t - 1, T, x - 1)}{P(t - 1, t, x - 1)} & \text{for } x = 1, \ldots, t. \end{cases}$$

This gives us, for example, the following table of values for $P(t, 4)$:

			t		
x	0	1	2	3	4
0	0.840\,000	0.914\,540	0.962\,583	0.988\,124	1.000\,000
1	—	0.845\,287	0.913\,355	0.962\,525	1.000\,000
2	—	—	0.866\,644	0.937\,589	1.000\,000
3	—	—	—	0.913\,299	1.000\,000
4	—	—	—	—	1.000\,000

The binomial lattice for prices and the corresponding spot rates

$$R(t, 4) = -\frac{\log(P(t, 4))}{4 - t}$$

and risk-free rates $r(t)$ are plotted in Figure 3.1. For illustration, the same sample path has been highlighted in each plot. In this figure we can observe the lattice structures, with convergence to $P(4, 4, x) = 1$ for all x. Also note that the risk-free rate of interest has a random walk structure with a time-varying drift and constant distance between nodes (that is, $r(t, x) - r(t, x - 1) = -\log k$; see Corollary 3.5).

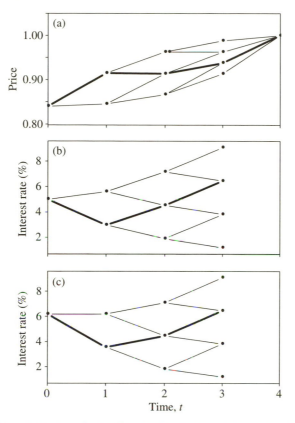

Figure 3.1. Binomial lattices for (a) $P(t, 4)$, (b) $R(t, 4)$ and (c) $r(t)$ for the Ho and Lee model and inputs given in Example 3.3. A single sample path is picked out in bold.

Remark 3.4. Under this model for $u(T)$, $d(T)$ let us consider the forward-rate curve. It can be shown that

$$F(t, T - 1, T) = \log \frac{P(t, T - 1)}{P(t, T)}$$

$$= F(0, T - 1, T) + \log \frac{u(T - t)}{u(T)} - D(t) \log k,$$

where

$$D(t) = \sum_{s=1}^{t} I(s)$$

is the number of down-steps and

$$I(s) = \begin{cases} 1 & \text{if there is a down-step at time } s, \\ 0 & \text{otherwise.} \end{cases}$$

Proof. We prove the result by induction.

The result is true for $t = 0$.

Suppose the result is true for t. Assume that $0 \leqslant D(t) \leqslant t$. First, consider the case $I(t + 1) = 0$. Then $D(t + 1) = D(t)$ and

$$
\begin{aligned}
&F(t + 1, T - 1, T, D(t + 1)) \\
&= \log \frac{P(t + 1, T - 1, D(t + 1))}{P(t + 1, T, D(t + 1))} \\
&= \log \frac{u(T - t - 1)P(t, T - 1, D(t))/P(t, t + 1, D(t))}{u(T - t)P(t, T, D(t))/P(t, t + 1, D(t))} \\
&= \log \frac{u(T - t - 1)}{u(T - t)} + F(t, T - 1, T, D(t)) \\
&= \log \frac{u(T - t - 1)}{u(T - t)} + F(0, T - 1, T) + \log \frac{u(T - t)}{u(T)} - D(t) \log k \\
&= F(0, T - 1, T) + \log \frac{u(T - t - 1)}{u(T)} - D(t + 1) \log k.
\end{aligned}
$$

So the result is true for $t + 1$ if $I(t + 1) = 0$.

Second, consider the case $I(t + 1) = 1$. Then $D(t + 1) = D(t) + 1$ and

$$
\begin{aligned}
&F(t + 1, T - 1, T, D(t + 1)) \\
&= \log \frac{d(T - t - 1)P(t, T - 1, D(t))/P(t, t + 1, D(t))}{d(T - t)P(t, T, D(t))/P(t, t + 1, D(t))} \\
&= \log \frac{d(T - t - 1)}{d(T - t)} + F(t, T - 1, T, D(t)) \\
&= \log \frac{u(T - t - 1)k^{T-t-2}}{u(T - t)k^{T-t-1}} + F(0, T - 1, T) + \log \frac{u(T - t)}{u(T)} - D(t) \log k \\
&= \log \frac{u(T - t - 1)}{u(T - t)} - \log k + F(0, T - 1, T) + \log \frac{u(T - t)}{u(T)} - D(t) \log k \\
&= F(0, T - 1, T) + \log \frac{u(T - t - 1)}{u(T)} - D(t + 1) \log k.
\end{aligned}
$$

So the result is true for $t + 1$ if $I(t + 1) = 1$.

Hence the result is true for $t + 1$, and the result follows by induction. $\qquad\square$

Thus, the size of the random component, $D(t) \log k$, is independent of T.

This model is the discrete-time version of the Ho and Lee (1986) model. We will see the continuous-time version of this model in Chapter 5. In other non-recombining binomial models the random component in $F(t, T, T + 1)$ typically decreases in size with T.

Corollary 3.5. *The risk-free rate of interest is then*

$$r(t) = F(t, t, t+1)$$

$$= F(0, t, t+1) + \log \frac{u(1)}{u(t+1)} - D(t) \log k$$

$$= F(0, t, t+1) - \log u(t+1) - D(t) \log k$$

$$= F(0, t, t+1) - \log d(t+1) + U(t) \log k,$$

where $U(t) = t - D(t)$ is the number of up-steps up to time t.

We observe that this model for $r(t)$ is a random walk with constant volatility but time-varying drift.

Remark 3.6. It is necessary to put constraints on q and k to ensure that the risk-free rate remains positive (that is, $P(t, t+1) < 1$) over a specified period of time. However, for any admissible (q, k), there will exist some t_0 which implies that, for all $t \geqslant t_0$, $P(t, t+1, i)$ will be greater than 1 for some i; that is, we cannot prevent interest rates from going negative eventually.

3.4 Models for the Risk-Free Rate of Interest

3.4.1 Time Homogeneity

In Sections 3.2 and 3.3 we considered first a very general binomial framework which was then refined to produce a model with certain features which are desirable from a computational point of view. In these models prices at time 0 form part of the input and this produces models which are usually time inhomogeneous. (That is, even if the risk-free rate of interest is the same at two different times, the term or price structures may be different.)

In some circumstances it is desirable to have a time-homogeneous model and this is best achieved using a time-homogeneous Markov model for the risk-free rate of interest under the equivalent martingale measure Q.

One way of achieving this is for the risk-free rate of interest to take values in some discrete-valued state space R in combination with a set of transition probabilities that determine how $r(t)$ changes from one period to the next. If the model is to form a complete market, then $r(t)$ should only be allowed to take one of two values one time step on.

Now consider the following binomial model for $r(t)$. Suppose $r(t) \in A$, where $A = \{\ldots, r_{-1}, r_0, r_1, r_2, \ldots\}$ and, under the real-world measure P,

$$\Pr_P[r(t+1) = r_{i-1} \text{ or } r_{i+1} \mid r(t) = r_i] = 1 \quad \text{for all } t \text{ and } i.$$

Suppose that, for all t,

$$P(t, t+2) \equiv P(t, t+2, r_i) = e^{-r_i}[q_i e^{-r_{i+1}} + (1 - q_i) e^{-r_{i-1}}] \tag{3.2}$$

for some set of constants q_i, $0 < q_i < 1$, for all $i \in \mathbb{Z}$. Then we can state formally the following theorem.

Theorem 3.7. *For all $T = t + 1, t + 2, \ldots,$*

$$P(t, T) = E_Q \left[\exp\left(-\int_t^T r(s) \, ds \right) \Bigg| \mathcal{F}_t \right]$$

$$= E_Q \left[\exp\left(-\sum_{s=t}^{T-1} r(s) \right) \Bigg| r(t) \right]$$

$$= P(t, t + 1) E_Q[P(t + 1, T) \mid r(t)],$$

where

$$\Pr_Q[r(t + 1) = r_{i+1} \mid r(t) = r_i] = q_i$$

and

$$\Pr_Q[r(t + 1) = r_{i-1} \mid r(t) = r_i] = 1 - q_i.$$

Proof. Let $Z(t, T) = P(t, T)/B(t)$ and define $D(t, T) = E_Q[B(T)^{-1} \mid \mathcal{F}_t]$. Note that $D(t, T)$ is a martingale under Q by the Tower Property for conditional expectation. We aim to show that $Z(t, T) = D(t, T)$. Now, by definition, if $r(t) = r_i$,

$$P(t, t + 2) = P(t, t + 2, r_i) = e^{-r_i}[q_i e^{-r_{i+1}} + (1 - q_i) e^{-r_{i-1}}],$$

$$Z(t, t + 2) = \frac{P(t, t + 2)}{B(t)} = \exp\left(-\sum_{s=0}^{t-1} r(s) \right) e^{-r(t)} E_Q[P(t + 1, t + 2) \mid \mathcal{F}_t],$$

$$= \exp\left(-\sum_{s=0}^{t} r(s) \right) E_Q[P(t + 1, t + 2) \mid \mathcal{F}_t]$$

$$= E_Q[Z(t + 1, t + 2) \mid \mathcal{F}_t].$$

That is, $Z(s, t + 2)$ is a martingale under Q from t to $t + 1$. By the Martingale Representation Theorem, there exists a previsible process $\phi(t, T)$ such that

$$D(t, T) = D(0, T) + \sum_{s=1}^{t} \phi(s, T) \Delta Z(s, s + 1),$$

where $\Delta Z(s, \tau) = Z(s, \tau) - Z(s - 1, \tau)$.

Let $\psi(t, T) = D(t - 1, T) - \phi(t, T) Z(t - 1, t + 1)$. Now consider the portfolio process which holds $\phi(t, T)$ units of the bond which matures at $t + 1$, $P(t - 1, t + 1)$, plus $\psi(t, T)$ units of the risk-free bond, $B(t - 1)$, from $t - 1$ to t. The value of this portfolio at time t *just after* rebalancing is

$$
\begin{aligned}
V(t, T) &= \phi(t + 1, T)P(t, t + 2) + \psi(t + 1, T)B(t) \\
&= B(t)[\phi(t + 1, T)Z(t, t + 2) + \psi(t + 1, T)] \\
&= B(t)D(t, T) \\
&= B(t)[D(t - 1, T) + \phi(t, T)\Delta Z(t, t + 1)] \\
&= B(t)[\phi(t, T)Z(t - 1, t + 1) + \psi(t, T) + \phi(t, T)\Delta Z(t, t + 1)] \\
&= B(t)[\phi(t, T)Z(t, t + 1) + \psi(t, T)] \\
&= \phi(t, T)P(t, t + 1) + \psi(t, T)B(t),
\end{aligned}
$$

which is the value of the portfolio at t *just before* rebalancing. Therefore the portfolio strategy is *self-financing*.

Furthermore, $V(T, T) = B(T)D(T, T) = 1$, so the portfolio strategy is replicating. The principle of no arbitrage (that is, the law of one price) indicates that $V(t, T)$ must, therefore, be the *unique* (given equation (3.2)) no-arbitrage price; that is,

$$
\begin{aligned}
P(t, T) = V(t, T) &= B(t)E_Q[B(T)^{-1} \mid \mathcal{F}_t] \\
&= E_Q\left[\exp\left(-\sum_{s=t}^{T-1} r(s)\right) \,\middle|\, r(t)\right].
\end{aligned}
$$

We can develop this further:

$$
\begin{aligned}
P(t, T) &= e^{-r(t)} E_Q\left[\exp\left(-\sum_{s=t+1}^{T-1} r(s)\right) \,\middle|\, r(t)\right] \\
&= e^{-r(t)} E_Q\left[E_Q\left\{\exp\left(-\sum_{s=t+1}^{T-1} r(s)\right) \,\middle|\, r(t+1)\right\} \,\middle|\, r(t)\right] \\
&= e^{-r(t)} E_Q[P(t + 1, T) \mid r(t)],
\end{aligned}
$$

where the relevant Q-probabilities are given in the statement of the theorem. $\qquad\square$

Example 3.8. The simplest example is the random-walk model for $r(t)$. The state space is then $R = \{r(0) + \delta n : n \in \mathbb{Z}\}$, where δ is the up- or down-step size.

For time homogeneity under Q we assume that the risk-neutral probabilities that $r(t)$ goes up and down (call these q and $1 - q$ respectively) are constant over time. The assumption of a random-walk model also indicates that these are independent of the state at any point in time. *However, the real-world probabilities of these events could be time or state dependent.*

Now recall Theorem 3.7. The risk-neutral probability q is determined most simply by considering at time 0 the price of the zero-coupon bond which matures at time 2. From the equation above,

$$P(0, 2) = P(0, 1)E_Q[P(1, 2) \mid r(0)]$$
$$= e^{-r(0)}(qe^{-(r(0)+\delta)} + (1 - q)e^{-(r(0)-\delta)}),$$
$$\Rightarrow \quad q = \frac{e^{-(r(0)-\delta)} - P(0, 2)e^{r(0)}}{e^{-(r(0)-\delta)} - e^{-(r(0)+\delta)}}.$$

Note that q is also the risk-neutral probability that prices go *down* relative to the risk-free cash account.

3.4.2 Calculation of Bond Prices

Suppose we are working with the random-walk model described in Example 3.8. It is sufficient that we know how many down-steps in bond prices there have been up to time t; we do not need to know their order of occurrence.

In general, prices can be calculated numerically by working backwards from the maturity date of the bond. As in Section 3.2, we extend our notation from $P(t, T)$ to $P(t, T, x)$, where x is the number of down-steps (in bond prices) from time 0 to time t. We take the following steps.

Step 1. For each state (t, x), let $r(t, x)$ be the risk-free rate of interest over the period t to $t + 1$ given x *down*-steps in bond prices (equivalently x *up*-steps in $r(s)$). For all $t \geqslant 0$ we have $P(t, t + 1) = P(t, t + 1, x) = \exp[-r(t, x)]$.

Step 2. Given the price $P(0, 2, 0)$, calculate

$$q = \frac{e^{-(r(0,0)-\delta)} - P(0, 2, 0)e^{r(0,0)}}{e^{-(r(0,0)-\delta)} - e^{-(r(0,0)+\delta)}}.$$

(Alternatively, q could be specified exogenously.)

Step 3. For $T = 2, 3, \dots$:

(a) Define $P(T, T, x) = 1$ for all $x = 0, 1, \dots, T$ and $P(T - 1, T, x) = \exp[-r(t, x)]$ for all $x = 0, 1, \dots, T - 1$.

(b) Suppose that we know the set of prices $P(s, T, x)$ for all $0 \leqslant x \leqslant s$ and for $s = t, t + 1, \dots, T$. We can then find the prices at time $t - 1$ in the following way. For each $x, 0 \leqslant x \leqslant t - 1$:

$$P(t - 1, T, x) = P(t - 1, t, x)E_Q[P(t, T) \mid r(t - 1) = r(t - 1, x)]$$
$$= e^{-r(t-1,x)}[qP(t, T, x + 1) + (1 - q)P(t, T, x)].$$

(c) Repeat step (b) until $t = 0$.

Example 3.9.

Step 1. Suppose that $r(0) = 0.05$, $r(t+1) = r(t) + (2I(t+1) - 1) \times 0.01$, where $I(t+1) = 1$ if the risk-free rate goes up at time $t+1$ and 0 otherwise.

Step 2. Suppose also that $P(0, 2) = P(0, 2, 0) = 0.909\,407$. Now calculate

$$q = \frac{e^{-0.04} - 0.909\,407 \times e^{0.05}}{e^{-0.04} - e^{-0.06}} = 0.25.$$

Step 3. For $T = 1$:

$$P(0, 1) = P(0, 1, 0) = e^{-0.05} = 0.951\,229.$$

For $T = 2$:

$$P(2, 2, x) = 1 \quad \text{for } x = 0, 1, 2,$$
$$P(1, 2, 1) = e^{-0.06} = 0.941\,765,$$
$$P(1, 2, 0) = e^{-0.04} = 0.960\,789,$$
$$P(0, 2, 0) = 0.909\,407 \quad \text{(given exogenously in step 2)}.$$

For $T = 3$:

$$P(3, 3, u) = 1 \quad \text{for } u = 0, 1, 2, 3,$$
$$P(2, 3, 2) = e^{-0.07} = 0.932\,394,$$
$$P(2, 3, 1) = e^{-0.05} = 0.951\,229,$$
$$P(2, 3, 0) = e^{-0.03} = 0.970\,446,$$
$$P(1, 3, 1) = P(1, 2, 1)[qP(2, 3, 2) + (1 - q)P(2, 3, 1)] = 0.891\,400,$$
$$P(1, 3, 0) = P(1, 2, 0)[qP(2, 3, 1) + (1 - q)P(2, 3, 0)] = 0.927\,778,$$
$$P(0, 3, 0) = P(0, 1, 0)[qP(1, 3, 1) + (1 - q)P(1, 3, 0)] = 0.873\,878,$$

and so on.

3.4.3 *Derivative Prices*

The prices of derivatives with payoffs that are contingent on bond prices at a given point in time can be calculated in a similar fashion.

Suppose that a derivative has a payoff Y at time T that is a function, for example, of the price at time T of the zero-coupon bond which matures at time $S > T$. Let this function be denoted by $f(p)$.

Again, assuming the recombining binomial tree in Sections 3.4.1 and 3.4.2, we denote by $V(t, x)$ the price at time t of the derivative, given that we have had x up-steps in the risk-free rate and $t - x$ down-steps up to time t.

Then $V(T, x) = f(P(T, S, x))$ and (in a similar fashion to the calculation of the underlying bond prices) for $t = T, T - 1, \ldots, 1$:

$$V(t - 1, x) = P(t - 1, t, x)[qV(t, x + 1) + (1 - q)V(t, x)].$$

Theorem 3.10. *Suppose a derivative contract pays $f(P(T, S))$ at time T $(T < S)$. Then the unique no-arbitrage price at time t for this contract is*

$$V(t) = E_Q\left[\exp\left(-\int_t^T r(s)\,\mathrm{d}s\right)f(P(T, S))\,\Big|\,\mathcal{F}_t\right]$$

$$= E_Q\left[\exp\left(-\sum_t^{T-1} r(s)\right)f(P(T, S))\,\Big|\,\mathcal{F}_t\right].$$

Proof. By Theorem 3.7, $Z(t, S) = P(t, S)/B(t)$ is a martingale under Q. Define

$$D(t) = E_Q\left[\frac{f(P(T, S))}{B(T)}\,\Big|\,\mathcal{F}_t\right].$$

This is also a martingale under Q. By the Martingale Representation Theorem, there exists a previsible process $\phi(t)$ (that is, for all t, $\phi(t)$ is known at time $t - 1$) such that $D(t) = D(0) + \sum_{u=1}^t \phi(u)\Delta Z(u, S)$.

Define $\psi(t) = D(t - 1) - \phi(t)Z(t - 1, S)$. Consider the portfolio strategy which holds $\phi(t)$ units of the S-bond and $\psi(t)$ units of the risk-free bond from $t - 1$ to t. This portfolio is self-financing (as in Theorem 3.7) and replicating. The value of this portfolio is, therefore, the unique no-arbitrage price for the derivative; that is,

$$V(t) = \phi(t + 1)P(t, S) + \psi(t + 1)B(t)$$

$$= B(t)D(t)$$

$$= E_Q\left[\exp\left(-\int_t^T r(s)\,\mathrm{d}s\right)f(P(T, S))\,\Big|\,\mathcal{F}_t\right].$$

$\square$

Example 3.11. Recall Example 3.9. Suppose that we have a call option on $P(t, 3)$ which matures at time 2 with a strike price of 0.95; that is, $Y = \max\{P(2, 3) - 0.95, 0\}$, or $f(p) = \max\{p - 0.95, 0\}$.

In Example 3.9 we found that $P(2, 3, 2) = 0.932\,394$, $P(2, 3, 1) = 0.951\,229$ and $P(2, 3, 0) = 0.970\,446$. It follows that $V(2, 2) = 0$, $V(2, 1) = 0.001\,229$ and $V(2, 0) = 0.020\,446$.

We can now apply Theorem 3.10 to calculate call option prices at earlier times. Thus,

$$V(1, 1) = P(1, 2, 1)[qV(2, 2) + (1 - q)V(2, 1)] = 0.000\,868,$$

$$V(1, 0) = P(1, 2, 0)[qV(2, 1) + (1 - q)V(2, 0)] = 0.015\,028,$$

$$V(0) = V(0, 0) = P(0, 1, 0)[qV(1, 1) + (1 - q)V(1, 0)] = 0.010\,928.$$

Example 3.12 (callable bonds). Suppose that $r(0) = 0.06$ and for all $t \geqslant 0$ we have risk-neutral probabilities

$$q = \Pr_Q(r(t+1) = r(t) + 0.01) = 0.5,$$
$$\Pr_Q(r(t+1) = r(t) - 0.01) = 0.5.$$

A zero-coupon, callable bond with a nominal value of 100 and a maximum term of four years is about to be sold. At each of times $t = 1, 2$ and 3, the bond may be redeemed early *at the option of the issuer*. The early redemption price at time t is $100 \exp[-0.055(4 - t)]$. At time 4 the bond will be redeemed at par (that is, 100) if this has not already happened.

Calculate the price for this bond at time 0 and for the equivalent zero-coupon bond with no early redemption option.

Solution. Let $X(t)$ be the number of up-steps in the risk-free rate of interest up to time t. The recombining binomial tree for the risk-free rate of interest is given in the table below, where $r(t, x)$ represents the risk-free rate of interest from t to $t + 1$ given $X(t) = x$:

			t		
x	0	1	2	3	4
4	—	—	—	—	0.10
3	—	—	—	0.09	0.08
2	—	—	0.08	0.07	0.06
1	—	0.07	0.06	0.05	0.04
0	0.06	0.05	0.04	0.03	0.02

The probability that $r(t)$ will go up is $q = 0.5$. Let us first calculate the prices $W(t, 4, x)$ of the conventional zero-coupon bond, where x is the number of steps up by time 4.

We start with $W(4, 4, x) = 100$ for $x = 0, 1, 2, 3, 4$. For all t and for all $0 \leqslant x \leqslant t$ we have

$$W(t, 4, x) = e^{-r(t,x)}[q W(t+1, 4, x+1) + (1-q)W(t+1, 4, x)].$$

Sample calculations are as follows:

$$W(3, 4, 3) = e^{-r(3,3)}[q W(4, 4, 4) + (1-q)W(4, 4, 3)]$$
$$= e^{-0.09}[0.5 \times 100 + 0.5 \times 100]$$
$$= 91.3931,$$
$$W(3, 4, 2) = e^{-r(3,2)}[q W(4, 4, 3) + (1-q)W(4, 4, 2)]$$
$$= e^{-0.07}[0.5 \times 100 + 0.5 \times 100]$$
$$= 93.2394,$$

$$W(2, 4, 2) = e^{-r(2,2)}[q\,W(3, 4, 3) + (1 - q)\,W(3, 4, 2)]$$
$$= e^{-0.08}[0.5 \times 91.3931 + 0.5 \times 93.2394]$$
$$= 85.2186,$$

and so on. The complete set of prices corresponding to the above table for $r(t)$ is given below:

			$W(t, 4, x)$		
			t		
x	0	1	2	3	4
4	—	—	—	—	100.0000
3	—	—	—	91.3931	100.0000
2	—	—	85.2186	93.2394	100.0000
1	—	81.0787	88.6965	95.1229	100.0000
0	78.7197	86.0923	92.3163	97.0446	100.0000

The calculation of prices for the callable bond is similar to that for the conventional bond, with the exception that, at each of times 1, 2 and 3, we compare two prices (one assuming early exercise and the other assuming the bond is not exercised). We assume that the issuer will redeem early if the exercise price is less than the price assuming no redemption. Thus, the price process $V(t, x)$ evolves according to the following recursive scheme:

$$V(4, x) = 100 \quad \text{for } x = 0, 1, 2, 3, 4.$$

For each $t = 3, 2, 1$ and $0 \leqslant x \leqslant t$:

$V(t, x)$
$$= \min\{100e^{-0.055(4-t)}, e^{-r(t,x)}(q\,V(t + 1, x + 1) + (1 - q)\,V(t + 1, x))\}.$$

Finally,
$$V(0, 0) = e^{-r(t,0)}(q\,V(1, 1) + (1 - q)\,V(1, 0)).$$

Sample calculations are as follows:

$$V(3, 3) = \min\{100e^{-0.055}, e^{-r(3,3)}(q\,V(4, 4) + (1 - q)\,V(4, 3))\}$$
$$= \min\{100e^{-0.055}, e^{-0.09}(\tfrac{1}{2} \times 100 + \tfrac{1}{2} \times 100)\}$$
$$= \min\{94.6485, 91.3931\}$$
$$= 91.3931,$$

$$V(3, 0) = \min\{100e^{-0.055}, e^{-r(3,0)}(qV(4, 1) + (1 - q)V(4, 0))\}$$
$$= \min\{100e^{-0.055}, e^{-0.03}(\tfrac{1}{2} \times 100 + \tfrac{1}{2} \times 100)\}$$
$$= \min\{94.6485, 97.0446\}$$
$$= 94.6485,$$

and so on. The complete set of prices corresponding to the above table for $r(t)$ is given below:

			$V(t, x)$		
			t		
x	0	1	2	3	4
4					100.0000
3				91.3931	100.0000
2			85.2186	93.2394	100.0000
1		80.9745	88.4731	**94.6485**	100.0000
0	78.0067	84.6863	**89.5834**	**94.6485**	100.0000

Those cells which have been typeset in bold indicate that early exercise is optimal; that is, $(t, x) = (2, 0)$, $(3, 0)$ and $(3, 1)$. The prices, $V(t, x)$, are generally *lower* than the prices of the corresponding conventional bond $W(t, 4, x)$ because the option characteristic favours the issuer rather than the holder of the bond. In some cases, though (for example, $(t, x) = (2, 2)$), $V(t, x) = W(t, 4, x)$, because early exercise will never be optimal from that particular position in the tree.

3.5 Futures Contracts

Let $f(t, S, T)$ be the futures price at time t for delivery at time S of the zero-coupon bond which matures at time T, where $S < T$.

Clearly, $f(S, S, T) = P(S, T)$, but what is the futures price at earlier times before the delivery date? In models for the equity market with a constant risk-free rate of interest we know that the forward and futures prices for an equity contract are equal (see, for example, Hull 2000). However, when the risk-free rate of interest is stochastic, forward and futures prices are not equal.

The futures price varies over time in such a way that immediately after the adjustment at time t the contract has value 0. Since the futures price varies over time, the futures exchange requires regular margin payments to pay for the adjustments. The mechanism employed by the exchange usually proceeds as follows. Consider an investor who has purchased one futures contract at time 0.

- At time 0, the net cashflow is 0. (There is no cost to set up the contract.)
- At time $t = 1, 2, \ldots, S$, the net cashflow to the investor is $f(t, S, T) - f(t - 1, S, T)$. This is called the margin payment.

Thus, for all $t = 0, 1, \ldots, S - 1$ we must set $f(t, S, T)$ in order that

$$E_Q\left[\sum_{n=t+1}^{S} \frac{B(t)}{B(n)}(f(n, S, T) - f(n-1, S, T)) \,\Big|\, \mathcal{F}_t\right] = 0.$$

This follows from Theorem 3.10 when we recognize that $f(n, S, T) - f(n-1, S, T)$ is an interest rate derivative payoff at time n. The sum of the expected discounted values under Q is then the unique no-arbitrage price for this *package* of derivative contracts with payoffs at times $t + 1$ up to T.

The problem is solved using a backwards recursion.

First, set $f(S, S, T) = P(S, T)$.

Now use backwards induction. Suppose the pricing structure, $f(m, S, T)$, is known for $m = t + 1, \ldots, S$. Thus, for each $n = t + 1, \ldots, S$, we already know that

$$E_Q\left[\sum_{m=n+1}^{S} \frac{B(n)}{B(m)}(f(m, S, T) - f(m-1, S, T)) \,\Big|\, \mathcal{F}_n\right] = 0.$$

Now consider what level to set $f(t, S, T)$ at. We require

$$E_Q\left[\sum_{n=t+1}^{S} \frac{B(t)}{B(n)}(f(n, S, T) - f(n-1, S, T)) \,\Big|\, \mathcal{F}_t\right] = 0. \qquad (3.3)$$

But

$$E_Q\left[\sum_{n=t+1}^{S} \frac{B(t)}{B(n)}(f(n, S, T) - f(n-1, S, T)) \,\Big|\, \mathcal{F}_t\right]$$

$$= E_Q\left[\frac{B(t)}{B(t+1)}(f(t+1, S, T) - f(t, S, T)) \,\Big|\, \mathcal{F}_t\right]$$

$$+ E_Q\left[\sum_{n=t+2}^{S} \frac{B(t)}{B(n)}(f(n, S, T) - f(n-1, S, T)) \,\Big|\, \mathcal{F}_t\right]$$

$$= P(t, t+1)E_Q[f(t+1, S, T) - f(t, S, T) \mid \mathcal{F}_t]$$

$$+ E_Q\left[\frac{B(t)}{B(t+1)}\right.$$

$$\left. \times E_Q\left[\sum_{n=t+2}^{S} \frac{B(t+1)}{B(n)}(f(n, S, T) - f(n-1, S, T)) \,\Big|\, \mathcal{F}_{t+1}\right] \,\Big|\, \mathcal{F}_t\right]$$

$$= P(t, t+1)E_Q[f(t+1, S, T) - f(t, S, T) \mid \mathcal{F}_t]$$
$$+ E_Q\left[\frac{B(t)}{B(t+1)} \times 0 \,\middle|\, \mathcal{F}_t\right] \qquad \text{(by equation (3.3))}$$
$$= P(t, t+1)E_Q[f(t+1, S, T) - f(t, S, T) \mid \mathcal{F}_t]$$
$$= 0.$$

Hence we solve

$$E_Q[f(t+1, S, T) - f(t, S, T) \mid \mathcal{F}_t] = 0,$$
$$\Rightarrow \quad f(t, S, T) = E_Q[f(t+1, S, T) \mid \mathcal{F}_t].$$

This formula is useful for recursive calculation of futures prices. However, we can also note the following important corollary.

Corollary 3.13.

$$f(t, S, T) = E_Q[P(S, T) \mid \mathcal{F}_t].$$

Proof. The result is true for $t = S$ since $f(S, S, T) = P(S, T)$ by definition. Suppose the result is true for $t+1, \ldots, S$. Then

$$f(t, S, T) = E_Q[E_Q(P(S, T) \mid \mathcal{F}_{t+1}) \mid \mathcal{F}_t] = E_Q[P(S, T) \mid \mathcal{F}_t].$$

Hence the result is true for all t by induction. $\qquad\square$

This corollary is in contrast to the forward contract under which, denoting the exercise price by K,

$$E_Q\left[\frac{B(t)}{B(S)}(P(S, T) - K) \,\middle|\, \mathcal{F}_t\right] = 0 \quad \Rightarrow \quad K = \frac{P(t, T)}{P(t, S)}.$$

The futures and forward prices are not equal because $P(S, T)$ and $B(t)/B(S)$ are not, in general, independent. (The exception to this in the discrete-time model is when $t = S - 1$.)

Example 3.14. Consider the following random-walk model for the risk-free rate of interest: $r(0) = 0.05$, $\Pr_Q(r(t+1) = r(t) + 0.01 \mid \mathcal{F}_t) = 0.6$ and $\Pr_Q(r(t+1) = r(t) - 0.01 \mid \mathcal{F}_t) = 0.4$.

Consider next the futures contract which delivers at time $S = 2$ the zero-coupon bond which matures at time $T = 3$. We will write $f(t, S, T, r)$ meaning $f(t, S, T)$ when $r(t) = r$ and, likewise, $P(t, T, r)$.

r	$P(2, 3, r)$	$f(2, 2, 3, r)$
0.07	0.932 394	0.932 394
0.05	0.951 229	0.951 229
0.03	0.970 446	0.970 446

Now consider $f(1, 2, 3, r)$. First take $r = 0.06$. We require

$$0 = E_Q[f(2, 2, 3, r(2)) - f(1, 2, 3, r(1)) \mid \mathcal{F}_1]$$
$$= (0.6 \times 0.932\,394 + 0.4 \times 0.951\,229) - f(1, 2, 3, 0.06)$$
$$= 0.939\,928 - f(1, 2, 3, 0.06)$$
$$\Rightarrow \quad f(1, 2, 3, 0.06) = 0.939\,928.$$

Similarly,

$$f(1, 2, 3, 0.04) = 0.6 \times 0.951\,229 + 0.4 \times 0.970\,446 = 0.958\,916,$$

and

$$f(0, 2, 3, 0.05) = 0.6 \times f(1, 2, 3, 0.06) + 0.4 \times f(1, 2, 3, 0.04)$$
$$= 0.947\,523.$$

As a check we can calculate $f(0, 2, 3, 0.05)$ directly using the relation

$$f(t, S, T) = E_Q[P(S, T) \mid \mathcal{F}_t]$$
$$\Rightarrow \quad f(0, 2, 3, 0.05) = 0.6^2 \times 0.932\,394 + 2 \times 0.6 \times 0.4 \times 0.951\,229$$
$$+ 0.4^2 \times 0.970\,446$$
$$= 0.947\,523.$$

Without much difficulty we can find that the prices at time 0 of the zero-coupon bonds are $P(0, 2) = 0.903\,073$ and $P(0, 3) = 0.855\,765$. It follows that the forward price at time 0 for delivery of $P(2, 3)$ at time 2 is $K = P(0, 3)/P(0, 2) = 0.947\,614$. This is slightly higher than the futures price because $B(0)/B(2)$ and $P(2, 3)$ are positively correlated.

3.6 Exercises

Exercise 3.1. In a certain bond market the prices at time 0 of the zero-coupon bonds maturing at times 1 to 6 are as follows:

T	1	2	3	4	5	6
$P(0, T)$	0.96	0.92	0.87	0.82	0.77	0.73

Furthermore, using the notation of Section 3.2, we have a recombining tree with $u(2) = 1.01$ and $d(2) = 0.96$. Throughout the binomial tree the risk-neutral probability, q, of an up-step is constant (that is, it does not depend upon time or the history of the process).

A put option which expires at time 2 has been issued on the zero-coupon bond which matures at time 3. The exercise price of the option is 0.96.

(a) Determine the values of $u(T)$ and $d(T)$ for $T = 3, 4, 5, 6$.

(b) Determine the risk-neutral probability, q, of an up-step at any point in the tree.

(c) For each of the outcomes at time 2 find the payoff on the option and the accumulation of cash up to time 2.

(d) Find the price of the put option at time 0 using risk-neutral expectations.

(e) Find the price of the put option by constructing a replicating strategy which uses (i) the zero-coupon bond maturing at time 3 and (ii) the zero-coupon bond maturing at time 6. Comment on the differences in the two replicating strategies.

(f) Write a program in C++ to solve this problem. Investigate the effects of using different values of $u(2)$ and $d(2)$.

Exercise 3.2. A recombining binomial model for the one-year interest rate r_n (which applies between times n and $n + 1$, with continuous compounding) is as follows.

- $r_0 = 0.06$.

- Given r_n, $r_{n+1} = r_n + 0.005(1 - 2I_{n+1})$, where I_n equals 0 with probability $q = 0.4$ (in the risk-neutral world) and equals 1 otherwise. Thus, $\sum_{i=1}^{n} I_i$ is the number of down-steps in the one-year rate.

(a) Find the prices at time 0 of the zero-coupon bonds that mature at times 1, 2, 3 and 4.

(b) Find the coupon rate ρ_4 (payable annually) for a coupon bond which matures at time 4 and which stands at par at time 0.

(c) An *interest rate swap* is a contract running for N years under which investor A pays a fixed rate of interest, R^*, to investor B at time $n + 1$ in return for a payment from B to A of the variable one-year rate of interest $R_n = e^{r_n} - 1$, with $n = 0, 1, \ldots, N - 1$. (R_n is commonly referred to as the 12-month LIBOR (the London Interbank Offer Rate). See Chapter 9.)

    (i) Show that this contract when $N = 4$ has zero value to each party when $R^* = \rho_4$.

    (ii) What is the value of the contract to investor A if $R^* = 0.06$?

(d) Another contract called a *swaption* gives A *the right but not the obligation* to enter into a swap agreement with B at time 1 under which A pays to B at time $n + 1$ a fixed rate of interest, R^*, in return for a payment from B to A of the variable one-year rate of interest $R_n = e^{r_n} - 1$, with $n = 1, \ldots, N - 1$. Here $N = 4$ and $R^* = 0.06$.

(i) Under what circumstances will A exercise the option at time 1?

(ii) What is the value of this contract to A at time 0?

(iii) Are there any circumstances under which this contract could have a negative value to A?

(e) Investor A holds a convertible zero-coupon bond which will, if no action is taken earlier, pay 1 at time 3. The bond offers A the option to convert the 1 unit of the convertible bond at time 2 into 1.06 units of the (conventional) zero-coupon bond which matures at time 4.

(i) Using the same model for r_t above calculate the three possible values that the underlying bonds and the convertible bond can have at time 2.

(ii) Hence calculate the value of the convertible bond at time 0.

(iii) Find the replicating strategies between times 0 and 2 using cash (that is, $P(t, t+1)$) and $P(t, 3)$.

(iv) Repeat part (iii) using cash and $P(t, 4)$.

Exercise 3.3. Consider the recombining binomial model in Section 3.3. Prove *from first principles* that

$$P(t, T) = \frac{P(0, T)}{P(0, t)} \exp\left[-\sum_{s=t+1}^{T} \log \frac{d(s + 1 - t)}{d(s + 1)} - (T - t)U_t \log k\right],$$

where U_t is the number of up-steps up to time t.

Exercise 3.4. Consider the recombining binomial model in Section 3.3. We have previously proved that

$$F(t, T - 1, T) = F(0, T - 1, T) + \log[u(T - t)/u(T)] - D(t) \log k.$$

Let $l_F(t) = \lim_{T \to \infty} F(t, T - 1, T)$ be the long forward rate.

Derive a formula for $l_F(t)$ in terms of $l_F(0)$, $D(t)$ and k.

Hence deduce that $l_F(t)$ satisfies the necessary condition in the Dybvig–Ingersoll–Ross result for an arbitrage-free term-structure model.

Exercise 3.5. Consider the following random-walk model for r_t:

$$r_{t+1} = r_t + \delta(2I_{t+1} - 1),$$

where $I_1, I_2, \ldots$ are independent and identically distributed under Q with $\Pr_Q(I_t = 1) = q$ and $\Pr_Q(I_t = 0) = 1 - q$.

(a) Use Remark 3.4 and Corollary 3.5 to determine the forward-rate curve at time 0 given r_0, δ and q.

Hence derive the theoretical zero-coupon prices at time 0 for this model.

(b) Verify your formula for zero-coupon bond prices at time 0 by direct calculation of prices using the formula

$$P(0, T) = E_Q\left[\exp\left(-\sum_{s=0}^{T-1} r_s\right) \,\Big|\, \mathcal{F}_0\right].$$

(Hint: prove the result by induction.)

4

Continuous-Time Interest Rate Models

4.1 One-Factor Models for the Risk-Free Rate

We will now consider one-factor models for the term structure of interest rates within a continuous-time framework. In particular, we will consider how to price bonds given a one-factor diffusion model for the risk-free rate, $r(t)$. We will assume that $r(t)$ is an Itô process with stochastic differential equation (SDE)

$$\mathrm{d}r(t) = a(t)\,\mathrm{d}t + b(t)\,\mathrm{d}W(t),$$

where $W(t)$ is a standard Brownian motion under the real-world measure P, $a(t)$ and $b(t)$ are previsible processes, and $\mathcal{F}_t = \sigma(\{W(s) : s \leqslant t\})$ is the sigma-algebra generated by the history of $W(s)$ up to time t.

For the one-factor models we will assume that $a(t) = a(r(t))$ and $b(t) = b(r(t))$ so that the process $r(t)$ is Markov and time homogeneous.

A range of time-homogeneous models have been proposed, including those given in Table 4.1.

There are three basic characteristics which, in varying degrees, are desirable but not essential for the development of a term-structure model.

- Interest rates should be positive. If a model does permit interest rates to become negative, then the technical development assumes that the *money-under-the-mattress* option is not available.

- $r(t)$ should be autoregressive. This feature assumes that $r(t)$ cannot drift off to plus or minus infinity or to zero, but will eventually be pulled back to some long-term target.

- We should get simple formulae for bond prices and for the prices of some derivatives. In contrast to the first two characteristics, this is a matter of computational convenience rather than of economic principle.

With regard to the final point, we should always remember that the existence of elegant formulae do not prove the worth of a model. We must always be prepared to demonstrate that a proposed model gives a good approximation to what we observe

Table 4.1. One-factor, time-homogeneous models for $r(t)$.

Model	$a(r)$	$b(r)$
Merton (1973) (M)	μ	σ
Dothan (1978) (D)	μr	σr
Vasicek (1977) (V)	$\alpha(\mu - r)$	σ
Cox–Ingersoll–Ross (1985) (CIR)	$\alpha(\mu - r)$	$\sigma\sqrt{r}$
Pearson–Sun (1994) (PS)	$\alpha(\mu - r)$	$\sigma\sqrt{r - \beta}$
Brennan–Schwartz (1979) (BS)	$\alpha(\mu - r)$	σr
Black–Karasinski (1991) (BK)	$\alpha r - \gamma r \log r$	σr

Table 4.2. Key characteristics of one-factor models.

Model	$r(t) \geqslant 0$?	Autoregressive?	Simple formulae?
M	N	N	Y
D	Y	N	N
V	N	Y	Y
CIR	Y	Y	Y
PS	Y if $\beta > 0$	Y	N
BS	Y	Y	N
BK	Y	Y ($\gamma > 0$)	N

in reality and that it is appropriate for the task in hand. All models are approximations to reality but, of course, some are better than others.

Table 4.2 shows which of these characteristics are displayed by each of the six models listed in Table 4.1.

With the Cox–Ingersoll–Ross model (see Section 4.6), analytical formulae do exist for bond prices and for European put and call options. However, these are rather more complex than those for other models and include the numerical calculation of percentage points for the non-central chi-squared distribution. This is a more complex procedure than that required for the Vasicek and other models which rely on computation of percentage points for the Normal distribution.

4.1.1 Other Characteristics

Other desirable characteristics of term-structure models include the following.

- Are bond and derivative prices straightforward and simple to calculate numerically? This relaxes the requirement that prices be available using analytical formulae and is a reflection of the existence of increasing computing power.

- Is the model flexible enough to cope with new and more complex derivative products?

- Does the model produce dynamics which are realistic? For example: does it produce features which we have observed in the past; does it produce a range of yield curves both now and in the future which are consistent with what we have observed in the past?

- Does the model fit historical data well (in the statistical sense) or at least adequately?

- If the model does keep interest rates positive, does it allow forward rates, spot rates and par yields to take values arbitrarily close to zero?

- Does the model have an equilibrium derivation? An equilibrium model attempts to mimic the characteristics of the whole of a particular market, including investors with varying appetites for risk. Bond price dynamics (in particular, risk premiums under P) depend upon the range of investors. The simplicity and elegance of equilibrium models can be compromised by the fact that, at any point in time, *theoretical* prices generally are not precisely equal to *observed* prices. No-arbitrage models avoid this problem by using initial observed prices as part of the input, but they lose the benefits of time homogeneity.

One-factor models generally fail many of these criteria because their dependence on a single factor (normally the risk-free rate of interest) makes the model insufficiently flexible and realistic. For this reason models which incorporate more than one factor are often used. These models are discussed in Chapters 6, 8 and 9.

In this chapter and Chapter 5 we concentrate on one-factor models because they form a solid foundation upon which we can build more complex models from a position of understanding. In the current chapter we will look at general approaches to bond pricing which allow us to bridge the gap between the SDE for the risk-free rate of interest and a coherent set of price dynamics that does not admit arbitrage. Sections 4.5–4.9 will apply these results to some of the models described above.

We will develop general pricing formulae in two different ways. First, we will describe pricing using the more modern martingale approach. Second, we will develop the results using the partial differential equation approach of Vasicek (1977).

4.2 The Martingale Approach

We will now describe the more modern approach to the pricing of bonds and interest rate derivatives that uses the theory of martingales to establish prices and hedging strategies. Some of the earliest descriptions of this approach can be found in the papers by Harrison and Kreps (1979) and Harrison and Pliska (1981, 1983). An accessible and more detailed account of the application to the equities market can be found in Baxter and Rennie (1996), amongst other places.

Consider a term-structure model with a one-dimensional Brownian motion as the only source of randomness. Suppose that we have the stochastic differential equations (SDEs) for the risk-free rate of interest, $r(t)$, and the price at time t of a zero-coupon bond which matures at time T, $P(t, T)$, then

$$dr(t) = a(t)\,dt + b(t)\,dW(t), \tag{4.1}$$

$$dP(t, T) = P(t, T)[m(t, T)\,dt + S(t, T)\,dW(t)], \tag{4.2}$$

where $a(t), b(t), m(t, T)$ and $S(t, T)$ are previsible functions (so they are potentially stochastic). We call $S(t, T)$ the volatility of the bond $P(t, T)$. Although we will treat $S(t, T)$ as an essentially arbitrary (but non-zero) function, its form is, in fact, dictated by the model for $r(t)$. A summary of the relevant theory of Brownian motion and SDEs is given in Appendix A.

Associated with these processes we have the risk-free cash account, $B(t)$, which satisfies the SDE

$$dB(t) = r(t)B(t)\,dt,$$

to which the solution is

$$B(t) = B(0) \exp\left[\int_0^t r(u)\,du\right].$$

The *risk premium* at time t on the risky bond $P(t, T)$ is defined as the excess expected rate of return on the bond, $m(t, T)$, over the risk-free rate of interest, $r(t)$. The risk premium represents the extra reward we get for investing in the risky asset rather than a risk-free cash account. Associated with this we define the *market price of risk* to be the previsible process $\gamma(t) = (m(t, T) - r(t))/S(t, T)$. The market price of risk represents the excess expected return per unit of volatility.

Consider, now, an interest rate derivative contract which pays X_S (which is $\mathcal{F}_S$-measurable) at some time $S < T$. What is the fair (no-arbitrage) price, $V(t)$, at time $t < S$ for this contract, given equations (4.1) and (4.2)?

We get the following result.

Theorem 4.1. *There exists a measure Q equivalent to P with*

$$V(t) = E_Q\left[\exp\left(-\int_t^S r(u)\,du\right)X_S \,\bigg|\, \mathcal{F}_t\right],$$

where $dr(t) = (a(t) - \gamma(t)b(t))\,dt + b(t)\,d\tilde{W}(t)$ and $\tilde{W}(t)$ is a standard Brownian motion under Q.

Proof. First, define the discounted price process

$$Z(t, T) = \frac{P(t, T)}{B(t)} = P(t, T)\exp\left(-\int_0^t r(u)\,du\right).$$

We now break the proof up into five steps.

Step 1. *Establish the probability measure Q equivalent to P under which the discounted price process, $Z(t, T)$, is a martingale.*

By the product rule for stochastic differential equations (see Corollary A.8) we have

$$\mathrm{d}Z(t, T) = B(t)^{-1}\,\mathrm{d}P(t, T) + P(t, T)\,\mathrm{d}(B(t)^{-1}) + \mathrm{d}\langle B^{-1}, P\rangle(t),$$

but

$$\mathrm{d}(B(t)^{-1}) = -\frac{1}{B(t)^2}\,\mathrm{d}B(t) + \frac{1}{2}\frac{2}{B(t)^3}\,\mathrm{d}\langle B\rangle(t) \qquad \text{(by Itô's formula)}$$

$$= -\frac{r(t)\,\mathrm{d}t}{B(t)} \qquad \text{(since } \mathrm{d}\langle B\rangle(t) = 0.\mathrm{d}t\text{)},$$

$$\Rightarrow \quad \mathrm{d}Z(t, T) = \frac{P(t, T)}{B(t)}(m(t, T)\,\mathrm{d}t + S(t, T)\,\mathrm{d}W(t)) - \frac{r(t)P(t, T)\,\mathrm{d}t}{B(t)} + 0.\mathrm{d}t$$

$$= Z(t, T)[(m(t, T) - r(t))\,\mathrm{d}t + S(t, T)\,\mathrm{d}W(t)].$$

Now recall that $\gamma(t) = (m(t, T) - r(t))/S(t, T)$. Let us also define a new process $\tilde{W}(t) = W(t) + \int_0^t \gamma(u)\,\mathrm{d}u$. Then we have

$$\mathrm{d}Z(t, T)$$
$$= Z(t, T)[(m(t, T) - r(t) - \gamma(t)S(t, T))\,\mathrm{d}t + S(t, T)(\mathrm{d}W(t) + \gamma(t)\,\mathrm{d}t)]$$
$$= Z(t, T)S(t, T)\,\mathrm{d}\tilde{W}(t). \tag{4.3}$$

Provided $\gamma(s)$ satisfies the *Novikov* condition

$$E_P\left[\exp\left(\frac{1}{2}\int_0^T \gamma(u)^2\,\mathrm{d}u\right)\right] < \infty,$$

we can apply the Girsanov Theorem (see Theorem A.12). This allows us to state that there exists a measure Q equivalent to P with Radon–Nikodým derivative

$$\frac{\mathrm{d}Q}{\mathrm{d}P} = \exp\left(-\int_0^T \gamma(u)\,\mathrm{d}W(u) - \frac{1}{2}\int_0^T \gamma(u)^2\,\mathrm{d}u\right)$$

and under which $\tilde{W}(t)$ is a standard Brownian motion.

It is important to note that under the same change of measure we have

$$\mathrm{d}P(t, T) = P(t, T)[r(t)\,\mathrm{d}t + S(t, T)\,\mathrm{d}\tilde{W}(t)]. \tag{4.4}$$

This is something which we always observe for diffusion models (one factor or multifactor) and for all tradable assets (not just zero-coupon bonds). In particular, under Q, the prices of all tradable assets have drift equal to the current price times the risk-free rate.

Now we note that the SDE for $Z(t, T)$ under Q (equation (4.3)) has zero drift (that is, no dt term). It follows that $Z(t, T)$ is a martingale under Q if one of the following (sufficient) technical conditions is satisfied:

$$E_Q\left[\left(\int_0^T S(t, T)^2 Z(t, T)^2 \, \mathrm{d}t\right)^{1/2}\right] < \infty$$

or

$$E_Q\left[\exp\left(\frac{1}{2}\int_0^T S(t, T)^2 \, \mathrm{d}t\right)\right] < \infty.$$

(A *necessary* condition for $Z(t, T)$ to be a martingale is that it has zero drift in the SDE.)

Step 2. *For $t < S < T$ define $D(t) = E_Q[B(S)^{-1}X_S \mid \mathcal{F}_t]$.* This is a martingale under Q by the Tower Property for conditional expectations.

Step 3. *Since $Z(t, T)$ and $D(t)$ are both Q-martingales, by the Martingale Representation Theorem (see Theorem A.10) there exists a previsible process $\phi(t)$ such that*

$$D(t) = D(0) + \int_0^t \phi(u) \, \mathrm{d}Z(u, T).$$

Note that this requires $S(t, T)$ to be non-zero for all $t < S$ almost surely.

Step 4. *Define $\psi(t) = D(t) - \phi(t)Z(t, T)$. Now suppose that we employ the portfolio strategy which holds $\phi(t)$ units of $P(t, T)$ and $\psi(t)$ units of $B(t)$ at time t. This portfolio can be shown to be self-financing.*

The value at time t of this portfolio is

$$V(t) = \phi(t)P(t, T) + \psi(t)B(t) = B(t)[\phi(t)Z(t, T) + \psi(t)] = B(t)D(t).$$

Now consider the infinitesimal interval $[t, t + \mathrm{d}t)$. Over this period we hold $\phi(t)$ units of $P(s, T)$ and $\psi(t)$ units of $B(s)$. The change in the value of the portfolio (the *instantaneous investment gain*) is then

$$\phi(t) \, \mathrm{d}P(t, T) + \psi(t) \, \mathrm{d}B(t).$$

The corresponding instantaneous change in the portfolio value from t to $t + \mathrm{d}t$ is

$$\begin{aligned}
\mathrm{d}V(t) &= \mathrm{d}[B(t)D(t)] \\
&= B(t) \, \mathrm{d}D(t) + D(t) \, \mathrm{d}B(t) + \mathrm{d}B(t) \, \mathrm{d}D(t) \\
&= B(t)\phi(t) \, \mathrm{d}Z(t, T) + D(t)r(t)B(t) \, \mathrm{d}t + 0.\mathrm{d}t \\
&= \phi(t)B(t)S(t, T)Z(t, T) \, \mathrm{d}\tilde{W}(t) + (\phi(t)Z(t, T) + \psi(t))r(t)B(t) \, \mathrm{d}t \\
&= \phi(t)P(t, T)(r(t) \, \mathrm{d}t + S(t, T) \, \mathrm{d}\tilde{W}(t)) + \psi(t)r(t)B(t) \, \mathrm{d}t \\
&= \phi(t) \, \mathrm{d}P(t, T) + \psi(t) \, \mathrm{d}B(t). \tag{4.5}
\end{aligned}$$

This is equal to the instantaneous investment gain over the same period, so the portfolio process (that is, the hedging strategy) is self-financing.

Step 5. *We also have* $V(S) = B(S)E_Q[B(S)^{-1}X_S \mid \mathcal{F}_S] = X_S$. *This implies not only that the portfolio process is self-financing but also that it replicates the derivative payoff. It follows that, for $t < S$, $V(t)$ is the unique no-arbitrage price at time t for X_S payable at S.*

Note finally that

$$V(t) = B(t)D(t) = E_Q\left[\frac{B(t)}{B(S)}X_S \,\middle|\, \mathcal{F}_t\right]$$

$$= E_Q\left[\exp\left(-\int_t^S r(u)\,\mathrm{d}u\right)X_S \,\middle|\, \mathcal{F}_t\right]$$

and the proof is complete. □

Note that the uniqueness of the price $V(t)$ stems from the following points.

- We are considering a one-factor model.
- We specified the real-world price dynamics for one zero-coupon bond maturing at time T. This allows us to establish the market price of risk.
- The use of Brownian motion ensures that we have a complete market. With other processes driving changes in $r(t)$ (e.g. jump processes), $V(t)$ may not be uniquely determined given the real-world SDEs for $r(t)$ and $P(t, T)$ only.

Corollary 4.2. *For all S such that $0 < S < T$,*

$$P(t, S) = E_Q\left[\exp\left(-\int_t^S r(u)\,\mathrm{d}u\right) \,\middle|\, \mathcal{F}_t\right].$$

Proof. From the previous theorem take $X_S = 1$ and the result follows immediately. □

Remark 4.3. The general theory extends simply to diffusion models driven by m independent Brownian motions $W_1(t), \ldots, W_m(t)$. In particular, a unique no-arbitrage price can still be established by constructing a suitable portfolio process involving m risky bonds, $P(t, T_k)$ for $k = 1, \ldots, m$ (and $S < T_1 < \cdots < T_m$), plus the risk-free cash account $B(t)$. The price for the derivative is the same as that given in Theorem 4.1.

Remark 4.4. Recall equation (4.5). From this we can see that

$$\begin{aligned}
\mathrm{d}V(t) &= \phi(t)P(t, T)[r(t)\,\mathrm{d}t + S(t, T)\,\mathrm{d}\tilde{W}(t)] + \psi(t)B(t)r(t)\,\mathrm{d}t \\
&= [\phi(t)P(t, T) + \psi(t)B(t)]r(t)\,\mathrm{d}t + \phi(t)P(t, T)S(t, T)\,\mathrm{d}\tilde{W}(t) \\
&= V(t)[r(t)\,\mathrm{d}t + \sigma_V(t)\,\mathrm{d}\tilde{W}(t)],
\end{aligned}$$

where
$$V(t)\sigma_V(t) = \phi(t)P(t, T)S(t, T).$$

So we see that, under Q, the prices of all tradable assets have the risk-free rate of interest as the expected growth rate. Now consider the price dynamics under the real-world measure P. For a general tradable asset we have

$$dV(t) = V(t)[r(t)\,dt + \sigma_V(t)(dW(t) + \gamma(t)\,dt)]$$
$$= V(t)[(r(t) + \gamma(t)\sigma_V(t))\,dt + \sigma_V(t)\,dW(t)].$$

Definition 4.5. The excess expected growth rate under P on the bond or derivative, $\gamma(t)\sigma_V(t)$, is called the *market risk premium* or simply *risk premium*.

Because of its dependence on the market price of risk, $\gamma(t)$, we can see that the risk premiums on different assets are closely linked. In particular, they can differ (in a one-factor model) only through the volatility in the tradable asset (for example, $\sigma_V(t)$ in the derivative or $S(t, T)$ for a zero-coupon bond). In general, we anticipate that zero-coupon bonds will have a positive risk premium (that is, $\gamma(t)S(t, T) > 0$ for all $T > t$) to reward investors for the extra risk they are taking on. It follows that derivatives, $V(t)$, for which $\sigma_V(t)$ has the same sign as $S(t, T)$ (for example, call options on $P(t, T)$) also have a positive risk premium. On the other hand, if $\sigma_V(t)$ has the opposite sign (for example, a put option on $P(t, T)$) then the derivative will have a negative risk premium.

4.3 The PDE Approach to Pricing

We will now present the general approach taken by Vasicek (1977). Vasicek's paper is, perhaps, better known for the specific model described in Section 4.5. However, the earlier stages of his paper present a general partial differential equation (PDE) approach which is similar to the equity-derivative approach developed by Black and Scholes (1973). Although the martingale approach is generally thought to be the more powerful and intuitive, the PDE approach still provides us with a useful tool for the development of numerical methods (see Chapter 10).

The general principles in this development are that

- $r(t)$ is Markov;
- prices $P(t, T)$ depend upon an assessment at time t of how $r(s)$ will vary between t and T;
- the market is efficient, without transaction costs and all investors are rational.

The first two principles ensure that $a(t) = a(t, r(t))$, $b(t) = b(t, r(t))$, and $P(t, T) = P(t, T, r(t))$. Thus, under a one-factor model, price changes for all bonds with different maturity dates are perfectly (but non-linearly) correlated.

Now let us consider $P(t, T)$. By Itô's formula

$$\begin{aligned}
dP &= \frac{\partial P}{\partial t}\,dt + \frac{\partial P}{\partial r}\,dr + \frac{1}{2}\frac{\partial^2 P}{\partial r^2}\,d\langle r\rangle \\
&= \frac{\partial P}{\partial t}\,dt + \frac{\partial P}{\partial r}(a\,dt + b\,dW) + \frac{1}{2}\frac{\partial^2 P}{\partial r^2}b^2\,dt \\
&= \left[\frac{\partial P}{\partial t} + a\frac{\partial P}{\partial r} + \tfrac{1}{2}b^2\frac{\partial^2 P}{\partial r^2}\right]dt + b\frac{\partial P}{\partial r}\,dW
\end{aligned}$$

or

$$dP = P(t, T, r)[m(t, T, r)\,dt + S(t, T, r)\,dW],$$

where

$$m(t, T, r) = \frac{1}{P}\left[\frac{\partial P}{\partial t} + a\frac{\partial P}{\partial r} + \tfrac{1}{2}b^2\frac{\partial^2 P}{\partial r^2}\right],$$

$$S(t, T, r) = \frac{1}{P}b\frac{\partial P}{\partial r} \qquad \text{(the volatility of the price).} \tag{4.6}$$

We can now begin to consider the *market price of risk*. Consider two bonds maturing at times T_1 and T_2 (with $T_1 < T_2$). At time t suppose that we hold amounts $-V_1(t)$ in the T_1-bond (a short position) and $V_2(t)$ in the T_2-bond (a long position). The total wealth is $V(t) = V_2(t) - V_1(t)$. We will vary $V_1(t)$ and $V_2(t)$ in such a way that the portfolio is risk-free. The instantaneous investment gain from t to $t + dt$ is

$$\begin{aligned}
-\frac{V_1(t)}{P(t, T_1)}\,dP(t, T_1) &+ \frac{V_2(t)}{P(t, T_2)}\,dP(t, T_2) \\
&= -V_1(t)(m_1\,dt + S_1\,dW) + V_2(t)(m_2\,dt + S_2\,dW) \\
&= (V_2 m_2 - V_1 m_1)\,dt + (V_2 S_2 - V_1 S_1)\,dW,
\end{aligned}$$

where, for notational compactness, we write

$$m_i = m(t, T_i, r(t)) \quad \text{and} \quad S_i = S(t, T_i, r(t)) \quad \text{for } i = 1, 2.$$

Furthermore, suppose that, for all t,

$$\frac{V_1(t)}{V_2(t)} = \frac{S(t, T_2, r(t))}{S(t, T_1, r(t))} = \frac{S_2}{S_1}.$$

Then

$$V_2 S_2 - V_1 S_1 = 0$$

and

$$V_2 m_2 - V_1 m_1 = \frac{S_1 V}{S_1 - S_2}m_2 - \frac{S_2 V}{S_1 - S_2}m_1.$$

Hence, the instantaneous investment gain is equal to

$$V\left(\frac{m_2 S_1 - m_1 S_2}{S_1 - S_2}\right) dt.$$

Thus, through our choice of portfolio strategy, we have a risk-free investment strategy. Furthermore, by varying $V_1(t)$ in an appropriate way, we can ensure that this portfolio is also self-financing: that is,

$$dV = V\left(\frac{m_2 S_1 - m_1 S_2}{S_1 - S_2}\right) dt.$$

Since this portfolio is risk-free, the principle of no arbitrage dictates that the portfolio growth rate must equal $r(t)$; that is,

$$\left(\frac{m_2 S_1 - m_1 S_2}{S_1 - S_2}\right) = r(t) \quad \text{or} \quad \frac{m_1 - r}{S_1} = \frac{m_2 - r}{S_2}.$$

This must be true for all maturities. Thus, for all $T > t$,

$$\frac{m(t, T, r(t)) - r(t)}{S(t, T, r(t))} = \gamma(t, r(t)),$$

where $\gamma(t, r(t))$ is the *market price of risk*; that is, the extra return over $r(t)$ per unit of risk. The key observation here is that $\gamma(\cdot)$ cannot depend on the maturity date T.

(Depending on how a model is parametrized, $\gamma(t, r(t))$ can often be negative! This follows from the observations that, first, $P(t, T, r(t))$ is usually a decreasing function of $r(t)$. Second, suppose the volatility, $b(t, r(t))$, of $r(t)$ is positive. Then we have $S(t, T, r(t)) < 0$. Thus, $\gamma(t, r(t))$ must be negative to ensure that expected returns under P, $m(t, T, r(t))$, are greater than the risk-free rate, $r(t)$.)

So we have

$$m(t, T, r) = r(t) + \gamma(t, r) S(t, T, r)$$

and (recapping equation (4.6))

$$m(t, T, r) = \frac{1}{P}\left[\frac{\partial P}{\partial t} + a\frac{\partial P}{\partial r} + \tfrac{1}{2}b^2\frac{\partial^2 P}{\partial r^2}\right],$$

$$S(t, T, r) = \frac{1}{P}b\frac{\partial P}{\partial r}.$$

If we equate the two expressions for $m(t, T, r)$, we find that

$$\frac{\partial P}{\partial t} + (a - b.\gamma)\frac{\partial P}{\partial r} + \tfrac{1}{2}b^2\frac{\partial^2 P}{\partial r^2} - rP = 0. \tag{4.7}$$

This is of a suitable form to allow us to apply the Feynman–Kac formula (Theorem A.9)

$$\frac{\partial P}{\partial t} + f(t, r)\frac{\partial P}{\partial r} + \tfrac{1}{2}\rho^2(t, r)\frac{\partial^2 P}{\partial r^2} - R(r)P + h(t, r) = 0,$$

where $f(t, r) = a(t, r) - \gamma(t, r)b(t, r)$, $\rho(t, r) = b(t, r)$, $R(r) = r$ and $h(t, r) = 0$. The boundary condition for this PDE is $P(T, T, r) = \psi(r) = 1$ for all T, r.

By the Feynman–Kac formula there exists a suitable probability triple $(\Omega, \mathcal{F}, Q)$ with filtration $\{\mathcal{F}_t : 0 \leqslant t < \infty\}$ under which

$$P(t, T, r(t)) = E_Q \left[\exp\left(-\int_t^T \tilde{r}(s)\, ds\right) \,\bigg|\, \mathcal{F}_t \right]. \tag{4.8}$$

The process $\tilde{r}(s)$ $(t \leqslant s \leqslant T)$ is a Markov diffusion process with $\tilde{r}(t) = r(t)$ and, under the measure Q, $\tilde{r}(u)$ satisfies the SDE

$$d\tilde{r}(u) = f(u, \tilde{r}(u))\, du + \rho(u, \tilde{r}(u))\, d\tilde{W}(u), \tag{4.9}$$

where $\tilde{W}(u)$ is a standard Brownian motion under Q.

Now we also have the original model $dr(t) = a(t)\, dt + b(t)\, dW(t)$, where $W(t)$ is a Brownian motion under P. We need to satisfy ourselves that the two measures P and Q are equivalent to complete the story. Suppose that

- $\gamma(s, r(s))$ satisfies the Novikov condition

$$E_Q \left[\exp\left(\frac{1}{2}\int_0^T \gamma(s, r(s))^2\, ds\right) \right] < \infty;$$

- we define $\tilde{W}(t) = W(t) + \int_0^t \gamma(s, r(s))\, ds$.

Then, by the Girsanov Theorem, there exists an equivalent measure Q under which $\tilde{W}(t)$ (for $0 \leqslant t \leqslant T$) is a Brownian motion and with Radon–Nikodým derivative

$$\frac{dQ}{dP} = \exp\left[-\int_0^T \gamma(t, r(t))\, dW(t) - \frac{1}{2}\int_0^T \gamma(t, r(t))^2\, dt \right].$$

Note that we then have

$$
\begin{aligned}
dP(t, T) &\\
&= P(t, T)(m(t, T, r(t))\, dt + S(t, T, r(t))\, dW) \\
&= P(t, T)[m(t, T, r(t))\, dt + S(t, T, r(t))\{d\tilde{W} - \gamma(t, r(t))\, dt\}] \\
&= P(t, T)[\{m(t, T, r(t)) - \gamma(t, r(t))S(t, T, r(t))\}\, dt + S(t, T, r(t))\, d\tilde{W}] \\
&= P(t, T)(r(t)\, dt + S(t, T, r(t))\, d\tilde{W}).
\end{aligned}
$$

Thus, under Q, the expected return on any bond is equal to the risk-free rate. Hence Q is the risk-neutral equivalent measure.

The Feynman–Kac formula can be applied to interest rate derivative contracts. We continue the argument as follows. Let $V(t)$ be the price at time t of a derivative which will have a payoff to the holder of $\psi(r(T))$ at time T (which could be described as a function of $P(T, S, r(T))$ if the underlying quantity is $P(t, S, r(t)))$. As above we will have

$$\frac{\partial V}{\partial t} + f(t, r)\frac{\partial V}{\partial r} + \frac{1}{2}\rho^2(t, r)\frac{\partial^2 V}{\partial r^2} - R(r)V = 0, \quad \text{subject to } V(T) = \psi(r),$$

(4.10)

where $f(t, r) = a(t, r) - \gamma(t, r)b(t, r)$, $\rho(t, r) = b(t, r)$ and $R(r) = r$. Again by the Feynman–Kac formula we have

$$V(t) = E_Q\left[\exp\left(-\int_t^T \tilde{r}(s)\,\mathrm{d}s\right)\psi(\tilde{r}(T))\ \middle|\ \mathcal{F}_t\right],$$

(4.11)

where $\tilde{r}(s)$ is as in equation (4.9). So the only difference in the PDE problem when compared with the zero-coupon-bond case is in the boundary condition.

These formulae are, of course, the same as those in Theorem 4.1 and Corollary 4.2.

4.4 Further Comment on the General Results

Remark 4.6. We have developed these results by specifying first the dynamics of the model under P before transferring to the equivalent measure Q. In practice, modellers typically start by specifying the dynamics under Q directly. This immediately gives us the relevant pricing formulae provided we know the parameter values.

Knowledge of the dynamics under P is not always required but, if they are, the market price of risk, $\gamma(t)$, can then be introduced at this stage. If this approach is taken, modellers must be confident that $\gamma(t)$ satisfies the Novikov condition.

4.5 The Vasicek Model

In the previous sections we have developed the general framework within which all one-factor models must fit. We will now look at some specific models, look at their properties individually and make some simple comparisons between them. We have noted before that for some applications (more complex or longer-dated derivatives) these one-factor models may not be suitable. However, our aim here is to give readers a good understanding of the basic principles of term-structure modelling. Furthermore, it is important to learn how to investigate the characteristics of individual models in order to come to a well-informed decision about a model's suitability for a particular application. These aims are best served by looking at the well-known, tractable one-factor models. Perhaps the best known of these models was proposed by Vasicek (1977).

Vasicek proposed the following model for the risk-free rate of interest, $r(t)$, based on the SDE

$$\mathrm{d}r(t) = \alpha(\mu - r(t))\,\mathrm{d}t + \sigma\,\mathrm{d}\tilde{W}(t),$$

where $\tilde{W}(t)$ is a standard Brownian motion under the risk-neutral measure Q, and α, μ and σ are all strictly positive constants. This process is well known to experts in stochastic differential equations under the name of the Ornstein–Uhlenbeck process (see, for example, Øksendal 1998). Its properties are well known but the key feature in the present context is its mean-reverting structure. In the model:

- μ represents the *risk-neutral* long-term mean risk-free rate;
- α represents the rate at which $r(t)$ reverts back to this long-term mean;
- σ represents the local volatility of short-term interest rates.

In more detail, for $s > 0$, $r(t + s)$ given $r(t)$ is Normally distributed under Q with mean $\mu + (r(t) - \mu)e^{-\alpha s}$ and variance $\sigma^2[1 - e^{-2\alpha s}]/2\alpha$. This implies that the long-term standard-deviation of $r(t)$ is $\sigma/\sqrt{2\alpha}$.

4.5.1 Bond and Option Prices

Let us now apply the results of Theorem 4.1 to derive prices for zero-coupon bonds and call options. We will state within this chapter the respective pricing formulae in Theorem 4.7. A detailed (and lengthy!) proof of this result is given in Appendix B. A key aspect of the proof is the establishment of the joint distribution under Q of $\int_t^T r(s)\,ds$ and $r(T)$ given $r(t)$ (Lemma B.1).

Theorem 4.7.

(a) *Prices for zero-coupon bonds are given by*

$$P(t, T) = \exp[A(t, T) - B(t, T)r(t)],$$

where

$$B(t, T) = \frac{1 - e^{-\alpha(T-t)}}{\alpha},$$

$$A(t, T) = (B(t, T) - (T - t))\left(\mu - \frac{\sigma^2}{2\alpha^2}\right) - \frac{\sigma^2}{4\alpha}B(t, T)^2.$$

(b) *The price of a European call option with the zero-coupon bond which matures at time S as the underlying security with strike price K and exercise date T (with $T < S$) is*

$$V(t) = P(t, S)\Phi(d_1) - KP(t, T)\Phi(d_2),$$

where

$$d_1 = \frac{1}{\sigma_p}\log\frac{P(t, S)}{KP(t, T)} + \frac{\sigma_p}{2}, \qquad d_2 = d_1 - \sigma_p,$$

$$\sigma_p = \frac{\sigma}{\alpha}(1 - e^{-\alpha(S-T)})\sqrt{\frac{1 - e^{-2\alpha(T-t)}}{2\alpha}},$$

and $\Phi(z)$ is the cumulative distribution function of a standard normal random variable.

Proof. See Section B.1.   □

4.6   The Cox–Ingersoll–Ross Model

4.6.1   Introduction

It is evident that the Vasicek model has a significant failing; namely, that the risk-free rate of interest can become negative. Indeed, all spot rates and forward rates for finite maturity can become negative. From a theoretical standpoint this is a rather unsatisfactory state of affairs: we seem to be sacrificing a key requirement for a term structure for the sake of mathematical convenience. In practice, one can argue that if the probabilities of $r(t)$ becoming negative are small (either because the timescale is short or because the volatility of $r(t)$ is small), then the effect of introducing a minimum barrier of zero into the Vasicek model will be minimal. However, under many circumstances and realistic parameter sets the probabilities of negative rates can be significant. An additional drawback of the Vasicek model is that empirical evidence (see, for example, Chan et al. 1992) suggests that the volatility of $r(t)$ is not constant but is, instead, an increasing function of $r(t)$.

The first tractable model for $r(t)$ which keeps rates of interest positive was proposed by Cox, Ingersoll and Ross (1985) (CIR). They considered the following one-factor model for the risk-free rate of interest:

$$dr(t) = \alpha(\mu - r(t))\,dt + \sigma\sqrt{r(t)}\,d\tilde{W}(t),$$

where $\alpha, \mu, \sigma > 0$ and $\tilde{W}(t)$ is a standard Brownian motion under the risk-neutral measure Q.

This section introduces a number of interesting new concepts which we will deal with shortly in order to prove the following results.

We include within the statement of the theorem the Laplace transform for the joint random variables $(\int_t^T r(s)\,ds, r(T))$ given $r(t)$:

$$P_{\mathrm{L}}(t, T, r, v, \omega) = E_Q\left[\exp\left(-v\int_t^T r(s)\,ds - \omega r(T)\right)\,\middle|\, r(t) = r\right].$$

When $v = 1$ and $\omega = 0$ we abbreviate this from $P_{\mathrm{L}}(t, T, r, 1, 0)$ to the zero-coupon-bond price $P(t, T, r)$. This is a key step in developing the subsequent formulae for both zero-coupon bond prices and prices of European call options on zero-coupon bonds.

Theorem 4.8.

(a) *The Laplace transform takes the affine form*

$$P_L(t, T, r, v, \omega) = \exp[A(t, T, v, \omega) - B(t, T, v, \omega)r],$$

where

$$A(t, T, v, \omega) = \frac{2\alpha\mu}{\sigma^2} \log\left(\frac{2\gamma(v)e^{(\gamma(v)+\alpha)(T-t)/2}}{(\sigma^2\omega + \gamma(v) + \alpha)(e^{\gamma(v)(T-t)} - 1) + 2\gamma(v)}\right),$$

$$\gamma(v) = \sqrt{\alpha^2 + 2\sigma^2 v},$$

$$B(t, T, v, \omega) = \frac{\omega(2\gamma(v) + (\gamma(v) - \alpha)(e^{\gamma(v)(T-t)} - 1)) + 2v(e^{\gamma(v)(T-t)} - 1)}{(\sigma^2\omega + \gamma(v) + \alpha)(e^{\gamma(v)(T-t)} - 1) + 2\gamma(v)}.$$

$$(4.12)$$

(b) *Hence (by taking $v = 1$ and $\omega = 0$)*

$$P(t, T, r) = \exp[\bar{A}(T - t) - \bar{B}(T - t)r],$$

where

$$\bar{A}(\tau) = \frac{2\alpha\mu}{\sigma^2} \log\left(\frac{2\gamma e^{(\gamma+\alpha)\tau/2}}{(\gamma + \alpha)(e^{\gamma\tau} - 1) + 2\gamma}\right),$$

$$\gamma = \sqrt{\alpha^2 + 2\sigma^2},$$

$$\bar{B}(\tau) = \frac{2(e^{\gamma\tau} - 1)}{(\gamma + \alpha)(e^{\gamma\tau} - 1) + 2\gamma}.$$

$$(4.13)$$

(c) *Given $r(0) = r$, $r(T)/k_Q$ has a non-central chi-squared distribution under Q with $d = 4\alpha\mu/\sigma^2$ degrees of freedom, and non-centrality parameter λ_Q, where*

$$\lambda_Q = \frac{4\alpha r(0)}{\sigma^2(e^{\alpha T} - 1)} \quad \text{and} \quad k_Q = \frac{\sigma^2(1 - e^{-\alpha T})}{4\alpha}. \quad (4.14)$$

(d) *Given $r(0) = r > 0$, let $U = \inf\{t : r(t) \leqslant 0\}$ (where $\inf \emptyset = \infty$). Then*

$$2\alpha\mu \geqslant \sigma^2 \Rightarrow \quad \Pr(U = \infty) = 1 \quad \text{and} \quad 2\alpha\mu < \sigma^2 \Rightarrow \quad \Pr(U < \infty) = 1.$$

$$(4.15)$$

(e) *Let C be the price at time 0 of a European call option on the zero-coupon bond maturing at time $U = T + \tau$ with an exercise date T and an exercise price K. Then, given $r(0) = r$,*

$$C = P(0, T + \tau, r)\chi^2(d, \lambda_1; y_1) - KP(0, T, r)\chi^2(d, \lambda_2; y_2), \quad (4.16)$$

where $\chi^2(d, \lambda; y)$ *is the cumulative distribution function of the non-central chi-squared distribution with d degrees of freedom and non-centrality parameter* λ. *The required inputs* d, λ_1, λ_2, y_1 *and* y_2 *are calculated as follows:*

$$d = \frac{4\alpha\mu}{\sigma^2},$$

$$\gamma = \sqrt{\alpha^2 + 2\sigma^2},$$

$$\lambda_1 = \frac{8\gamma^2 e^{\gamma T} r}{\sigma^2 (e^{\gamma T} - 1)(2\gamma + (\gamma + \alpha + \sigma^2 \bar{B}(U - T))(e^{\gamma T} - 1))},$$

$$\lambda_2 = \frac{8\gamma^2 e^{\gamma T} r}{\sigma^2 (e^{\gamma T} - 1)(2\gamma + (\gamma + \alpha)(e^{\gamma T} - 1))},$$

$$k_1 = \frac{\sigma^2 (e^{\gamma T} - 1)}{2(2\gamma + (\gamma + \alpha + \sigma^2 \bar{B}(U - T))(e^{\gamma T} - 1))},$$

$$k_2 = \frac{\sigma^2 (e^{\gamma T} - 1)}{2(2\gamma + (\gamma + \alpha)(e^{\gamma T} - 1))},$$

$$r^* = \frac{\bar{A}(U - T) - \log K}{\bar{B}(U - T)},$$

$$y_1 = r^*/k_1,$$

$$y_2 = r^*/k_2.$$

These results are proved in detail in Section B.2. However, it is useful for readers to consider some related probability and statistics before moving onto the detailed proofs in the appendix. This will give the reader more of an intuitive feel for the CIR model.

4.6.2 *Multi-Dimensional Ornstein–Uhlenbeck Processes*

Suppose that $X_1(t)$, $X_2(t)$, ..., $X_d(t)$ are d independent Ornstein–Uhlenbeck processes with

$$dX_i(t) = -\tfrac{1}{2}\alpha X_i(t)\, dt + \sqrt{\alpha}\, dW_i(t),$$

where the $W_i(t)$ are independent standard Brownian motions. The use here of a volatility of $\sqrt{\alpha}$ and a mean-reversion rate of $-\tfrac{1}{2}\alpha$ will seem unusual at this point. We will see shortly, though, why this is a helpful parametrization.

The properties of the $X_i(t)$ are well known; that is, $X_i(t)$ has a Normal distribution with mean $X_i(0)e^{-\alpha t/2}$ and variance $1 - e^{-\alpha t}$.

Consider the vector $X(t) = (X_1(t), \ldots, X_d(t))'$. The squared radius of this vector process is

$$R(t) = \sum_{i=1}^{d} X_i(t)^2. \tag{4.17}$$

Then

$$dR(t) = \sum_{i=1}^{d} (2X_i(t)\, dX_i(t) + d\langle X_i \rangle(t))$$

$$= -\alpha \sum_{i=1}^{d} X_i(t)^2\, dt + 2 \sum_{i=1}^{d} X_i(t)\sqrt{\alpha}\, dW_i(t) + d\alpha\, dt$$

$$\stackrel{\mathcal{D}}{=} \alpha(d - R(t))\, dt + \sqrt{4\alpha R(t)}\, d\tilde{W}(t)$$

for another standard, one-dimensional Brownian motion $\tilde{W}(t)$.

The CIR process is defined as

$$dr(t) = \alpha(\mu - r(t))\, dt + \sigma\sqrt{r(t)}\, d\tilde{W}(t).$$

Thus, if we take $\theta = 4\alpha/\sigma^2$ and $d = 4\alpha\mu/\sigma^2$, we have

$$r(t) = R(t)/\theta. \tag{4.18}$$

We note, of course, that this representation of $r(t)$ is only valid whenever $d = 4\alpha\mu/\sigma^2$ is a positive integer. However, the representation gives us a useful geometric interpretation of the CIR process.

We need to think next about the distribution of $r(t)$.

4.6.3 The Non-Central Chi-Squared Distribution

Here we consider the definition and relevant properties of the non-central chi-squared distribution. A more comprehensive discussion of this distribution including computational issues can be found in Johnson, Kotz and Balakrishnan (1995).

We start by considering the case where there is an integer number of degrees of freedom. Let $W_1, W_2, \ldots, W_d$ be d independent and identically distributed standard normal random variables and $\delta_1, \delta_2, \ldots, \delta_d$ be d real numbers of any size.

It is well known that $\sum_{i=1}^{d} W_i^2$ has a chi-squared distribution with d degrees of freedom. Now consider

$$R = \sum_{i=1}^{d} (W_i + \delta_i)^2.$$

Then we say that R has a non-central chi-squared distribution with d degrees of freedom and non-centrality parameter

$$\lambda = \sum_{i=1}^{d} \delta_i^2.$$

It is important to note that the distribution of R depends upon the δ_i only through λ. Given λ we will now show that the precise values of the δ_i are irrelevant.

Consider the Laplace transform of R

$$E[\exp(-kR)] = \prod_{i=1}^{d} E[e^{-k(W_i+\delta_i)^2}]$$

$$= \prod_{i=1}^{d} \int_{-\infty}^{\infty} \frac{1}{\sqrt{2\pi}} \exp(-k(w^2 + 2\delta_i w + \delta_i^2) - \tfrac{1}{2}w^2)\,dw$$

$$= \prod_{i=1}^{d} (1+2k)^{-1/2} \exp\left(-\frac{k}{1+2k}\delta_j^2\right)$$

$$= (1+2k)^{-d/2} e^{-\lambda/2} \exp\left(\frac{\lambda}{2(1+2k)}\right), \tag{4.19}$$

where $\lambda = \sum_{i=1}^{d} \delta_i^2$. (The Laplace transform in equation (4.19) is defined for $k > -\tfrac{1}{2}$.)

We can note two points. First, when $\lambda = 0$, we get the Laplace transform of the chi-squared distribution. Second, the Laplace transform depends only upon λ and not on the δ_i individually. This confirms the claim made earlier that the distribution of R depends only on the summary statistic λ rather than individually on the δ_i.

The definition of the non-central chi-squared distribution can be extended (as can the chi-squared distribution) to non-integer values of d. Quite simply, we assume that the Laplace transform of the non-central chi-squared distribution with a non-integer number of degrees of freedom, d, is that given in equation (4.19).

We can now return to the multi-dimensional Ornstein–Uhlenbeck process and the CIR process. In particular, recall equations (4.17) and (4.18). Since the $X_i(t)$ are all Normally distributed with the same variance we can see that $R(t)/[1 - e^{-\alpha t}]$ has a non-central chi-squared distribution. Similarly, we can note that, for integer $d = 4\alpha\mu/\sigma^2$, $4\alpha r(t)/[\sigma^2(1 - e^{-\alpha t})]$ has a non-central chi-squared distribution with d degrees of freedom and non-centrality parameter $\lambda = 4\alpha r(0)/[\sigma^2(e^{\alpha t} - 1)]$.

We have now introduced the essential elements of the CIR model, and interested readers should now work through the proof of Theorem 4.8 in Appendix B.

4.7 A Comparison of the Vasicek and Cox–Ingersoll–Ross Models

4.7.1 Introduction

We have seen that the Vasicek and Cox–Ingersoll–Ross models have some elements of their form which are essentially the same and some which are different. In particular, the processes for the risk-free rate of interest, $r(t)$, have the same form for the drift, $\alpha(\mu - r(t))$, but different volatility functions σ versus $\sigma\sqrt{r(t)}$. In addition, some or all of the parameter values will, necessarily, be different.

Table 4.3.   Sample parameter values for the Vasicek and CIR models.

Model	μ	α	σ
Vasicek	0.06	0.25	0.02
Cox–Ingersoll–Ross	0.060 15	0.232	0.082

From a mathematical, qualitative point of view, these are two quite different models that merit separate analyses. However, the question arises as to whether the two models produce results which are significantly different from a quantitative point of view.

4.7.2   Forward-Rate Curves

The parameter values we will use to compare the two models are given in Table 4.3.

This choice of parameters ensures, first, that the volatility of $r(t)$ under the CIR model is approximately equal to that under the Vasicek model when $r(t) = \mu_{\mathrm{CIR}}$; that is, $\sigma_{\mathrm{CIR}}\sqrt{\mu_{\mathrm{CIR}}} \approx \sigma_{\mathrm{Vas}}$. Second, the limiting forward and spot rates are $\mu - \sigma^2/2\alpha^2$ under the Vasicek model and $\mu\alpha(\gamma - \alpha)/\sigma^2$ under the CIR model (where $\gamma = \sqrt{\alpha^2 + 2\sigma^2}$). The parameters have been chosen so that these limiting rates are equal. Third, α was chosen so that the shapes of forward curves roughly match over a range of values for $r(0)$.

The second and third criteria are reflected in Figures 4.1 and 4.2. We see that the differences between the two sets of forward-rate curves are quite small (normally much less than 0.1% or 10 *basis points*). Figures 4.1 and 4.2 also give us a good indication of the range of shapes of forward-rate curves which the two models can produce. Thus, we can have rising, falling or slightly humped forward-rate curves. The models cannot produce curves which have troughs, more than one turning point or with more than just a slight hump.

Figure 4.2 uses the three initial conditions $r(0) = 0.02$, 0.06 and 0.15 which we will now investigate further.

4.7.3   Probabilistic Differences and Option Prices

In Figure 4.3 we look at the cost of a European call option (see Theorems 4.7 and 4.8) as a function of the strike price K. In the four cases plotted we have $r(0) = 0.02$, 0.06, 0.15 and 0.15 with exercise dates $T = 5, 5, 5$ and 20 respectively. In each case the maturity date of the underlying zero-coupon bond is $U = T + 3$. We observe the following general points. The price falls as the strike price rises, reflecting the lower expected payouts. As $r(0)$ increases from 0.02 to 0.15 the price of the option falls. As $r(0)$ increases, the current price of the underlying bond decreases so again the dependence of the price of the call option on $r(0)$ is clear.

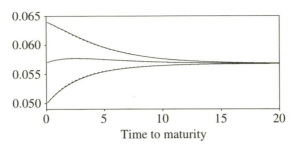

Figure 4.1. Possible shapes for forward-rate curves under the Vasicek model (dotted lines, $\mu = 0.06, \alpha = 0.25, \sigma = 0.02$) and the Cox–Ingersoll–Ross model (solid lines, $\mu = 0.060\,15, \alpha = 0.232, \sigma = 0.082$). Initial values for $r(0)$ are 0.05, 0.057 and 0.064.

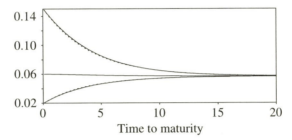

Figure 4.2. Possible shapes for forward-rate curves under the Vasicek model (dotted lines, $\mu = 0.06, \alpha = 0.25, \sigma = 0.02$) and the Cox–Ingersoll–Ross model (solid lines, $\mu = 0.060\,15, \alpha = 0.232, \sigma = 0.082$). Initial values for $r(0)$ are 0.02, 0.06 and 0.15.

Now concentrate on the differences between the call option prices indicated by the Vasicek and CIR models. Prices under the two models are similar for some values of $r(0)$ and K. However, often the prices are quite different, particularly in relative terms.

The most obvious feature is that the Vasicek model almost always gives substantially higher option prices for high strike prices, K. One reason for this is clear: the Vasicek model allows $r(t)$ to become negative and for bond prices to take on any value greater than zero. The CIR model constrains $r(t)$ to remain positive with the price of the underlying bond at maturity remaining strictly less than $\exp \bar{A}(3) < 1$. It follows that if $1 > K > \exp \bar{A}(3) < 1$ the price of the call option must be zero. Equally, we can argue that if K is close to 1 then the Vasicek model will overvalue the call option (for example, when $K = 1$ the price will be greater than zero).

When $r(0) = 0.02$ Vasicek prices are generally higher than CIR and when $r(0) = 0.15$ Vasicek prices are generally lower than CIR. This reflects the differences in volatility between the two models. It is well known that the price of a call option is, on the whole, an increasing function of the volatility of the price of the underlying. The form of the price functions under the two models indicates that price volatility

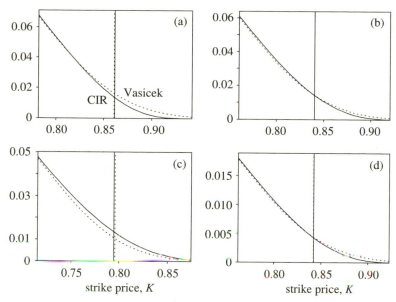

Figure 4.3. Call option prices as a function of the strike price K under the Vasicek model (dotted lines, $\mu = 0.06, \alpha = 0.25, \sigma = 0.02$) and the Cox–Ingersoll–Ross model (solid lines, $\mu = 0.060\,15, \alpha = 0.232, \sigma = 0.082$). (a) $T = 5, U = 8, r(0) = 0.02$; (b) $T = 5, U = 8, r(0) = 0.06$; (c) $T = 5, U = 8, r(0) = 0.15$; (d) $T = 20, U = 23, r(0) = 0.15$. Vertical lines give the forward price for delivery at T of the zero-coupon bond maturing at time U.

is closely linked to volatility in $r(t)$. When $r(0) = 0.02$ the volatility of the CIR model will be low relative to the Vasicek model, giving rise to lower call-option prices. The converse is true if $r(0) = 0.15$.

We can get more of a feel for this by considering the distributions plotted in Figure 4.4. Here we plot the probability density function of $r(T)$ for the four sets of initial conditions as before. First consider the distribution of $r(T)$ under the Vasicek model for $T = 5$. The spread of this distribution is not affected by $r(0)$ but the location is. Now consider the distribution of $r(T)$ under the CIR model. It is thinnest when $r(0) = 0.02$ and fattest when $r(0) = 0.15$. This confirms the earlier comments about the volatility of the CIR process affecting the price of the call option.

If we continue a comparison of the Vasicek and CIR models, we see that the Vasicek model produces a symmetric distribution while the CIR model produces a skewed distribution. The latter is constrained to the positive real numbers and has a fatter right-hand tail than the Vasicek model. The left-hand constraint means that CIR is less likely to produce small $r(T)$ than the Vasicek model. This explains why the Vasicek model typically produces higher call-option prices when K is high.

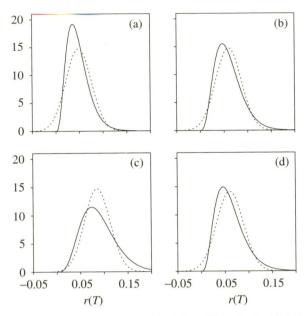

Figure 4.4. Probability density function under Q for $r(T)$ under the Vasicek model (dotted lines, $\mu = 0.06, \alpha = 0.25, \sigma = 0.02$) and the Cox–Ingersoll–Ross model (solid lines, $\mu = 0.060\,15, \alpha = 0.232, \sigma = 0.082$). (a) $T = 5, r(0) = 0.02$; (b) $T = 5, r(0) = 0.06$; (c) $T = 5, r(0) = 0.15$; (d) $T = 20, r(0) = 0.15$.

Conversely, the CIR model will produce higher put-option prices when K is low because it has a fatter right-hand tail.

4.8 Affine Short-Rate Models

We have seen in the preceding sections that both the Vasicek and CIR models have zero-coupon bond prices which are of the *affine* form $P(t, T) = \exp[A(t, T) - B(t, T)r(t)]$ for functions A and B which are specific to each model. It is interesting to ask the question: are there any other models which give rise to similar affine forms for $P(t, T)$?

Consider the general SDE for $r(t)$

$$dr(t) = m(t, r(t))\,dt + s(t, r(t))\,d\tilde{W}(t),$$

where $\tilde{W}(t)$ is a standard Brownian motion under the risk-neutral measure Q. Suppose that $P(t, T) = \exp[A(t, T) - B(t, T)r(t)]$. Then by application of Itô's formula we have

$$dP(t, T) = P(t, T)\left[\left(\frac{\partial A}{\partial t} - \frac{\partial B}{\partial t}r(t) - Bm + \tfrac{1}{2}Bs^2\right)dt - Bs\,d\tilde{W}(t)\right] \quad (4.20)$$

(where $m \equiv m(t, r(t))$, etc.). But we also know that, under Q,

$$dP(t, T) = P(t, T)[r(t)\,dt + S(t, T, r(t))\,d\tilde{W}(t)], \tag{4.21}$$

where $S(t, T, r(t))$ is the volatility of $P(t, T)$. This equality comes from the requirement that all tradable assets must have expected growth at the risk-free rate under Q. It follows that if we define

$$g(t, r) = \frac{\partial A}{\partial t} - \frac{\partial B}{\partial t}r - Bm(t, r) + \tfrac{1}{2}Bs(t, r)^2 - r,$$

then $g(t, r) = 0$ for all t and r. Now differentiate twice with respect to r

$$\frac{\partial^2 g}{\partial r^2} = -B(t, T)\frac{\partial^2 m(t, r)}{\partial r^2} + \tfrac{1}{2}B(t, T)^2\frac{\partial^2 (s(t, r)^2)}{\partial r^2} = 0$$

$$\Rightarrow \quad -\frac{\partial^2 m(t, r)}{\partial r^2} + \tfrac{1}{2}B(t, T)\frac{\partial^2 (s(t, r)^2)}{\partial r^2} = 0.$$

Since $B(t, T)$ is a function of T as well as t, this identity can only hold if both

$$\frac{\partial^2 (s(t, r)^2)}{\partial r^2} = 0 \quad \text{and} \quad \frac{\partial^2 m(t, r)}{\partial r^2} = 0.$$

Corollary 4.9. *It is a necessary condition for bond-pricing formulae to be of the form $P(t, T) = \exp[A(t, T) - B(t, T)r(t)]$ that the risk-neutral drift and volatility of $r(t)$ are of the form*

$$m(t, r(t)) = a(t) + b(t)r(t) \quad \text{and} \quad s(t, r(t)) = \sqrt{\gamma(t)r(t) + \delta(t)},$$

where $a(t)$, $b(t)$, $\gamma(t)$ and $\delta(t)$ are deterministic functions.

For general, time-dependent $a(t)$, $b(t)$, $\gamma(t)$ and $\delta(t)$, analytical solutions for $A(t, T)$ and $B(t, T)$ are not normally available. However, when they are constant we can derive formulae for the functions A and B. In particular, we have the following cases.

Vasicek (1977). $\gamma = 0$, $\delta = \sigma^2$, $b = -\alpha$ and $a = \alpha\mu$, which implies that $dr(t) = \alpha(\mu - r(t))\,dt + \sigma\,d\tilde{W}(t)$. Earlier, for this model, we found that

$$B(t, T) = (1 - e^{-\alpha(T-t)})/\alpha$$

and

$$A(t, T) = (B(t, T) - (T - t))(\mu - \sigma^2/2\alpha^2) - \frac{\sigma^2}{4\alpha}B(t, T)^2.$$

Cox, Ingersoll and Ross (1985). $\delta = 0$, $\gamma = \sigma^2$, $b = -\alpha$ and $a = \alpha\mu$, which implies that $dr(t) = \alpha(\mu - r(t))\,dt + \sigma\sqrt{r(t)}\,d\tilde{W}(t)$. This gave us

$$A(t, T) = \frac{2\alpha\mu}{\sigma^2}\log\left(\frac{2\gamma e^{(\gamma + \alpha)(T-t)/2}}{(\gamma + \alpha)(e^{\gamma(T-t)} - 1) + 2\gamma}\right),$$

$$\gamma = \sqrt{\alpha^2 + 2\sigma^2},$$

$$B(t, T) = \frac{2(e^{\gamma(T-t)} - 1)}{(\gamma + \alpha)(e^{\gamma(T-t)} - 1) + 2\gamma}. \tag{4.22}$$

Merton (1973). $\gamma = 0$, $\delta = \sigma^2$ and $b = 0$, which implies that $dr(t) = a\,dt + \sigma\,d\tilde{W}(t)$. This results in

$$B(t, T) = T - t,$$

$$A(t, T) = \tfrac{1}{6}\sigma^2(T - t)^3 - \tfrac{1}{2}a(T - t)^2.$$

(Note that this implies $P(t, T) \to +\infty$ as $T \to \infty$ even if the drift a is positive!)

Pearson and Sun (1994). $\delta = \sigma^2$, $\gamma = -\beta/\sigma^2$, $b = -\alpha$ and $a = \alpha(\mu + \beta)$, which implies that $dr(t) = \alpha(\mu - r(t))\,dt + \sigma\sqrt{r(t) - \beta}\,d\tilde{W}(t)$. In a similar way to the CIR model, this model has a minimum value for $r(t)$ at β. Let $X(t) = r(t) - \beta$. Then we have

$$dX(t) = \alpha(\mu - \beta - X(t))\,dt + \sigma\sqrt{X(t)}\,d\tilde{W}(t).$$

Also we have

$$
\begin{aligned}
P(t, T) &= E_Q\left[\exp\left(-\int_t^T r(s)\,ds\right) \,\middle|\, \mathcal{F}_t\right] \\
&= e^{-\beta(T-t)}E_Q\left[\exp\left(-\int_t^T X(s)\,ds\right) \,\middle|\, \mathcal{F}_t\right] \\
&= e^{-\beta(T-t)}e^{A(t,T)-B(t,T)X(t)} \\
&= e^{-\beta(T-t)}e^{A(t,T)-B(t,T)(X(t)+\beta)+\beta B(t,T)} \\
&= e^{A_P(t,T)-B(t,T)r(t)},
\end{aligned}
$$

where $A(t, T)$ and $B(t, T)$ are given in equations (4.22) (replacing μ by $\mu - \beta$) and $A_P(t, T) = A(t, T) - \beta(T - t) + \beta B(t, T)$.

Two further cases where $a(t)$ is permitted to be time dependent have been investigated by Ho and Lee (1986) and Hull and White (1990), corresponding to the Merton and Vasicek models respectively. Their models are investigated in the next chapter (Sections 5.2.1 and 5.2.2). In both cases analytical solutions can be found for both bond prices and option prices.

4.9 Other Short-Rate Models

A number of other one-factor models have been developed which incorporate volatility in $r(t)$ of the form $\sigma r(t)^{\gamma}$, where $\gamma > 0.5$. From a statistical perspective, such models have gained credibility in recent years since Chan et al. (1992) found that $\gamma = 1$ fitted historical data significantly better, and $\gamma = 1.5$, even better still.[1]

This was preceded by the Brennan and Schwartz (1979) model where $dr(t) = \alpha(\mu - r(t))\,dt + \sigma r(t)\,d\tilde{W}(t)$. Although this is a potentially attractive model, it does not give rise to analytical formulae.

Another popular model is the lognormal model proposed by Black, Derman and Toy (1990) and later generalized by Black and Karasinski (1991). We will describe briefly here the time-homogeneous version of the Black–Karasinski model and we will return to consider the time-inhomogeneous version in Chapter 5. We start with a state-variable $L(t)$ with SDE under Q:

$$dL(t) = \alpha(\mu - L(t))\,dt + \sigma\,d\tilde{W}(t).$$

Then we know from the properties of the Vasicek model that $L(T)$, given $L(t)$, is Normally distributed with $E_Q[L(T) \mid L(t)] = \mu + (L(t) - \mu)\exp(-\alpha(T - t))$ and $\mathrm{Var}_Q[L(T) \mid L(t)] = \sigma^2\{1 - \exp(-2\alpha(T - t))\}/2\alpha$. We then define $r(t) = e^{L(t)}$, so that $r(T)$ is lognormal. The Black–Karasinski model does not give rise to closed-form bond and option prices but these can be calculated straightforwardly using numerical methods (see Chapter 10).

A further group of time-homogeneous, one-factor models is described later in this book (Chapter 8) once we have developed some further theory.

4.10 Options on Coupon-Paying Securities

The options we have considered so far are options on zero-coupon bonds. Jamshidian (1989) proposed the following simple means of extending these formulae to deal with options on fixed-interest contracts with more than one payment, provided we are dealing with a *one-factor* interest rate model.

Suppose we have a bond which pays the fixed amount $c_j > 0$ at times T_j for $j = 1, \ldots, m$ and where $T_1 < T_2 < \cdots < T_m$. Suppose also that we have a European call option on this bond with exercise date $T < T_1$ and exercise price K. The price at time t for the bond is

$$C(t, r(t)) = \sum_{j=1}^{m} c_j P(t, T_j, r(t)).$$

---

[1]Chan et al. (1992) use the Generalized Method of Moments and obtain an estimate for γ of about 1.5. However, the alternative use of Maximum Likelihood Estimation yields an estimate for γ of around 1. Therefore, we need to act with caution.

(We have included $r(t)$ as a variable here to assist with the arguments that follow.)

For any one-factor model assume (reasonably) that zero-coupon bond prices $P(t, S, r(t))$ are decreasing functions of $r(t)$ and denote by $V_j(t, k)$ the price at time t of a European call option on the zero-coupon bond $P(t, T_j, r(t))$ with exercise date T (the same as the bond $C(t, r(t))$) and exercise price k. Since the zero-coupon bond prices decrease with $r(t)$ and since the c_j are all positive it follows that $C(t, r(t))$ is also a decreasing function of $r(t)$. There is therefore a critical value, r_0, of $r(T)$ at which $C(t, r_0) = K$.

Now let $K_j = P(T, T_j, r_0)$ so that $\sum_{j=1}^{m} c_j K_j = K$. Then the call option payoff on the coupon-bond $C(T, r(T))$ is

$$V(T) = \max\{C(T, r(T)) - K, 0\}$$

$$= \max\left\{\sum_{j=1}^{m} c_j[P(T, T_j, r(T)) - K_j], 0\right\}$$

$$= \sum_{j=1}^{m} c_j \max\{P(T, T_j, r(T)) - K_j, 0\}$$

$$= \sum_{j=1}^{m} c_j V_j(T, K_j).$$

It follows by the law of one price that the price at time t of the call option is then

$$V(t) = \sum_{j=1}^{m} c_j V_j(t, K_j).$$

This formula holds for any one-factor interest rate model. If we are using, in particular, a model which yields analytical formulae for the $V_j(t, K)$, then we obtain an analytical formula for $V(t)$.

4.11   Exercises

General

Exercise 4.1. Consider a derivative which has the following payoff at time t: $X = \max\{R(T, T + 1) - R(T, T + 10), 0\}$. Why might it be inappropriate to use a *one-factor* interest rate model to value this option?

Exercise 4.2. Explain why the PDE in equation (4.7) involves the risk-neutral drift for $r(t)$, $a(t, r) - b(t, r)\gamma(t, r)$, and not the drift under the real-world measure.

Exercise 4.3. Suppose the market price of risk ensures that investments in risky bonds have a suitable positive risk premium over investments in risk-free cash.

What does this imply for the dynamics of the risk-free rate of interest; that is, will $r(t)$ be generally higher or lower when we switch from the risk-neutral measure to the real-world measure?

Exercise 4.4. Recall that $dP(t, T) = P(t, T)[r(t)\,dt + S(t, T)\,d\tilde{W}(t)]$. Consider the spot-rate process $R(t, T) = -(T - t)^{-1}\log P(t, T)$. Use Itô's formula to show that

$$dR(t, T) = \frac{1}{T - t}[(R(t, T) + \tfrac{1}{2}S(t, T)^2 - r(t))\,dt - S(t, T)\,d\tilde{W}(t)].$$

Exercise 4.5. Let $Y(t, T) = -\log P(t, T)$. Assume that

$$d\left(\frac{\partial Y(t, T)}{\partial T}\right) = \frac{\partial}{\partial T}\,dY(t, T).$$

Show that the instantaneous forward-rate curve $f(t, T)$ has SDE

$$df(t, T) = -\sigma(t, T)S(t, T)\,dt + \sigma(t, T)\,d\tilde{W}(t),$$

where $\sigma(t, T) = -\partial S(t, T)/\partial T$. (See the Heath–Jarrow–Morton framework in Section 5.3.)

Exercise 4.6. In equation (4.4) it was noted that

$$dP(t, T) = P(t, T)[r(t)\,dt + S(t, T)\,d\tilde{W}(t)].$$

Now consider a derivative with $P(t, T)$ as the underlying which pays $g(P(S, T))$ at time $S < T$. The *delta-hedge* for this contract holds $\phi(t) = \phi(t, T)$ units of $P(t, T)$ at time t along with the appropriate number of units of the risk-free cash account, $B(t)$, if we wish to replicate the derivative payoff.

(a) Use the product rule for Itô processes to show that

$$dV(t) = V(t)\left[r(t)\,dt + \frac{B(t)}{V(t)}\,dD(t)\right].$$

(b) Show that

$$dV(t) = V(t)[r(t)\,dt + \sigma_V(t)\,d\tilde{W}(t)],$$

where $\sigma_V(t) = \phi(t, T)P(t, T)S(t, T)/V(t)$.

(c) What conclusion do you draw from this SDE?

(d) Suppose we wish to hedge the same derivative contract using the bond which matures at time S. Find an expression for $\phi(t, S)$.

(e) It is known that $S(t, \tau) \to 0$ as $t \to \tau$. What problems do you foresee if we choose to hedge the derivative using $P(t, S)$?

Exercise 4.7 (futures prices). Suppose $dr(t) = a(t)\,dt + b(t)\,dW(t)$ for some suitable previsible functions $a(t)$ and $b(t)$, where $W(t)$ is a Brownian motion under the real-world measure P.

Consider a futures contract for delivery of the underlying $P(t, S)$ at time $T < S$. The futures price at time t for this contract is $F(t)$. The variation margin payable to the holder of the futures contract at the end of the time interval $[t, t + dt)$ is $dF(t) = F(t + dt) - F(t)$ for infinitesimally small dt. Immediately after the payment the price for the contract is always 0.

Prove that for all $t < T$: $F(t) = E_Q[P(T, S) \mid \mathcal{F}_t]$.

Two possible methods are as follows.

Method A. Start by assuming that $P(T, S) = E_Q[B(T)/B(S) \mid \mathcal{F}_T]$ as usual. However, then assume that $P(t, S)$ is not tradable between 0 and T. Then establish how you can replicate $P(T, S)$ using cash plus futures contracts.

Method B. (a) Let $\pi(t)$ be the accumulation in cash of the net cashflows $dF(t)$ to the holder of the long position; that is, $d\pi(t) = r(t)\pi(t)\,dt + dF(t)$. (b) Also note that $B(t)E_Q[\int_t^T B(s)^{-1}\,dF(s) \mid \mathcal{F}_t] = 0$ (state why!). (c) Define $\tilde{\pi}(t) = \pi(t)/B(t)$. Show that $d\tilde{\pi}(t) = B(t)^{-1}\,dF(t)$. (d) Show that $\tilde{\pi}(t)$ is a martingale under Q. (e) Hence show that $F(t)$ is a martingale under Q and complete the result.

Exercise 4.8. In practice, futures contracts on exchanges incorporate additional option characteristics. Identify what these option features are (for example, go to www.liffe.com or www.cbot.com).

Specific Models

Exercise 4.9. Derive the form of the forward-rate curve, $f(t, T)$, for the Vasicek model in terms of α, μ, σ and $r(t)$, given the pricing formula in Theorem 4.7.

Exercise 4.10. Consider a European call option on a zero-coupon bond with strike price $K < 1$. Why is it inappropriate to use the Cox–Ingersoll–Ross model if K is too close to 1?

Exercise 4.11. Suppose that X and Y are normal random variables with mean zero, variance 1 and correlation ρ under a measure Q.

(a) For any real numbers ω and ν, write down

$$E_Q[e^{-\omega Y - \nu X}].$$

(b) Let P be another measure equivalent to Q defined by the Radon–Nikodým derivative

$$\frac{dP}{dQ} = \frac{e^{-\omega Y}}{E_Q[e^{-\omega Y}]}.$$

Find an expression for $E_Q[\exp(-\omega Y - \nu X)]$ in terms of $E_Q[\exp(-\omega Y)]$ and $E_P[\exp(-\nu X)]$.

(c) Hence show that under P, X has a Normal distribution with mean $-\omega\rho$ and variance 1.

(d) Use your answers to parts (b) and (c) to show that

$$E_Q[e^{-Y}I(X < x)] = e^{1/2}\Phi(x + \rho),$$

where $I(X < x) = 1$ if $X < x$ and 0 otherwise and $\Phi(z)$ is the cumulative distribution function of the standard Normal distribution.

(e) Use the fact that $E_Q[e^{-Y}I(X < x)] = E_Q[I(X < x)E_Q[e^{-Y} \mid X]]$ to verify directly (that is, without using the change of measure) the result in part (d).

Exercise 4.12. Make use of the following facts to investigate the problems below relating to the Vasicek model.

- If $X = \int_0^T g(s)\,dW_s$ for some deterministic function $g(s)$, then X has a Normal distribution with mean 0 and variance $\int_0^T g(s)^2\,ds$.

- If $\sigma(t, u)$ is such that

$$E\left[\int_0^T \left|\int_0^u \sigma(t, u)\,dW_t\right| du\right] < \infty,$$

  then

$$\int_0^t \int_0^u \sigma(s, u)\,dW_s\,du = \int_0^t \int_s^t \sigma(s, u)\,du\,dW_s.$$

- If Y has a Normal distribution with mean μ and variance σ^2, then $E[e^Y] = e^{\mu+\sigma^2/2}$.

Under the Vasicek model $dr(t) = \alpha(\mu - r(t))\,dt + \sigma\,d\tilde{W}_t$, where $\tilde{W}_t$ is a Brownian motion under the equivalent martingale measure Q.

(a) Show that $r(t) = \mu + (r(0) - \mu)e^{-\alpha t} + \sigma \int_0^t e^{-\alpha(t-s)}\,d\tilde{W}_s$.

(b) Show that

$$X(T) = \int_0^T r(t)\,dt$$

$$= \mu T + (r(0) - \mu)\frac{1 - e^{-\alpha T}}{\alpha} + \sigma \int_0^T \frac{1 - e^{-\alpha(T-s)}}{\alpha}\,d\tilde{W}_s.$$

(c) Find the joint distribution for $(r(T), X(T))$.

(d) Using your answers to (b) and (c) find an expression for

$$P(t, T) = E_Q[\exp\{-(X(T) - X(t))\} \mid \mathcal{F}_t],$$

expressing this in the form $\exp[A(t, T) - B(t, T)r(t)]$.

(e) Using part (b), find the joint distribution for $(r(T), X(S))$, where $T < S$.

(f) Consider a European call option with exercise date T and strike price K with the zero-coupon bond maturing at time S (where $T < S$) as the underlying. The price at time 0 of this bond is

$$C(0) = E_Q[e^{-X(T)} \max\{P(T, S) - K, 0\}].$$

Let P_1 and P_2 be two measures equivalent to Q defined by the Radon–Nikodým derivatives

$$\frac{dP_1}{dQ} = \frac{e^{-X(S)}}{E_Q[e^{-X(S)}]} \quad \text{and} \quad \frac{dP_2}{dQ} = \frac{e^{-X(T)}}{E_Q[e^{-X(T)}]}.$$

  (i) Show that

$$C(0) = P(0, S)E_{P_1}[I(r(T) < r^*)] - KP(0, T)E_{P_2}[I(r(T) < r^*)],$$

where $r^* = (A(T, S) - \log K)/B(T, S)$ and $I(r(T) < r^*) = 1$ if $r(T) < r^*$ and 0 otherwise.

 (ii) Use your answers to Exercise 4.11 to show that

$$C(0) = P(0, S)\Phi(d_1) - KP(0, T)\Phi(d_2),$$

where

$$d_1 = \frac{1}{\sigma_p} \log \frac{P(0, S)}{KP(0, T)} + \frac{\sigma_p}{2},$$

$$d_2 = d_1 - \sigma_p,$$

$$\sigma_p = \frac{\sigma}{\alpha}(1 - e^{-\alpha(S-T)})\sqrt{\frac{1 - e^{-2\alpha T}}{2\alpha}}.$$

Exercise 4.13.

(a) Under the CIR model we have

$$dr(t) = \alpha(\mu - r(t))\,dt + \sigma\sqrt{r(t)}\,d\tilde{W}(t).$$

It is required that $2\alpha\mu \geqslant \sigma^2$ to keep the process strictly positive. By considering the process $X(t) = \log r(t)$, discuss why this requirement might be necessary (a rigorous proof is not required).

(b) Under CIR we have prices of the form

$$P(t, T) = \exp[A(t, T) - B(t, T)r(t)],$$

where

$$B(t, T) = \frac{2(e^{\gamma(T-t)} - 1)}{(\gamma + \alpha)(e^{\gamma(T-t)} - 1) + 2\gamma},$$

$$A(t, T) = \frac{2\alpha\mu}{\sigma^2} \log\left[\frac{2\gamma e^{(\alpha+\gamma)(T-t)/2}}{(\gamma + \alpha)(e^{\gamma(T-t)} - 1) + 2\gamma}\right],$$

$$\gamma = \sqrt{\alpha^2 + 2\sigma^2}.$$

The Pearson–Sun model uses

$$dr(t) = \alpha(\mu - r(t)) \, dt + \sigma\sqrt{r(t) - \beta} \, d\tilde{W}(t).$$

Make use of the CIR pricing formula to derive prices for the PS model.

(c) Under the Brennan–Schwartz model we have $dr(t) = \alpha(\mu - r(t)) \, dt + \sigma r(t) \, d\tilde{W}(t)$. Other than $\alpha > 0$, $\mu > 0$ and $\sigma > 0$, do we need any special conditions to ensure that $r(t)$ does not hit 0?

Exercise 4.14. Consider the CIR model

$$dr(t) = \alpha(\mu - r(t)) \, dt + \sigma\sqrt{r(t)} \, d\tilde{W}(t),$$

where $\tilde{W}(t)$ is standard Brownian motion under the equivalent martingale measure Q and the parameter values are $\alpha = 0.0125$, $\mu = 0.05$, $\sigma = 0.05$, with $r(0) = 0.1$.

Prices can be expressed in the form $P(0, T) = \exp[\bar{A}(T) - \bar{B}(T)r(0)]$. For the above parameters:

u	$\bar{A}(u)$	$\bar{B}(u)$
1	−0.000 311 1	0.993 365
10	−0.029 416 1	9.048 922
11	−0.035 314 0	9.819 592

(a) Calculate the prices $P(0, 1)$ and $P(0, 11)$.

(b) Suppose $Z \sim N(0, 1)$ and $Y = (Z + \sqrt{\lambda})^2$.

Show that $\Pr(Y < y) = \Phi(\sqrt{y} - \sqrt{\lambda}) - \Phi(-\sqrt{y} - \sqrt{\lambda})$.

(c) A European call option written on $P(t, 11)$ has an exercise date of $T = 1$ and a strike price of $K = e^{-0.8}$.

Calculate the price of this call option at time 0.

5

No-Arbitrage Models

5.1 Introduction

In Section 2.7 we discussed the two main classes of interest rate model: short-rate models and no-arbitrage models. In Chapter 4 we looked at the general theory behind arbitrage-free models before focusing on specific time-homogeneous, short-rate models. In this chapter we will focus on no-arbitrage models.

The reason for the development of no-arbitrage models is simple. When we use a short-rate model such as the Vasicek or CIR model we will usually find that the theoretical prices, $\hat{P}(t, T)$, output by the model do not match precisely the bond prices we observe, $P_{\text{obs}}(t, T)$, in the market place. For example, model parameters may have been calibrated using historical data and this may result on a given date with differences between the $\hat{P}(t, T)$ and $P_{\text{obs}}(t, T)$. These differences can be reduced by ignoring historical data and instead calibrating the (fixed) model parameters to achieve the best possible match between the $\hat{P}(t, T)$ and $P_{\text{obs}}(t, T)$ (for example, using non-linear regression). However, since we typically have many more bonds than parameters in the model we will still find that the $\hat{P}(t, T)$ differ from the $P_{\text{obs}}(t, T)$, albeit to a smaller extent. This approach is somewhat unsatisfactory, since we would need to recalibrate the model daily (if not more frequently), resulting in parameter values which fluctuate over time. But the model parameters are supposed to be fixed rather than time varying (and stochastic), which means that the very process of calibration is violating the basic assumptions of the model.

While we have focused above on the differences between the underlying bond prices, we have ignored the market prices of traded, interest-rate derivatives. Again, the theoretical model prices will differ from those we observe on the derivatives exchanges.

These differences can be dealt with in one of two ways. First, we can develop multifactor, time-homogeneous models where parameters are calibrated to historical data. Let θ be the fixed parameter set and $X(t)$ be the vector of stochastic state variables. Suppose that we have estimated θ from past data. We will have a satisfactory model when we are happy that

- we can always get a *close* fit between theoretical and observed prices of bonds and derivatives by regular (continuous) calibration of $X(t)$ (note that theoretical and observed prices can be allowed to differ to the extent that transaction costs prevent arbitrage),

- the quality of fit can be achieved without the need to recalibrate the parameters θ, and

- the time series of calibrated values of $X(t)$ (for example, at times $t = t_1, t_2, \ldots, t_N$) is consistent with the model dynamics for $X(t)$.

General multifactor models are discussed further in Chapters 6, 8 and 9.

Second, we can develop time-inhomogeneous, no-arbitrage models where observed prices of bonds and derivatives form part of the input, so that, by design, the $\hat{P}(t, T)$ precisely match the $P_{obs}(t, T)$ at the time of calibration, t.

In general, market practitioners prefer the second, no-arbitrage, of these approaches for giving price quotations for the following reasons.

- In the traded bonds and options markets, market makers must be prepared to both buy and sell at the quoted prices. If the market-maker's model gives prices which are different from the rest of the market, then this will give rise to arbitrage opportunities which would probably be to the disadvantage of the market maker.

- When new information comes into the public domain (for example, a change in the short-term rate of interest) the market maker wants to be able to anticipate how the other market makers are likely to change their prices.

- Market makers are typically concerned with the pricing and hedging of short-term derivatives contracts. Concern might be expressed about the potential impact of using an approach to calibration which is inconsistent with the assumptions of the model; for example, treating what are constant parameters in the model as if they are stochastic. However, the impact of such errors in the modelling process is felt to be much smaller for short-term contracts than long-term contracts.

- Investment banks offering over-the-counter derivatives want to price these contracts in a way which is consistent with the closest-matching traded derivatives.

5.2 Markov Models

Here we consider models for which

- the prices are input, and

- the probability law of $P(s, T) \mid \mathcal{F}_t$, where $t < s < T$, is equal to that of $P(s, T) \mid X(t)$, where $X(t)$ is some finite-dimensional Itô process. (Thus, it is sufficient to know the value of $X(t)$ to describe the future dynamics of $P(s, T)$. Any additional information about the past does not refine our description of its future dynamics.)

In particular, we will consider two models under which $X(t) = r(t)$ (so is one dimensional).

5.2.1 The Ho and Lee Model

Ho and Lee (1986) considered the following model for the risk-free rate:

$$dr(t) = \theta(t)\, dt + \sigma\, d\tilde{W}(t),$$

where $\tilde{W}(t)$ is a Brownian motion under the equivalent martingale measure Q. This is a more general version of the Merton random-walk model under which $\theta(t)$ is a constant.

Suppose that we have as input $P(0, T)$ for all $T > 0$. Let

$$f(0, T) = -\frac{\partial}{\partial T} \log P(0, T)$$

be the initial forward-rate curve. If

$$\theta(T) = \frac{\partial}{\partial T} f(0, T) + \sigma^2 T,$$

then it can be shown that

$$E_Q\left[\exp\left(-\int_0^T r(t)\, dt \right) \,\middle|\, r(0) \right] = P(0, T)$$

and

$$P(t, T) = \exp[A(t, T) - (T - t)r(t)],$$

where

$$A(t, T) = \log \frac{P(0, T)}{P(0, t)} + (T - t)f(0, t) - \tfrac{1}{2}\sigma^2 t(T - t)^2.$$

Note that the $P(t, T)$ have the affine form characteristic of the Vasicek and CIR models.

It follows that

$$f(t, T) = -\frac{\partial}{\partial T} \log P(t, T) = r(t) + f(0, T) - f(0, t) + \sigma^2 t(T - t).$$

Now it is straightforward to find a solution for $r(t)$:

$$
\begin{aligned}
r(t) &= r(0) + \int_0^t \theta(s)\, ds + \sigma \tilde{W}(t) \\
&= r(0) + f(0, t) - f(0, 0) + \tfrac{1}{2}\sigma^2 t^2 + \sigma \tilde{W}(t) \\
&= f(0, t) + \tfrac{1}{2}\sigma^2 t^2 + \sigma \tilde{W}(t).
\end{aligned}
$$

Hence

$$f(t, T) = f(0, t) + \tfrac{1}{2}\sigma^2 t^2 + \sigma \tilde{W}(t) + f(0, T) - f(0, t) + \sigma^2 t(T - t)$$
$$= f(0, T) + \tfrac{1}{2}\sigma^2 T^2 - \tfrac{1}{2}\sigma^2(T - t)^2 + \sigma \tilde{W}(t).$$

(Recall, in the discrete-time version of this model (Section 3.2) we had

$$F(t, T, T + 1) = F(0, T, T + 1) + \log \frac{d(T - t)}{d(T)} + \log k \sum_{s=1}^{t} I_s,$$

which has a very similar form.)

The Ho and Lee model can be generalized quite easily to have $\sigma(t)$ deterministic but time dependent.

5.2.2 The Hull and White Model

Hull and White (1990) proposed a simple but powerful generalization of the Vasicek model in which

$$dr(t) = \alpha(\mu(t) - r(t)) \, dt + \sigma \, d\tilde{W}(t),$$

where $\tilde{W}(t)$ is a Brownian motion under Q and $\mu(t)$ is a deterministic function of time. Often this is expressed in the form $dr(t) = (\theta(t) - \alpha r(t)) \, dt + \sigma \, d\tilde{W}(t)$, but the deterministic function $\theta(t)$ is much less meaningful than $\mu(t)$, which has the straightforward interpretation of a local mean-reversion level.

In the Vasicek model $\mu(t) = \mu$ is constant. The Ho and Lee model is a special case (provided we restrict ourselves to finite maturities) where $\alpha \to 0$ and $\alpha\mu(t) \to \theta(t)$ as $\alpha \to 0$.

In order for initial theoretical and observed prices to match we require

$$\left.\begin{aligned}
\mu(t) &= \frac{1}{\alpha}\frac{\partial}{\partial t} f(0, t) + f(0, t) + \frac{\sigma^2}{2\alpha^2}(1 - e^{-2\alpha t}), \\
P(t, T) &= \exp[A(t, T) - B(t, T)r(t)],
\end{aligned}\right\} \tag{5.1}$$

where

$$B(t, T) = \frac{1 - e^{-\alpha(T-t)}}{\alpha},$$

$$A(t, T) = \log \frac{P(0, T)}{P(0, t)} + B(t, T)f(0, t) - \frac{\sigma^2}{4\alpha^3}(1 - e^{-\alpha(T-t)})^2(1 - e^{-2\alpha t}).$$

The properties of the Ornstein–Uhlenbeck process result in the expression

$$r(t) = e^{-\alpha t} r(0) + \alpha \int_0^t e^{-\alpha(t-s)}\mu(s) \, ds + \sigma \int_0^t e^{-\alpha(t-s)} \, d\tilde{W}(s).$$

From the earlier expression for $\mu(t)$ we have

$$\alpha \int_0^t e^{-\alpha(t-s)} \mu(s)\,ds = f(0,t) - e^{-\alpha t} r(0) + \frac{\sigma^2}{2\alpha^2}(1 - e^{-\alpha t})^2$$

$$\Rightarrow \quad r(t) = f(0,t) + \frac{\sigma^2}{2\alpha^2}(1 - e^{-\alpha t})^2 + \sigma \int_0^t e^{-\alpha(t-s)}\,d\tilde{W}(s).$$

The distributional properties of the Hull and White model are very similar to those of the Vasicek model, making derivative pricing equally straightforward. In particular, the price at time t of a European call option with $P(u, S)$ as the underlying, T as the maturity date and K as the strike price is (see Theorem 4.7)

$$V(t) = P(t, S)\Phi(d_1) - KP(t, T)\Phi(d_2), \tag{5.2}$$

where

$$d_1 = \frac{1}{\sigma_p} \log \frac{P(t, S)}{KP(t, T)} + \frac{\sigma_p}{2}, \qquad d_2 = d_1 - \sigma_p,$$

$$\sigma_p = \frac{\sigma}{\alpha}(1 - e^{-\alpha(S-T)}) \sqrt{\frac{1 - e^{-2\alpha(T-t)}}{2\alpha}}.$$

The Hull and White model can be generalized quite easily to have $\alpha(t)$ and $\sigma(t)$ deterministic but time dependent (see Exercise 5.8).

5.2.3 The Black–Karasinski Model

In Section 4.9 we introduced the time-homogeneous version of the Black–Karasinski model. In their original paper, Black and Karasinski (1991) used a more general, time-inhomogeneous version. We start with the state variable $Y(t) = \log r(t)$ with SDE

$$dY(t) = \alpha(t)(\log \mu(t) - Y(t))\,dt + \sigma(t)\,d\tilde{W}(t),$$

where $\tilde{W}(t)$ is a standard Brownian motion under Q and $\alpha(t)$, $\mu(t)$ and $\sigma(t)$ are deterministic functions of time. Application of Itô's formula gives us the SDE for $r(t)$:

$$dr(t) = \alpha(t)r(t)\left[\log\mu(t) + \frac{\sigma(t)^2}{2\alpha(t)} - \log r(t)\right]dt + \sigma(t)r(t)\,d\tilde{W}(t). \tag{5.3}$$

This implies that $r(t)$ has a local mean-reversion level of $\mu(t)\exp[\sigma(t)^2/2\alpha(t)]$.

In an earlier paper, Black, Derman and Toy (1990) introduced a simpler, discrete-time version of the model. This model is written as $r(t) = m(t)\exp[\sigma(t)\tilde{W}(t)]$ in continuous time. If we take $Y(t) = \log r(t)$ as before, then

$$dY(t) = \alpha(t)(\log\mu(t) - Y(t))\,dt + \sigma(r)\,d\tilde{W}(t),$$

where

$$\alpha(t) = -\sigma'(t)/\sigma(t) \quad \text{and} \quad \log\mu(t) = \log m(t) - \frac{\sigma(t)}{\sigma'(t)}\frac{m'(t)}{m(t)},$$

provided $\sigma'(t) \neq 0$ for all t. This model is overly restrictive compared with the later Black–Karasinski model. Consequently, the latter tends to be used in practice.

Return now to the Black–Karasinski model. Let $A(t) = \int_0^t \alpha(u)\,\mathrm{d}u$. Then we can show that

$$Y(T) = e^{A(t)-A(T)}Y(t)$$
$$+ \int_t^T \alpha(u)e^{A(u)-A(T)} \log \mu(u)\,\mathrm{d}u + \int_t^T \sigma(u)e^{A(u)-A(T)}\,\mathrm{d}\tilde{W}(u).$$

Since $\sigma(u)\exp[A(u) - A(T)]$ is deterministic, it follows that $r(T)$ given $r(t)$ is lognormal with

$$E_Q[\log r(T) \mid \mathcal{F}_t] = e^{A(t)-A(T)}Y(t) + \int_t^T \alpha(u)e^{A(u)-A(T)} \log \mu(u)\,\mathrm{d}u$$

and

$$\mathrm{Var}_Q[r(T) \mid \mathcal{F}_t] = \int_t^T \sigma(u)^2 e^{2(A(u)-A(T))}\,\mathrm{d}u.$$

Some other models (see, for example, the market models described in Chapter 9) have the same volatility, $\sigma(t)r(t)$, but different forms for the drift. These models are typically also described as *lognormal* models even though the distribution of $r(t)$ is not lognormal.

The model does not give us closed-form solutions for bond and option prices. However, the model is straightforward to implement using lattice-based numerical methods (see Chapter 10). Numerical procedures in a no-arbitrage framework require us first to calibrate the functions $\alpha(t)$, $\mu(t)$ and $\sigma(t)$. Black and Karasinski suggest that this is done using three inputs:

- the zero-coupon yield curve (that is, the spot rates $R(t, T)$);

- the curve of the volatilities of each of the $R(t, T)$;

- the *cap curve* (that is, the prices of at-the-money interest rate caps).

The lattice framework discretizes time into n intervals $t = t_0, t_1, \ldots, t_n = T$, where $t_j = t + (T - t)j/n$. We assume that each of the $\alpha(u)$, $\mu(u)$ and $\sigma(u)$ is constant over each interval $[t_j, t_{j+1})$ and calculate each of these values recursively for $j = 0, 1, \ldots$.

A drawback of the Black–Karasinski model is that the expected accumulation of cash over any positive time interval, $E_Q[B(T)/B(t) \mid \mathcal{F}_t]$, is infinite (Hogan and Weintraub 1993; Sandmann and Sondermann 1997). This means that the model cannot be used to price Eurodollar futures contracts. This particular problem with lognormal rates of interest was subsequently avoided by shifting attention away from the continuously compounding risk-free rate to effective rates of interest. This new approach is described in Chapter 9.

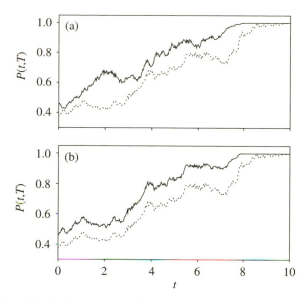

Figure 5.1. Possible simultaneous sample paths for $P(t, 8)$ and $P(t, 10)$. (a) Independent sources of randomness affecting prices for different maturities. (b) The same source of randomness affects both prices.

5.3 The Heath–Jarrow–Morton (HJM) Framework

This approach gives us a very general framework within which we can work to develop more specific no-arbitrage models; in particular, we find out how a general model for the term structure should evolve in a way that is arbitrage free and in which the initial forward-rate curve is part of the input.

In such a setting we find ourselves considering infinitely many processes, $f(t, T)$ for $0 \leqslant t < T$: one process for each $T \in \mathbb{R}$. The same level of complexity applies to the prices of zero-coupon bonds (see Figure 5.1(a)). Fortunately, our models need not be so complex. In particular, if a model depends upon, say, three sources of randomness, then we are able to consider the problem as three dimensional rather than infinite dimensional. Thus, given $f(t, T)$ for all T it is sufficient over the small interval $(t, t + \mathrm{d}t]$ that we know the changes in the forward-rate curve at only three maturity dates in order for us to specify the changes over the interval at all other maturities.

In this chapter we will consider one-factor models only. The multifactor development is described in Chapter 6.

We take the initial forward-rate curve $f(0, T)$ as our starting point. For a fixed maturity, T, $f(t, T)$ is an Itô process satisfying

$$\mathrm{d}f(t, T) = \alpha(t, T) \,\mathrm{d}t + \sigma(t, T) \,\mathrm{d}W(t)$$

or

$$f(t, T) = f(0, T) + \int_0^t \alpha(s, T)\, ds + \int_0^t \sigma(s, T)\, dW(s)$$

for each $T > t$, where $\alpha(t, T)$ and $\sigma(t, T)$ may depend upon $f(t, T)$, or the whole forward-rate curve at time t, or, even more generally, upon $\mathcal{F}_t = \sigma(\{W(s) : s \leqslant t\})$.

In a one-factor model we define, for all maturities T, Itô processes that are dependent upon the same one-dimensional source of uncertainty $W(t)$. It follows that changes over the whole of the forward-rate curve are perfectly but non-linearly correlated.

Technical conditions

(i) For all T, $\sigma(t, T)$ and $\alpha(t, T)$ are previsible and depend upon the history of $W(s)$ up to time t.

(ii) $\displaystyle\int_0^T \sigma^2(t, T)\, dt$ and $\displaystyle\int_0^T |\alpha(t, T)|\, dt$ are finite almost surely.

(iii) $\displaystyle\int_0^T \int_0^u |\alpha(t, u)|\, dt\, du$ is finite.

(iv) $f(0, T)$ is deterministic and satisfies $\displaystyle\int_0^T |f(0, u)|\, du < \infty$.

(v) $\displaystyle E\left[\left| \int_0^u \sigma(t, u)\, dW(u) \right| du \right] < \infty$.

5.3.1   The Risk-Free Asset

Since $d f(t, T) = \alpha(t, T)\, dt + \sigma(t, T)\, dW(t)$ we get

$$r(T) = \lim_{t \to T^-} f(t, T) = f(0, T) + \int_0^T \sigma(s, T)\, dW(s) + \int_0^T \alpha(s, T)\, ds.$$

$r(T)$ may or may not be Markov depending upon the form of $\sigma(s, T)$.

The cash account has value $B(t)$ with the stochastic differential equation

$$dB(t) = r(t)B(t)\, dt$$

$$\Rightarrow \quad B(t) = B(0) \exp\left[\int_0^t r(u)\, du \right]$$

$$= B(0) \exp\left[\int_0^t f(0, u)\, du + \int_0^t \int_s^t \alpha(s, u)\, du\, ds \right.$$

$$\left. + \int_0^t \left(\int_s^t \sigma(s, u)\, du \right) dW(s) \right].$$

In arriving at the third term in the exponential we made use of the technical condition (v) above which allows us to change the order of integration taking us from $\int_0^t \int_0^u (\sigma(s, u)\, dW(s))\, du$ to $\int_0^t (\int_s^t \sigma(s, u)\, du)\, dW(s)$.

5.3.2 Tradable Assets

The zero-coupon bonds in the market are priced in the obvious way:

$$P(t, T) = \exp\left[-\int_t^T f(t, u)\, du\right]$$

$$= \exp\left[-\int_0^t \left(\int_t^T \sigma(s, u)\, du\right) dW(s)\right.$$

$$\left. -\int_t^T f(0, u)\, du - \int_0^t \int_t^T \alpha(s, u)\, du\, ds\right].$$

Define the discounted asset price

$$Z(t, T) = \frac{P(t, T)}{B(t)}$$

$$= \exp\left[\int_0^t S(s, T)\, dW(s) - \int_0^T f(0, u)\, du - \int_0^t \int_s^T \alpha(s, u)\, du\, ds\right],$$

where $S(s, T) = -\int_s^T \sigma(s, u)\, du$. Using Itô's formula we then have

$$dZ(t, T) = Z(t, T)\left[\left(\tfrac{1}{2}S^2(t, T) - \int_t^T \alpha(t, u)\, du\right) dt + S(t, T)\, dW(t)\right].$$

Since the cash account, $B(t)$, has zero volatility, $S(t, T)$ is interpreted as the volatility of $P(t, T)$. Note that the volatility of the bond price is negative, since prices go down when interest rates go up.

5.3.3 Change of Measure

Now we want to make the discounted asset price into a martingale, which we can do by changing measure. The required change-of-measure drift (that is, the market price of risk) for the bond maturing at time T is

$$\gamma(t) = \tfrac{1}{2}S(t, T) - \frac{1}{S(t, T)}\int_t^T \alpha(t, u)\, du$$

(which is therefore a previsible process).

Referring to the Girsanov Theorem we note that $\gamma(t)$ must satisfy the Novikov condition $E_P[\exp(\tfrac{1}{2}\int_0^T \gamma(t)^2\, dt)] < \infty$. Then there will be a new measure Q equivalent to P such that

$$\tilde{W}(t) = W(t) + \int_0^t \gamma(s)\, ds$$

is a Brownian motion under Q. Under Q we have

$$dZ(t, T) = Z(t, T)S(t, T)\, d\tilde{W}(t).$$

Thus, $Z(t, T)$ is a martingale under Q, subject to the technical condition that

$$E_Q\left[\exp\left(\frac{1}{2}\int_0^T S^2(t, T)\,\mathrm{d}t\right)\right] < \infty.$$

It follows that

$$\mathrm{d}P(t, T) = P(t, T)(r(t)\,\mathrm{d}t + S(t, T)\,\mathrm{d}\tilde{W}(t)). \qquad (5.4)$$

Note that, as usual, the drift term under Q is equal to the risk-free rate times the price.

Here we have established what the appropriate change of measure should be, given the dynamics of a specific zero-coupon bond maturing at time T. In the next two subsections we will consider how no-arbitrage arguments can be used to determine what the relationship between $\alpha(t, T)$, $\sigma(t, T)$ and $\gamma(t)$ should be for all t and T.

5.3.4 Replicating Strategies

Suppose that we have a derivative which pays X (contingent upon $\mathcal{F}_S$) at time S ($S < T$). Let us see if we can construct a hedging strategy for X using cash and the T-bond, $P(t, T)$.

The five steps involved in constructing the replicating strategy are as follows (see also Section 4.2 on the martingale approach).

- Find the equivalent measure Q under which $Z(t, T)$ is a martingale.
- Define the Q-martingale $D(t) = E_Q[B(S)^{-1}X \mid \mathcal{F}_t]$.
- Find the previsible process $\phi(t)$ such that $D(t) = D(0) + \int_0^t \phi(s)\,\mathrm{d}Z(s, T)$.
- Define $\psi(t) = D(t) - \phi(t)Z(t, T)$.
- The trading strategy $(\psi(t), \phi(t))$ representing the number of units of $B(t)$ and $P(t, T)$ respectively is a self-financing, replicating strategy for the derivative payment X at time S.

We will now work through these five steps in more detail.

We have already established the equivalent martingale measure Q.

Let

$$D(t) = E_Q[B(S)^{-1}X \mid \mathcal{F}_t] \quad \text{for } t < S.$$

This is a martingale under Q because of the Tower Property for conditional expectation.

$Z(t, T)$ is also a Q-martingale. Thus, we can call on the Martingale Representation Theorem which tells us that there is some previsible process $\phi(t)$ such that

$$\mathrm{d}D(t) = \phi(t)\,\mathrm{d}Z(t, T) \quad \text{or} \quad D(t) = D(0) + \int_0^t \phi(s)\,\mathrm{d}Z(s, T).$$

Suppose that we have a trading strategy which holds $\phi(t)$ units of the T-bond, $P(t, T)$, at time t and $\psi(t) = D(t) - \phi(t)Z(t, T)$ units of the cash account, $B(t)$, at time t.

The value of this portfolio at time t is

$$V(t) = B(t)D(t) = B(t)E_Q[B(S)^{-1}X \mid \mathcal{F}_t].$$

It follows that the instantaneous change in the portfolio value is

$$
\begin{aligned}
dV(t) &= d(B(t)D(t)) \\
&= D(t)\,dB(t) + B(t)\,dD(t) + dB(t)\,dD(t) \\
&= r(t)B(t)D(t)\,dt + B(t)\phi(t)\,dZ(t, T) + 0.dt.
\end{aligned}
$$

Now $dP(t, T) = d(B(t)Z(t, T)) = r(t)B(t)Z(t, T)\,dt + B(t)\,dZ(t, T)$, so the instantaneous investment gain is

$$
\begin{aligned}
\psi(t)\,&dB(t) + \phi(t)\,dP(t, T) \\
&= (D(t) - \phi(t)Z(t, T))B(t)r(t)\,dt + \phi(t)[r(t)B(t)Z(t, T)\,dt + B(t)\,dZ(t, T)] \\
&= r(t)B(t)D(t)\,dt + \phi(t)B(t)\,dZ(t, T) \\
&= dV(t).
\end{aligned}
$$

So the investment strategy is self-financing.

Thus, $V(t)$ is the value at time t of the derivative which pays off X at time S.

5.3.5 The Arbitrage-Free Market

Suppose that $X = 1$ in the above problem; that is, the derivative is the zero-coupon bond which matures at time S. We have established that the fair price for this bond in an arbitrage-free bond market, given that it can be replicated using cash and the T-bond, must be

$$P(t, S) = B(t)E_Q[B(S)^{-1} \mid \mathcal{F}_t] = E_Q\left[\exp\left(-\int_t^S r(u)\,du\right) \bigg| \mathcal{F}_t\right].$$

The discounted S-bond is

$$Z(t, S) = \frac{P(t, S)}{B(t)} = E_Q[B(S)^{-1} \mid \mathcal{F}_t],$$

so that $Z(t, S)$ is a Q-martingale. This, of course, must be true for all bonds. It follows that they must all be turned into martingales by making the same change of measure; that is, they all have the same market price of risk $\gamma(t)$. Thus, for all maturities T,

$$\tfrac{1}{2}S(t, T) - \frac{1}{S(t, T)}\int_t^T \alpha(t, u)\,du = \gamma(t)$$

or

$$\int_t^T \alpha(t, u) \, du = \tfrac{1}{2} S(t, T)^2 - \gamma(t) S(t, T).$$

We differentiate with respect to T to find that

$$\alpha(t, T) = \sigma(t, T)(\gamma(t) - S(t, T))$$

since

$$\frac{\partial}{\partial T} S(t, T) = -\sigma(t, T).$$

Now we return to the original model

$$\begin{aligned}
\mathrm{d} f(t, T) &= \alpha(t, T) \, \mathrm{d}t + \sigma(t, T) \, \mathrm{d}W(t) \\
&= \alpha(t, T) \, \mathrm{d}t + \sigma(t, T)(\mathrm{d}\tilde{W}(t) - \gamma(t) \, \mathrm{d}t) \\
&= -\sigma(t, T) S(t, T) \, \mathrm{d}t + \sigma(t, T) \, \mathrm{d}\tilde{W}(t)
\end{aligned}$$

and consequently

$$r(t) = f(0, t) - \int_0^t \sigma(s, t) S(s, t) \, \mathrm{d}s + \int_0^t \sigma(s, t) \, \mathrm{d}\tilde{W}(s).$$

In general, this means that $r(t)$ is not a Markov process.

5.4   Relationship between HJM and Markov Models

5.4.1   Ho and Lee

Suppose that we set $\sigma(s, t) = \sigma$ under HJM for all s and t, so that $S(s, t) = -(t-s)\sigma$ and we find that

$$r(t) = f(0, t) + \tfrac{1}{2}\sigma^2 t^2 + \sigma \tilde{W}(t).$$

This is of the form of $r(t)$ under the Ho and Lee model.

5.4.2   Hull and White

Suppose that $\sigma(s, t) = \sigma e^{-\alpha(t-s)}$. Then

$$S(s, t) = -\frac{\sigma}{\alpha}(1 - e^{-\alpha(t-s)})$$

$$\Rightarrow \quad -\int_0^t \sigma(s, t) S(s, t) \, \mathrm{d}s = \frac{\sigma^2}{\alpha} \int_0^t e^{-\alpha(t-s)}(1 - e^{-\alpha(t-s)}) \, \mathrm{d}s$$

$$= \frac{\sigma^2}{2\alpha^2}(1 - e^{-\alpha t})^2.$$

Hence

$$r(t) = f(0, t) + \frac{\sigma^2}{2\alpha^2}(1 - e^{-\alpha t})^2 + \sigma \int_0^t e^{-\alpha(t-s)} \, \mathrm{d}\tilde{W}(s),$$

which is of the form of the Hull and White model.

5.5 Exercises

Exercise 5.1. Consider the Ho and Lee model. Let $X(T) = \int_0^T \tilde{W}(t)\,dt$.

(a) What is the distribution of $X(T)$?

(b) Find $E_Q[\exp(-X(T))]$.

(c) Let $r(T) = r(0) + \int_0^T \theta(t)\,dt + \sigma\tilde{W}(T)$ for some deterministic function $\theta(t)$. Find an expression for $P(t, T)$.

(d) Hence show that if the initial forward-rate curve $f(0, T)$ is given, then

$$\theta(T) = \frac{\partial}{\partial T} f(0, T) + \sigma^2 T.$$

(e) Show that there exists a similar expression to that under the Vasicek model for the price of a European call option on a zero-coupon bond.

Exercise 5.2. Under the Ho and Lee model, show that the spot rate

$$R(t, T) = F(0, t, T) + \tfrac{1}{2}\sigma^2 t T(T - t) + \sigma\tilde{W}(t).$$

Exercise 5.3. Using the Ho and Lee model, derive a formula for the price at time t of a European put option with $P(t, S)$ as the underlying, $T < S$ as the exercise date and K as the strike price.

Exercise 5.4. Consider the Ho and Lee model. Given that the model was calibrated at time 0, show that for all $t > 0$, $f(t, T) \to +\infty$ as $T \to \infty$.

Exercise 5.5. Consider the Hull and White model. Suppose that $f(0, t) = 0.06 + 0.01e^{-0.2t}$. Suppose also that we know that

$$\lim_{t\to\infty} \text{Var}[r(t)] = \frac{\sigma^2}{2\alpha} = 0.02^2$$

is fixed.

(a) Investigate the form of $\mu(t)$ in the Hull and White model for various choices of α.

(b) For what value of α does $\mu(0) = \mu(\infty)$?

Exercise 5.6. Consider the zero-coupon bond pricing formula (5.1) using the Hull and White model. Reorganize this formula in a way which highlights the forward price at time 0 for the bond and the difference between the forward rate $f(0, t)$ and $r(t)$. Comment on your formula.

Exercise 5.7. Under the Hull and White model, suppose that $\alpha = 0.24$, $\sigma = 0.02$ and $f(0, t) = 0.06 + 0.01e^{-0.2t}$.

(a) Calculate the price of a 3-month European call option written on a zero-coupon bond which will mature in 10 years time with a nominal value of £100 and a strike price of £53.50.

(b) What is the minimum amount of information required to make the calculation in (a)?

Exercise 5.8. Let us generalize the Hull and White model to

$$dr(t) = \alpha(\mu(t) - r(t)) \, dt + \sigma(t) \, d\tilde{W}(t)$$

for deterministic functions $\mu(t)$ and $\sigma(t)$.

Show from first principles that

(a) $r(t) = e^{-\alpha t} r(0) + \alpha \int_0^t e^{-\alpha(t-s)} \mu(s) \, ds + \int_0^t e^{-\alpha(t-s)} \sigma(s) \, d\tilde{W}(s).$

(b) $\mu(t) = \dfrac{1}{\alpha} \dfrac{\partial}{\partial t} f(0, t) + f(0, t) + \dfrac{1}{\alpha} \int_0^t \sigma(s)^2 e^{-2\alpha(t-s)} \, ds.$

Exercise 5.9. Under the Heath–Jarrow–Morton framework we have

$$df(t, T) = \alpha(t, T) \, dt + \sigma(t, T) \, dW(t).$$

(a) Under what circumstances is such a model arbitrage free?

(b) Under what circumstances is this model Markov? (You should identify separately an 'impractical' and a 'practical/useful' definition.)

(c) Which of the following models are Markov under the equivalent martingale measure Q for a suitable drift term $\alpha(t, T)$.

    (i) $\sigma(t, T) = \sigma$ for all t, T.

    (ii) $\sigma(t, T) = (1 - e^{-\delta t})\sigma$ for all t, T.

    (iii) $\sigma(t, T) = \sigma(t)$ for all t, T, where $\sigma(t)$ is an Itô process satisfying the SDE $d\sigma(t) = a(m - \sigma(t)) \, dt + b\sqrt{\sigma(t)} \, d\hat{W}(t)$ and $\hat{W}(t)$ is a Brownian motion under Q which is independent of $\tilde{W}(t)$.

    (iv) $\sigma(t, T) = \sigma/(T - t + \delta)$ for all t, T and for some $\delta > 0$.

    (v) $\sigma(t, T) = \sigma e^{-\alpha(T-t)}$ for all t, T.

    (vi) $\sigma(t, T) = \sigma_1 e^{-\alpha_1(T-t)} + \sigma_2 e^{-\alpha_2(T-t)}$ for all t, T?

Exercise 5.10. Suppose

$$f(0, T) = \lambda_0 + \lambda_1 e^{-\alpha T} - \frac{\sigma^2}{2\alpha^2}(1 - e^{-\alpha T})^2,$$

$$\sigma(t, T) = \sigma e^{-\alpha(T-t)},$$

$$df(t, T) = \theta(t, T) \, dt + \sigma(t, T) \, d\tilde{W}(t),$$

where

$$\theta(t, T) = -\sigma(t, T)S(t, T), \qquad S(t, T) = -\int_t^T \sigma(t, u) \, du.$$

Derive a formula for $r(t)$ of the form

$$r(t) = g(t, r(0)) + \int_0^t h(s, t) \, d\tilde{W}(s)$$

for suitable deterministic functions g and h.

What name is given to this model?

Exercise 5.11. Suppose that the model $df(t, T) = \alpha(t, T) \, dt + \sigma(t, T) \, dW(t)$, where $W(t)$ is a Brownian Motion under the real-world measure P, is arbitrage free and where $\sigma(t, T)$ is deterministic. The initial forward-rate curve $f(0, u)$ is given.

(a) Why is $f(t, T)$ not necessarily Gaussian?

(b) Suppose that the market price of risk $\gamma(t)$ is deterministic. Prove that $f(t, T)$ is now Gaussian.

(c) Under the equivalent martingale measure Q we have

$$df(t, T) = -\sigma(t, T)S(t, T) \, dt + \sigma(t, T) \, d\tilde{W}(t),$$

where $\tilde{W}(t)$ is a Brownian motion under Q and $S(t, T) = -\int_t^T \sigma(t, u) \, du$.

Given $P(0, \tau)$ for all τ, show that for any $0 < t < T < \infty$ $P(t, T)$ is lognormally distributed under Q.

Exercise 5.12. The dynamics of zero-coupon prices are defined by

$$dP(t, T) = P(t, T)(r(t) \, dt + S(t, T) \, d\tilde{W}(t))$$

for all T, where $\tilde{W}(t)$ is Brownian motion under the equivalent martingale measure Q.

A coupon bond pays a coupon rate of g per annum continuously until the maturity date T when the nominal capital of 100 is repaid. The price at time t of this bond is denoted by $V(t)$.

(a) Show that for some functions a_V and b_V

$$dV(t) = a_V(t, r(t), V(t)) \, dt + b_V(t, r(t), \mathscr{P}(t)) \, d\tilde{W}(t),$$

where $\mathscr{P}(t) = \{P(t, u) : t \leqslant u \leqslant T\}$.

(b) Suppose that

$$
\begin{aligned}
P(0, u) &= e^{-0.1u} & &\text{for all } u, \\
S(t, u) &= -10\sigma(1 - e^{-0.1(u-t)}) & &\text{for all } t, u, \\
g &= 10.
\end{aligned}
$$

(i) What is $V(0)$ as a function of T?

(ii) What is the volatility of $V(t)$ at time 0 (that is, the $d\tilde{W}$ component of $dV(t)/V(t)$)?

(iii) Hence deduce that the irredeemable bond ($T = \infty$) has the highest volatility amongst all bonds with a coupon of 10%.

(iv) Give an example of a bond which has a higher volatility than the irredeemable 10% coupon bond.

Exercise 5.13. In this exercise we look at the pricing of equity derivatives in the presence of interest rate risk. We assume a simple model where equity volatility is independent of interest rate volatility.

Suppose that the risk-free rate of interest $r(t)$ follows the Hull and White interest rate model

$$dr(t) = \alpha(\theta(t) - r(t))\,dt + \sigma\,d\tilde{W}(t),$$

where $\tilde{W}(t)$ is standard Brownian motion under the *risk-neutral measure* Q, and $\theta(t)$ is a deterministic function which is determined by observed bond prices $P(0, T)$ at time 0.

A non-dividend-paying stock has price $R(t)$ at time t with

$$dR(t) = R(t)(\mu(t)\,dt + \sigma_R\,dW^R(t))$$

and $W^R(t)$ is a standard Brownian motion under the *real-world measure* P. We denote $\tilde{W}^R(t)$ for the equivalent Brownian motion to $W^R(t)$ driving stock prices under Q. $\tilde{W}^R(t)$ and $\tilde{W}(t)$ are independent.

A binary call option on the stock pays £1 at time T if $R(T) \geqslant K$ and £0 otherwise.

(a) Use the forward-measure approach (Chapter 7) to determine a value for this option at time $t < T$.

(b) Without developing formulae, discuss briefly how you would hedge this option in order to replicate the payoff.

(c) Suppose now that $\alpha = 0.5$, $\sigma = 0.03$, $\theta(t) = \theta = 0.06$, $r(0) = 0.03$, $\mu(t) = r(t) + 0.04$, $\sigma_R = 0.25$, $R(0) = 95$, $K = 100$ and $T = 0.5$.

Calculate the price at time 0 of the binary option,

(i) using the formula derived in part (a),

(ii) using the standard Black–Scholes model with a constant deterministic risk-free rate of $r = -(\log P(0, T))/T$,

and compare the results.

(d) Discuss whether or not the comparison in part (c) would be different if $R(0) = 105$.

6

Multifactor Models

6.1 Introduction

In previous chapters we have focused on one-factor models with the aim of getting across the main elements of the arbitrage-free theory of interest rate models. However, for a variety of reasons it is necessary or desirable to use models which include more than one source of randomness.

We need only to look at historical interest rate data to see that changes in interest rates with different maturities are not perfectly correlated as predicted by one-factor models. This can be seen to some extent in Figures 6.1–6.3 where we plot UK short-term interest rates (the Bank of England base rate) and consol (perpetual bond) yields. A plot of the raw rates (Figure 6.1) suggests a high degree of correlation between long and short rates and the possibility of a one-factor model. However, when we do some additional (but simplistic) analysis (Figures 6.2 and 6.3) we see that there is sufficient evidence to suggest that there is more than one random factor at play. These graphs do not prove the point, but they do add substance to the argument that a multifactor model would be appropriate. Certainly we can say that a one-factor, *time-homogeneous* model is inappropriate.

Data for the United States[1] (Figure 6.4) demonstrate similar features, again suggesting a multifactor model should be used.

A final reason for using a multifactor model is to deal with more-complex interest rates options which refer to two or more stochastic underlying quantities. For example, an option may be defined in terms of the difference between the one- and five-year spot rates. A one-factor model would possibly overprice this contract because of its assumption that the underlying rates are perfectly, non-linearly correlated.

---

[1] Source: www.federalreserve.gov. Data from 1986 to 1993 for the 20-year maturity date are missing. Actual 30-year bond yields have been substituted for this period. Where published data for 20-year and 30-year rates overlap, yields are very close, although 20-year yields tend to be slightly higher (see Brown and Schaefer 2000).

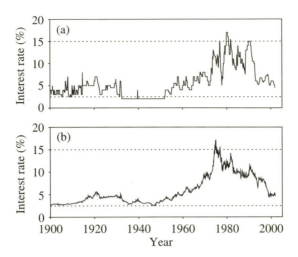

Figure 6.1. UK interest rates from 1900: (a) UK short-term Bank of England base rate; (b) UK consols yield (perpetual bonds).

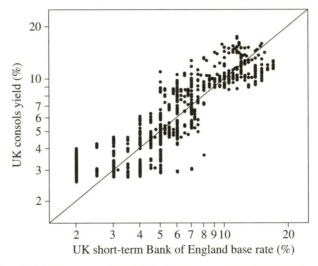

Figure 6.2. UK interest rates from 1900. Short-term versus long-term interest rates.

6.2 Affine Models

In Chapters 4 and 5 we investigated, amongst other things, the one-factor affine models by Vasicek (1977), Cox, Ingersoll and Ross (1985), Ho and Lee (1986) and Hull and White (1990). We will now look at multifactor models which have an affine form. Consider, then, a diffusion model with state variables $X_1(t), X_2(t), \ldots, X_n(t)$ (or, using vector notation, $X(t) = (X_1(t), X_2(t), \ldots, X_n(t))'$). The model is said

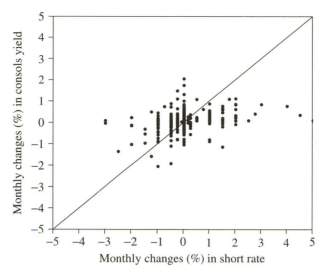

Figure 6.3. UK interest rates from 1900. Monthly changes in short-term interest rates versus monthly changes in long-term interest rates.

to be *affine* if the zero-coupon bond prices can be written in the form

$$P(t, T) = \exp\left[A(t, T) + \sum_{j=1}^{n} B_j(t, T)X_j(t)\right]$$

$$= \exp[A(t, T) + B(t, T)'X(t)], \tag{6.1}$$

where

$$B(t, T) = (B_1(t, T), \ldots, B_n(t, T))'. \tag{6.2}$$

The model is time homogeneous if both the state variables, $X(t)$, are time homogeneous and the functions $A(t, T)$ and $B(t, T)$ are functions of $T - t$ only. (These conditions, of course, are not independent.) Here we will restrict ourselves to time-homogeneous models for simplicity and use $A(T - t)$ and $B(T - t)$. In Chapter 4 we noted (Corollary 4.9) that in order for a time-homogeneous, one-factor model to have the affine form we would require $r(t)$ to have the SDE:

$$dr(t) = (a + br(t))\, dt + \sqrt{\gamma r(t) + \delta}\, d\tilde{W}(t).$$

We now ask the same question in the multifactor case, with the answer given in the following theorem.

Theorem 6.1 (Duffie and Kan 1996). *Suppose the $P(t, t + \tau)$ have the form* $\exp[A(\tau) + B(\tau)'X(t)]$. *Then $X(t)$ must have the SDE (using the n-dimensional Brownian motion $\tilde{W}(t)$ under Q)*

$$dX(t) = (\alpha + \beta X(t))\, dt + S\, D(X(t))\, d\tilde{W}(t), \tag{6.3}$$

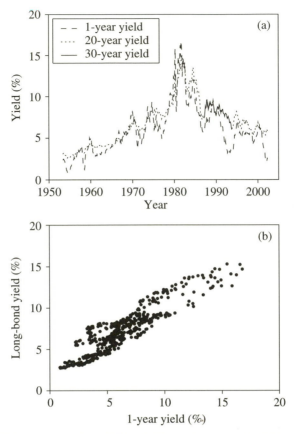

Figure 6.4. US interest rates from 1953 to 2002. (a) 1-year yields (thick solid curve) over the period 1953–2002. 20-year yields (dotted line) are plotted over the periods 1953–1986 and 1993–2002. 30-year yields (thin solid line) are plotted over the period 1986–1993. (b) Short-term versus long-term interest rates.

where $\alpha = (\alpha_1, \ldots, \alpha_n)'$ is a constant vector, $\beta = (\beta_{ij})_{i,j=1}^{n}$ is a constant matrix, $S = (\sigma_{ij})_{i,j=1}^{n}$ is a constant matrix, and, finally, $D(X(t))$ is the diagonal matrix:

$$
D(X(t)) = \begin{pmatrix}
\sqrt{\gamma_1' X(t) + \delta_1} & 0 & \cdots & \cdots & 0 \\
0 & \sqrt{\gamma_2' X(t) + \delta_2} & 0 & \cdots & 0 \\
\vdots & & \ddots & & \vdots \\
\vdots & & & \ddots & \vdots \\
0 & \cdots & \cdots & 0 & \sqrt{\gamma_n' X(t) + \delta_n}
\end{pmatrix},
$$

$$(6.4)$$

where $\delta_1, \ldots, \delta_n$ are constants and each $\gamma_i = (\gamma_{i1}, \ldots, \gamma_{in})'$ is a constant vector.

Proof. See Duffie and Kan (1996). □

Within this general framework it is necessary to ensure that each of the volatility processes $\gamma_i' X(t) + \delta_i$ remains positive (preferably strictly positive). Duffie and Kan (1996) provide conditions on the parameters for this to be the case corresponding to those in Theorem 4.8(d) for the one-factor CIR model.

Let us now look at the spot rates

$$R(t, t + \tau_j) = -\frac{1}{\tau_j}[A(\tau_j) + B(\tau_j)'X(t)] \quad \text{for } j = 1, \ldots, n,$$

for terms to maturity $\tau_1 < \tau_2 < \cdots < \tau_n$. In vector notation this is

$$R(t) = (R(t, t + \tau_1), \ldots, R(t, t + \tau_n))' = A_{\mathrm{R}} + B_{\mathrm{R}}X(t)$$

for a constant vector A_{R} and constant matrix B_{R} (given $\tau_1, \ldots, \tau_n$). This implies that

$$X(t) = B_{\mathrm{R}}^{-1}(R(t) - A_{\mathrm{R}}),$$

assuming that B_{R} is invertible. Hence

$$\begin{aligned} P(t, t + \tau) &= \exp[A(\tau) + B(\tau)'B_{\mathrm{R}}^{-1}(R(t) - A_{\mathrm{R}})] \\ &= \exp[\tilde{A}(\tau) + \tilde{B}(\tau)R(t)] \end{aligned}$$

for suitable functions $\tilde{A}(\tau)$ and $\tilde{B}(\tau)$. It follows that if the model is affine in a general process $X(t)$, then it can be reformulated in a way which is affine in $R(t)$. $R(t)$ must therefore have the same general form as $X(t)$ in Theorem 6.1.

These transformations make the n-factor model particularly simple to calibrate if we take as input n spot rates at each time t.

Models that are time homogeneous under Q do not need to be time homogeneous under the real-world measure P. In general, let $\lambda(t) = (\lambda_1(t), \ldots, \lambda_n(t))'$ be the vector of market prices of risk, which we assume to be previsible and to satisfy the Novikov condition. The SDE for $X(t)$ (equation (6.3) under Q) then becomes, under P,

$$\begin{aligned} \mathrm{d}X(t) &= (\alpha + \beta X(t))\,\mathrm{d}t + S\,D(X(t))(\mathrm{d}W(t) + \lambda(t)\,\mathrm{d}t) \\ &= (\alpha + \beta X(t) + S\,D(X(t))\lambda(t))\,\mathrm{d}t + S\,D(X(t))\,\mathrm{d}W(t). \end{aligned}$$

We will now describe briefly some specific multifactor affine models.

6.2.1 Gaussian Multifactor Models

Gaussian models have received little attention in the literature because of their drawback in allowing interest rates to become negative. The first multifactor extension of the Vasicek (1977) model was developed by Langetieg (1980). Later works include those by Beaglehole and Tenney (1991) (general theory) and Babbs and Nowman (1999) (parameter estimation using the Kalman filter). All time-homogeneous Gaussian models are based upon the following general model.

Let $X(t) = (X_1(t), \ldots, X_n(t))'$ be a diffusion process with SDE

$$dX(t) = BX(t)\,dt + K\,d\tilde{W}(t),$$

where B and K are some real-valued, constant $n \times n$ matrices and $\tilde{W}(t)$ is a standard n-dimensional Brownian motion under Q.

The risk-free rate of interest is

$$r(t) = \mu + \theta' X(t),$$

where $\theta = (\theta_1, \ldots, \theta_n)'$ is some vector of constants. If all of the θ_i are non-zero, then we can rescale the $X_i(t)$ and assume that all of the $\theta_i = 1$ without loss of generality. This is not possible where some of the θ_i equal 0.

Now the matrix B will have a spectral decomposition $B = B_R \Lambda B_L$, where

- $\Lambda = \text{diag}(\lambda_1, \ldots, \lambda_n)$ is the diagonal matrix of eigenvalues of B. Some of these eigenvalues may be complex. For $X(t)$ to be stationary we require that the real parts of all eigenvalues are negative.

- B_L and B_R are the matrices of left and right eigenvectors respectively of B; that is, column i of B_R is the right eigenvector of B corresponding to λ_i.

- The columns of B_R are scaled in a way which ensures that $B_R B_L = I$. It follows that $B^k = B_R \Lambda^k B_L$.

This decomposition is not unique.

Now define

$$Y(t) = e^{-\Lambda t} B_L X(t),$$

where, for any real-valued matrix A, $e^A = I + \sum_{k=1}^{\infty} A^k$. Application of Itô's formula then gives us

$$dY(t) = e^{-\Lambda t} B_L K\,d\tilde{W}(t).$$

Hence

$$Y(t) = Y(0) + \int_0^t e^{-\Lambda u} B_L K\,d\tilde{W}(u)$$

$$\Rightarrow \quad X(t) = B_R e^{\Lambda t} B_L X(0) + B_R e^{\Lambda t} \int_0^t e^{-\Lambda u} B_L K\,d\tilde{W}(u)$$

$$= e^{Bt} X(0) + \int_0^t e^{B(t-u)} K\,d\tilde{W}(u).$$

Note that the earlier requirement that the real parts of the eigenvalues of B are negative ensures that $\exp(\Lambda t)$ and e^{Bt} both tend to zero as t tends to infinity. Now let $R(T) = \int_0^T r(t)\,dt = \mu T + \int_0^T \theta' X(t)\,dt$. This is Normally distributed with

$$E_Q[R(T)] = \mu T + \theta' B_R \Lambda^{-1}(e^{\Lambda T} - I) B_L X(0)$$

and

$$\text{Var}_Q[R(T)] = \int_0^T \theta' B_R \Lambda^{-1}(e^{\Lambda t} - I) B_L K K' B_L'(e^{\Lambda t} - I)\Lambda^{-1} B_R' \theta\,dt.$$

We then have

$$P(0, T) = E_Q[\exp(-R(T)) \mid X(0)]$$
$$= \exp\{-E_Q[R(T) \mid X(0)] + \tfrac{1}{2}\text{Var}_Q[R(T) \mid X(0)]\}.$$

Example 6.2 (Beaglehole and Tenney 1991). We write $r(t) = (X_1(t) + \mu_1) + (X_2(t) + \mu_2)$, where $(X_1(t) + \mu_1)$ is the instantaneous rate of price inflation and $(X_2(t) + \mu_2)$ is the instantaneous real rate of interest with

$$dX_1(t) = -\alpha_1 X_2(t)\,dt + \sigma_{11}\,d\tilde{W}_1(t),$$
$$dX_2(t) = -\alpha_2 X_1(t)\,dt + \sigma_{21}\,d\tilde{W}_1(t) + \sigma_{22}\,d\tilde{W}_2(t).$$

Example 6.3.

$$r(t) = \mu + X_1(t),$$
$$dX_1(t) = \alpha_1(X_2(t) - X_1(t))\,dt + \sigma_{11}\,d\tilde{W}_1(t),$$
$$dX_2(t) = -\alpha_2 X_2(t)\,dt + \sigma_{21}\,d\tilde{W}_1(t) + \sigma_{22}\,d\tilde{W}_2(t).$$

In this model we treat $\mu + X_2(t)$ as a stochastic, *local* mean reversion level for $r(t)$.

6.2.2 Generalized CIR Models

A number of flexible multifactor CIR models have been proposed. In terms of equation (6.4) we have $\delta_i = 0$ for all i. Duffie (1996) describes the case where $X_1(t), \ldots, X_n(t)$ are independent processes of the one-factor CIR type. Thus, for $i = 1, \ldots, n$,

$$dX_i(t) = \alpha_i(\mu_i - X_i(t))\,dt + \sigma_i\sqrt{X_i(t)}\,d\tilde{W}_i(t),$$

where $\tilde{W}_1(t), \ldots, \tilde{W}_n(t)$ are independent and identically distributed (i.i.d.), standard Brownian motions under Q. The risk-free rate of interest is then defined as

$$r(t) = \sum_{i=1}^n X_i(t).$$

Now the $X_i(t)$ are independent, so we have

$$P(t, T) = E_Q\left[\exp\left(-\int_t^T r(u)\,du\right) \,\bigg|\, \mathcal{F}_t\right]$$

$$= E_Q\left[\exp\left(-\int_t^T \sum_{i=1}^n X_i(u)\,du\right) \,\bigg|\, \mathcal{F}_t\right]$$

$$= \prod_{i=1}^n E_Q\left[\exp\left(-\int_t^T X_i(u)\,du\right) \,\bigg|\, \mathcal{F}_t\right]$$

$$= \exp\left[\sum_{i=1}^n A_i(T-t) - \sum_{i=1}^n B_i(T-t)X_i(t)\right], \qquad (6.5)$$

where

$$A_i(\tau) = \frac{2\alpha_i\mu_i}{\sigma_i^2} \log\left(\frac{2\gamma_i e^{(\gamma_i+\alpha_i)\tau/2}}{(\gamma_i+\alpha_i)(e^{\gamma_i\tau}-1)+2\gamma_i}\right),$$

$$B_i(\tau) = \frac{2(e^{\gamma_i\tau}-1)}{(\gamma_i+\alpha_i)(e^{\gamma_i\tau}-1)+2\gamma_i}, \qquad \gamma_i = \sqrt{\alpha_i^2+2\sigma_i^2}.$$

Provided, for each $i = 1, \ldots, n$, we have $2\alpha_i\mu_i/\sigma_i^2 > 1$, the probability that any of the $X_i(t)$ will hit zero is zero.

Equation (6.5) follows from that fact that the $X_i(t)$ qualitatively all need the same treatment as $r(t)$ in the one-factor CIR model.

Remark 6.4. Consider the case $n = 2$. Suppose we choose one of the mean reversion rates, α_2 say, to be relatively small with σ_2 also small, while α_1 has a larger value similar to that in a typical one-factor version of the model. Over short timescales this will act rather like the one-factor Pearson and Sun (1994) model, since $X_2(t)$ will be almost constant. Over longer timescales $X_2(t)$ can vary substantially, giving us potentially long periods of both high and low interest rates.

An example of this is given in Figure 6.5. Here we combine two independent CIR-type processes, $X_1(t)$ and $X_2(t)$, with very different mean reversion rates, $\alpha_1 = 1$ and $\alpha_2 = 0.05$. Thus, we see that $X_1(t)$ has very short cycles of high and low values while $X_2(t)$ has very long cycles: typically of 15–20 years' duration. We then define $r(t) = X_1(t) + X_2(t)$ (see Figure 6.5(c)). On the one hand, we can see that the short-term volatility in $r(t)$ is mainly due to $X_1(t)$. On the other hand, there are sustained periods of both high and low $r(t)$, which are dictated by $X_2(t)$.

Thus, besides giving us the ability to price more-complex short-term derivatives, the two-factor model can be put to good use in the long-term risk management of bond portfolios. Models with such characteristics are, therefore, popular with life-insurers and pension plans that have very-long-term, fixed liabilities.

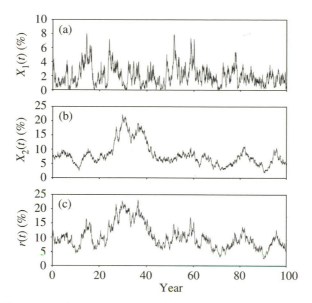

Figure 6.5. Simulation of a two-factor CIR model over 100 years. $r(t) = X_1(t) + X_2(t)$. $dX_i(t) = \alpha_i(\mu_i - X_i(t))\,dt + \sigma_i\sqrt{X_i(t)}\,d\tilde{W}_i(t)$. (a) $X_1(t)$ with $\mu_1 = 0.0225$, $\alpha_1 = 1$, $\sigma_1 = 0.15$ ($\Rightarrow d_1 = 4$ degrees of freedom). (b) $X_2(t)$ with $\mu_2 = 0.072$, $\alpha_2 = 0.05$, $\sigma_2 = 0.06$ ($\Rightarrow d_2 = 4$ degrees of freedom). (c) $r(t) = X_1(t) + X_2(t)$.

Countering this advantage, multifactor CIR models have the same disadvantage as the one-factor model; namely, that all interest rates other than $r(t)$ have a minimum value greater than 0. This means that we should be cautious when we are using the model to value derivatives that come into the money when interest rates get very low.

For the same set of parameter values as Figure 6.5, we plot a range of possible spot-rate curves in Figure 6.6. For this set of parameters we can see that the major innovation over the one-factor model is the ability to produce dipped curves as well as those with a slightly more pronounced hump. A feature of Figure 6.6 is the relatively wide range of values for the 30-year spot rate even though all curves ultimately converge to the same long-term value as $T \to \infty$. This spread of values is primarily dependent on the low value of α_2 (0.05). Finally, in the graph we can highlight the lowest curve (dashed line), which shows the lowest attainable values for the spot rate, $R(0, T)$.

6.2.3 *The Longstaff and Schwartz Model*

A variation on the two-factor version of the CIR model was proposed by Longstaff and Schwartz (1992). We take

$$dY_i(t) = \alpha_i(\mu_i - Y_i(t))\,dt + \sqrt{Y_i(t)}\,d\tilde{W}_i(t) \quad \text{for } i = 1, 2,$$

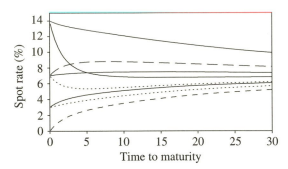

Figure 6.6. Spot-rate curves $R(0, T)$ for a two-factor CIR model with different initial conditions. $r(t) = X_1(t) + X_2(t)$. $dX_i(t) = \alpha_i(\mu_i - X_i(t))\,dt + \sigma_i\sqrt{X_i(t)}\,d\tilde{W}_i(t)$. $\mu_1 = 0.0225$, $\alpha_1 = 1$, $\sigma_1 = 0.15$. $\mu_2 = 0.072$, $\alpha_2 = 0.05$, $\sigma_2 = 0.06$. The lowest, dashed curve gives the minimum attainable spot-rate curve.

where $\tilde{W}_1(t)$ and $\tilde{W}_2(t)$ are independent. For positive constants c_1 and c_2, Longstaff and Schwartz define $r(t) = c_1 Y_1(t) + c_2 Y_2(t)$ and a second process $V(t) = c_1^2 Y_1(t) + c_2^2 Y_2(t)$.

Now consider

$$
\begin{aligned}
dr(t) &= c_1\,dY_1(t) + c_2\,dY_2(t) \\
&= \alpha_1 c_1(\mu_1 - Y_1(t))\,dt + c_1\sqrt{Y_1(t)}\,d\tilde{W}_1(t) \\
&\quad + \alpha_2 c_2(\mu_2 - Y_2(t))\,dt + c_2\sqrt{Y_2(t)}\,d\tilde{W}_2(t) \\
&= \left[\alpha_1 c_1 \mu_1 + \alpha_2 c_2 \mu_2 - \frac{(\alpha_1 c_2 - \alpha_2 c_1)r(t) + (\alpha_2 - \alpha_1)V(t)}{c_2 - c_1}\right]dt \\
&\quad + \sqrt{\frac{c_1(c_2 r(t) - V(t))}{c_2 - c_1}}\,d\tilde{W}_1(t) + \sqrt{\frac{c_2(V(t) - c_1 r(t))}{c_2 - c_1}}\,d\tilde{W}_2(t).
\end{aligned}
$$

It is straightforward to verify that the instantaneous variance of $r(t)$ is $V(t)\,dt$, which was the reason for formulating the model in this way. The use of $r(t)$ and $V(t)$ rather than $Y_1(t)$ and $Y_2(t)$ allows us to focus attention on variables that are believed to influence the pricing of derivatives; that is, level and volatility. In contrast, when we investigate models that concentrate on, say, level and slope or level and spread, it is often felt that knowledge of the second variable has little effect on prices.

Note that if $c_1 < c_2$ then the forms of $r(t)$ and $V(t)$ mean that $V(t)$ is limited at any point in time t to the range $(c_1 r(t), c_2 r(t))$. It follows that the richness of the model is enhanced when c_1 and c_2 are far apart.

A simulation example of the Longstaff and Schwartz model corresponding to the earlier example in Figure 6.5 is shown in Figures 6.7 and 6.8. In Figure 6.7 we plot the volatility $\sqrt{V(t)}$. In Figure 6.8 we plot $r(t)$ versus $V(t)$, showing how $V(t)$ always lies in the range $(c_1 r(t), c_2 r(t))$.

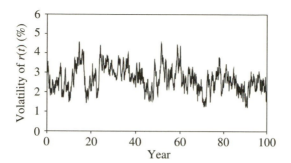

Figure 6.7. Simulation of Longstaff and Schwartz model over 100 years. $dY_i(t) = \alpha_i(\mu_i - Y_i(t)) dt + \sqrt{Y_i(t)} d\tilde{W}_i(t)$ with $\alpha_1 = 1$, $\mu_1 = 1$ and $\alpha_2 = 0.05$, $\mu_2 = 20$. $r(t) = c_1 Y_1(t) + c_2 Y_2(t)$, where $c_1 = 0.0225$ and $c_2 = 0.0036$. Instantaneous variance $V(t) = c_1^2 Y_1(t) + c_2^2 Y_2(t)$. The graph plots the instantaneous volatility $\sqrt{V(t)}$.

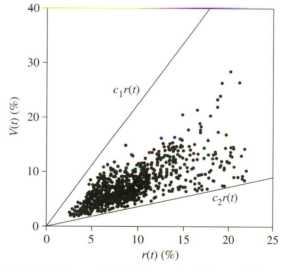

Figure 6.8. Simulation of Longstaff and Schwartz model over 100 years. Scatterplot of quarterly values of $r(t)$ versus $V(t)$. Solid lines: upper and lower bounds for $V(t)$ given $r(t)$.

A range of other multifactor affine models which are neither Gaussian nor multi-factor CIR are described in James and Webber (2000) and Bolder (2001) and in the references listed therein. However, we highlight here the paper of Dai and Single-ton (2000). They consider the full range of three-factor affine models. These are classified according to the number of state variables, $X_i(t)$, which are square-root processes (that is, $\gamma_i \neq 0$ in equation (6.4)). They carry out a careful statistical analysis of historical data before settling finally on a model with two square-root processes.

6.3   Consols Models

A number of two-factor models have been proposed which make use, either directly or indirectly, of the consols yield, $l(t)$, as well as $r(t)$. We assume that the consol is a perpetual bond which pays coupons continuously at the rate of 1 per annum and that it has no early redemption option.[2] If the yield (continuously compounding) is $l(t)$, then the price for the bond is

$$C(t) = \begin{cases} 1/l(t) & \text{if } l(t) > 0, \\ \infty & \text{if } l(t) \leqslant 0. \end{cases} \tag{6.6}$$

The general model proposed by Brennan and Schwartz (1979) has

$$dr(t) = \mu_r(r(t), l(t))\, dt + \sigma_r(r(t), l(t))\, dW_r(t),$$
$$dl(t) = \mu_l(r(t), l(t))\, dt + \sigma_l(r(t), l(t))\, dW_l(t),$$

where $W_r(t)$ and $W_l(t)$ are Brownian motions under the real-world measure P with instantaneous covariance $\rho\, dt$. For notational convenience we will use $\mu_r(r, l) \equiv \mu_r(r(t), l(t))$, etc. Given (6.6) the model needs to be formulated in a way which excludes the possibility that $l(t)$ can ever hit 0 or become negative.

Now consider the dynamics of $C(t)$. Under P we have

$$dC(t) = \left[-\frac{\mu_l(r, l)}{l(t)^2 + \sigma_l(r, l)^2} l(t)^3 \right] dt - \frac{\sigma_l(r, l)}{l(t)^2}\, dW_l(t). \tag{6.7}$$

Suppose that we wish to derive a full, arbitrage-free term structure. This requires us to establish the market prices of risk associated with each source of randomness. We look at the consol first. Under the risk-neutral measure Q we need the expected instantaneous rate of return on the asset from t to $t + dt$ including the coupon payment of $1\, dt$ to be equal to $C(t)r(t)\, dt$. This means that

$$dC(t) = (C(t)r(t) - 1)\, dt - \frac{\sigma_l(r, l)}{l(t)} C(t)\, d\tilde{W}_l(t), \tag{6.8}$$

where $d\tilde{W}_l(t) = dW_l(t) + \lambda_l(r, l)\, dt$ and $\lambda_l(r, l) \equiv \lambda_l(r(t), l(t))$ is the market price of risk associated with $W_l(t)$. Equating (6.7) with (6.8) we find that

$$\lambda_l(r(t), l(t)) = \frac{\mu_l(r(t), l(t)) + r(t)l(t) - l(t)^2}{\sigma_l(r(t), l(t))} - \frac{\sigma_l(r(t), l(t))}{l(t)}. \tag{6.9}$$

This can be reformulated as

$$\mu_l(r(t), l(t)) = \lambda_l(r(t), l(t))\sigma_l(r(t), l(t)) + l(t)^2 - r(t)l(t) + \frac{\sigma_l(r(t), l(t))^2}{l(t)}. \tag{6.10}$$

---

[2]Often, in fact, such bonds allow the issuer to redeem the bond for a specified price, K, at any given time (that is, if the prices ever reach or exceed K). However, this early redemption option typically, but not always, has a very low value because of the low coupon rate.

To some extent the market price of risk associated with $dW_r(t)$ is arbitrary until a further risky asset is introduced.

6.3.1 The Brennan and Schwartz Consols Model

In their paper, Brennan and Schwartz (1979) propose, for simplicity, that $\lambda_l(r, l) = \lambda_2$ is constant, along with the model formulation

$$\sigma_r(r, l) = \sigma_1 r, \qquad \sigma_l(r, l) = \sigma_2 l, \qquad \mu_r(r, l) = r\left[\alpha \log \frac{l}{pr} + \tfrac{1}{2}\sigma_1^2\right]$$

for some constant p. The form of $\mu_r(r, l)$ means that $\log r(t)$ has a local mean reversion level equal to $\log l(t)$ plus a constant depending on σ_1 and p.[3]

It is sometimes useful to work with independent Brownian motions. Therefore, let $W_1(t)$ and $W_2(t)$ be independent Brownian motions under P with $W_l(t) \equiv W_2(t)$ and $dW_r(t) = \sqrt{1 - \rho^2}\,dW_1(t) + \rho\,dW_2(t)$. Consequently, define $\sigma_{12} = \rho\sigma_1$, $\sigma_{11} = \sqrt{1 - \rho^2}\sigma_1$, $\sigma_{21} = 0$ and $\sigma_{22} = \sigma_2$. Then we have

$$dr(t) = \mu_r(r, l)\,dt + r(t)(\sigma_{11}\,dW_1(t) + \sigma_{12}\,dW_2(t))$$
$$= \tilde{\mu}_r(r, l)\,dt + \sigma_{11}r(t)\,d\tilde{W}_1(t) + \sigma_{12}r(t)\,d\tilde{W}_2(t), \qquad (6.11)$$
$$dl(t) = \mu_l(r, l)\,dt + \sigma_{22}l(t)\,dW_2(t)$$
$$= \tilde{\mu}_l(r, l)\,dt + \sigma_{22}l(t)\,d\tilde{W}_2(t), \qquad (6.12)$$

where

$$\tilde{\mu}_r(r, l) = \mu_r(r, l) - \sigma_{11}r(t)\lambda_1(r, l) - \sigma_{12}r(t)\lambda_2, \qquad (6.13)$$
$$\tilde{\mu}_l(r, l) = l(t)(\sigma_{22}^2 + l(t) - r(t)). \qquad (6.14)$$

(Note that the form of $\tilde{\mu}_l(r, l)$ is dictated by λ_2 and $\mu_l(r, l)$ in equation (6.10).) Finally, $\tilde{W}_1(t)$ and $\tilde{W}_2(t)$ are independent Brownian motions under the risk-neutral measure Q and $\lambda_1(r, l)$ is the market price of risk associated with $W_1(t)$.

Users of the Brennan and Schwartz model gradually identified some problems with the structure of the model and its parametrization. For example, the measures P and Q are not necessarily equivalent because of the possibility that $r(t)$ or $l(t)$ might hit zero in finite time under one measure and not the other. This endangers the arbitrage-free nature of the model. Of more concern was the observation that $l(t)$ could explode in finite time[4] (see, for example, Rebonato 1998): a consequence of the $l(t)^2$ term in the drift. This issue was formally resolved and further generalized by Hogan (1993) who proved that the process does indeed explode.

---

[3]We will see shortly that this model has a fundamental problem with stability. However, it is useful from an educational point of view to include models which are known to be flawed for one reason or another, provided these flaws are clearly highlighted.

[4]That is, tend to infinity in finite time.

Theorem 6.5 (Hogan 1993). *Assume that under Q we have*

$$dC(t) = (r(t)C(t) - 1)\,dt - C(t)^2 \sigma_l(r, l)\,d\tilde{W}_l(t)$$

and that the general stochastic differential equation for $r(t)$ is

$$dr(t) = \tilde{\mu}_r(r, l)\,dt + \sigma_r(r, l)\,d\tilde{W}_r(t),$$

with $d\langle \tilde{W}_r, \tilde{W}_l \rangle(t) = \rho\,dt$, $-1 < \rho < 1$. Also assume that $\sigma_r(r, l) = \eta_r(r)$ and $\sigma_l(r, l) = \eta_l(l)$ with $\eta_r(r)$ and $\eta_l(l)$ satisfying certain Lipschitz conditions (see Hogan 1993). If one of the following expressions for $\tilde{\mu}_r(r, l)$ holds,

(i) $\tilde{\mu}_r(r, l) = \alpha + \beta(l - r)$;

(ii) $\tilde{\mu}_r(r, l) = \alpha + \beta r(l - r)$; *or*

(iii) $\tilde{\mu}_r(r, l) = \alpha r + \beta r \log(l/r)$;

then the model explodes; that is, either $r(t)$ or $l(t)$ reaches ∞ in finite time almost surely.

Proof. See Hogan (1993).[5]   □

Note that case (iii) corresponds to the Brennan and Schwartz model.

The form of the Hogan (1993) theorem indicates that there is a certain amount of danger associated with taking this approach to multifactor models. Thus, we have seen that while the general Brennan and Schwartz (1979) formulation is sound (in particular, equation (6.10)), we have to be careful when we propose specific models to check for stability. There are, of course, plenty of stable multifactor models (formulated within other frameworks) under which the implied yields on consols are also stable. This indicates that it should be possible, if we are careful, to construct directly a model for $l(t)$ and $r(t)$ which is stable.

6.3.2 The Rebonato Lognormal Model

An alternative consols model to the Brennan and Schwartz (1979) model was proposed by Rebonato (1998). We define $k(t) = r(t)/l(t)$ and use the lognormal models [6]

$$dl(t) = l(t)[m_2(k, l)\,dt + \sigma_2\,d\tilde{W}_2(t)],$$

$$dk(t) = k(t)[m_1(t, k, l)\,dt + \sigma_1\,d\tilde{W}_1(t)],$$

---

[5] Note that there are several misprints in this paper including the statement of the theorem.

[6] As we have noted elsewhere in this book the name *lognormal model* is often applied to models in which the distribution of $r(t)$ is in fact not lognormal but where its volatility, $\sigma_r(r, l)$, is proportional to $r(t)$.

where σ_1 and σ_2 are constants and $\tilde{W}_1(t)$ and $\tilde{W}_2(t)$ are independent Brownian motions under Q. From the original Brennan and Schwartz (1979) result, no arbitrage indicates that

$$l(t)m_2(k(t), l(t)) \equiv \tilde{\mu}_l(k(t)l(t), l(t)).$$

Thus, we have

$$m_2(k, l) = \sigma_2^2 + l(t) - r(t) = \sigma_2^2 + l(t)(1 - k(t)).$$

Since $r(t) = k(t)l(t)$ we also have

$$dr(t) = r(t)[(\sigma_2^2 + (l(t) - r(t)) + m_1(t, k, l))\,dt + \sigma_1\,d\tilde{W}_1(t) + \sigma_2\,d\tilde{W}_2(t)]. \quad (6.15)$$

Rebonato (1998) suggests that

$$m_1(t, k, l) = \theta_0(t) + \theta_1 k(t)$$

for some deterministic function $\theta_0(t) > 0$, to be calibrated using current market data. The other parameter θ_1 is negative to ensure that $k(t)$ is mean reverting locally to $-\theta_0(t)/\theta_1$, thereby preventing $k(t)$ from exploding in finite time.

Although the form of $m_2(k, l)$ suggests that the process might be explosive, Rebonato (1998) reports that the model appears to be stable during simulations. However, the model cannot prevent $k(t)$ from spending periods of time between 0 and 1. During such periods the drift of $l(t)$ is always positive and leads to the potential for explosions. The model should, therefore, be treated with a degree of caution while acknowledging that the model is far less unstable than the Brennan and Schwartz (1979) model.

Other consols models have been proposed by Schaefer and Schwartz (1984) and Rhee (1999). Additional theoretical considerations can be found in Duffie, Ma and Yong (1995). They discuss the critical link between the consols price process $C(t) = 1/l(t)$ and $r(t)$.

Overall it seems that modelling $l(t)$ is fraught with difficulty, thereby explaining the relative scarcity of papers taking this approach.

6.4 Multifactor Heath–Jarrow–Morton Models

Multifactor models which operate within the no-arbitrage setting rather than the time-homogeneous setting are also straightforward to develop from their single-factor counterparts.

The one-factor Heath–Jarrow–Morton framework described in Section 5.3 is easily extended to cover more than one factor.

We will assume the existence of a given model under the risk-neutral measure Q. Thus we have

$$dP(t, T) = P(t, T)[r(t)\,dt + S(t, T)'\,d\tilde{W}(t)],$$

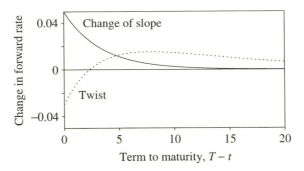

Figure 6.9. Volatility term structure for the two-factor Heath–Jarrow–Morton model. Solid curve is $\sigma_1(t, T) = 0.05e^{-0.3(T-t)}$. Dotted curve is $\sigma_2(t, T) = -0.08e^{-0.3(T-t)} + 0.05e^{-0.1(T-t)}$.

where the volatility function $S(t, T) = (S_1(t, T), \ldots, S_n(t, T))'$ is previsible and $\tilde{W}(t)$ is a standard n-dimensional Brownian motion under Q. If we next look at forward rates, then we get

$$\mathrm{d}f(t, T) = -\sigma(t, T)'S(t, T)\, \mathrm{d}t + \sigma(t, T)'\, \mathrm{d}\tilde{W}(t),$$

where

$$\sigma(t, T) = (\sigma_1(t, T), \ldots, \sigma_n(t, T))',$$

$$\sigma_i(t, T) = -\frac{\partial}{\partial T} S_i(t, T).$$

Example 6.6. A popular version of this framework is based upon the results of principal components analyses. These conclude that the primary source of randomness in the forward-rate curve is a change in the overall slope of the curve, followed, secondly, by a twist in the curve (see, for example, Rebonato 1996). An example of this is given in Figure 6.9. A simple implementation of this within a Gaussian setting takes

$$\sigma_1(t, T) = \sigma_{11}e^{-\alpha_1(T-t)},$$

$$\sigma_2(t, T) = \sigma_{21}e^{-\alpha_1(T-t)} + \sigma_{22}e^{-\alpha_2(T-t)}.$$

If, for example, $0 < \alpha_2 < \alpha_1$, the twist is achieved by adding the constraints that $\sigma_{22} > 0$ and $\sigma_{21} < -\sigma_{22}$.

6.5 Options on Coupon-Paying Securities

In Section 4.10, we presented the simple formula developed by Jamshidian (1989) for pricing options on coupon bonds or other securities with a series of fixed payments. Recall then that we have zero-coupon bonds $P(t, T, X(t))$, where $X(t)$ is the multifactor state variable, an underlying coupon bond $C(t) = \sum_{i=1}^{n} c_i P(t, T_i, X(t))$

(with $c_i > 0$ for $i = 1, \ldots, n$), and a call option on $C(t)$ with exercise date T and exercise price K.

Unfortunately, this result does not extend to a multifactor setting. There we exploited the fact that the prices of all zero-coupon bonds are strictly decreasing functions of the single state variable $r(t)$. This allowed us to determine a unique set of strike prices $K_1, \ldots, K_n$ for each of the zero-coupon bonds and a threshold risk-free rate r_0 at which the zero-coupon-bond prices $P(T, T_i, r_0) = K_i$ for all i and the coupon-bond price $C(T, r_0) = \sum_{i=1}^{n} c_i K_i = K$.

In a multifactor setting this no longer holds. There is a range of values for $X(T)$ for which $C(T, X(T)) = K$. In general, however, the $P(T, T_i, X(T))$ are not constant over the same set of values for $X(T)$.

What is, perhaps, less obvious is that this means that the Jamshidian approach will always overvalue the option on a coupon bond because of its assumption that the prices of all zero-coupon bonds are all perfectly non-linearly correlated. More formally, this is expressed in the following results.

Lemma 6.7. *For any set of numbers $y_1, \ldots, y_n \in \mathbb{R}$:*

$$\max \left\{ \sum_{i=1}^{n} y_i, 0 \right\} \leqslant \sum_{i=1}^{n} \max\{y_i, 0\}.$$

Theorem 6.8. *For any proposed set of zero-coupon-bond strike prices $K_1, \ldots, K_n$, with $\sum_{i=1}^{n} c_i K_i = K$, let $Y_i(X(T)) = c_i (P(T, T_i, X(T)) - K_i)$ so that*

$$C(T, X(T)) - K = \sum_{i=1}^{n} Y_i(X(T)).$$

Then, for all $X(T)$,

$$\max\{C(T, X(T)) - K, 0\} \leqslant \sum_{i=1}^{n} c_i \max\{P(T, T_i, X(T)) - K_i, 0\}. \qquad (6.16)$$

It follows that the price for the call option on the coupon bond is less than or equal to the sum of the prices of the call options on the zero-coupon bonds multiplied by the c_i.

The extent of this inequality can vary. In some cases the error might be quite significant. In other cases the error might in fact be quite small. For example, Pelsser (2003) argues that where $C(T)$ (an annuity in Pelsser's case) is made up mainly of payments far into the future, then the majority of the $P(T, T_i, X(T))$ are highly correlated with $C(T, X(T))$. This high correlation means that the Jamshidian approach is still a good approximation, provided we have minimized over the K_i on the right-hand side of (6.16).

6.6 Quadratic Term-Structure Models (QTSMs)

This general class of model is discussed in full generality by Ahn, Dittmar and Gallant (2002). Their work significantly generalized specific cases analysed by Longstaff (1989), Constantinides (1992) and Beaglehole and Tenney (1992). Additionally, the theoretical description of the general model contained in Ahn, Dittmar and Gallant was preceded by Rogers (1997) (see also the special case described later in Section 8.5, Example 3).

Ahn, Dittmar and Gallant (2002) start with an N-factor Gaussian diffusion, $Y(t)$, governed by the stochastic differential equation:

$$dY(t) = (\mu + \xi Y(t)) \, dt + \Sigma \, dW(t),$$

where μ is an $N \times 1$ vector, ξ and Σ are $N \times N$ matrices, and $W(t)$ is a standard N-dimensional Brownian motion under P. The risk-free rate of interest is defined as a quadratic function of $Y(t)$:

$$r(t) = \alpha + \beta' Y(t) + Y(t)' \Psi Y(t),$$

where $\alpha - \frac{1}{4} \beta' \Psi \beta \geqslant 0$ to ensure almost sure positivity of $r(t)$.

Prices are determined via the use of a state-price density process, as in Constantinides (1992), which has a specific functional form involving $Y(t)$ (see Chapter 8). Specifically, we find that

$$P(t, T) = \exp[A(T - t) + B(T - t)' Y(t) + Y(t)' C(T - t) Y(t)]$$

where $A(\tau)$ is a scalar function, $B(\tau)$ is $N \times 1$, and $C(\tau)$ is $N \times N$, and each function can be found be solving a straightforward set of ordinary differential equations (normally using numerical methods).

Ahn, Dittmar and Gallant (2002) continue with an extensive statistical analysis of the three-factor version of the model. They report that it fits relevant historical data significantly better than the best of the three-factor affine models fitted by Dai and Singleton (2000) (see Section 6.2 above).

6.7 Other Multifactor Models

The multifactor models we have discussed in this chapter have all (with the exception of the QTSMs above) carried out pricing calculations using the risk-neutral measure Q. In subsequent chapters (8 and 9) we will discuss two relatively new approaches (the positive interest framework and market models respectively). These use a pricing measure which is different from both P and Q. An introduction to this further change of measure is given in the next chapter. Both of these new approaches immediately set themselves up for the development of multifactor models.

6.8 Exercises

Exercise 6.1. Consider the multifactor, affine term-structure model in Section 6.2.

(a) Show that

$$r(t) = -\frac{dA(\tau)}{d\tau}\bigg|_{\tau=0} - \left(\frac{dB(\tau)}{d\tau}\bigg|_{\tau=0}\right)' X(t).$$

(b) Based on this model derive the HJM term structure of volatility $\sigma(t, T)$.

Exercise 6.2. Derive pricing formulae for the models described in Examples 6.2 and 6.3.

Exercise 6.3. Show that the two-factor model in Example 6.6 can be written down as a Gaussian model which is a generalization of the one-factor Hull and White model (Section 5.2.2).

7

The Forward-Measure Approach

In the problems that we have considered so far, it has been convenient to express prices in terms of expectations under the risk-neutral equivalent martingale measure Q. Sometimes it may be straightforward to evaluate an expected value (for example, pricing *bonds* under the Vasicek model). In other cases the calculation can be quite complex. Sometimes the calculation can be made much easier by making a second change of measure from the risk-neutral measure, Q, to what we call a *forward* measure. This is the case for the pricing of *bond options* under the Vasicek model.

In the present chapter, then, the change of measure is applied in circumstances where the underlying model has already been established under Q.

In later chapters we will see how the introduction of the forward-measure concept has been critical in recent years in the development of totally new approaches to interest rate modelling. In contrast to the previous chapters these new approaches allow the development of models under the forward measure rather than Q.

7.1 A New Numeraire

We will use the same notation as before. Thus, for a bond maturing at time T, we have the stochastic differential equation

$$dP(t, T) = P(t, T)[r(t)\, dt + S(t, T)\, d\tilde{W}(t)],$$

where $\tilde{W}(t)$ is a Brownian motion under Q.

Previously, we considered the discounted price process

$$Z(t, U) = \frac{P(t, U)}{B(t)},$$

where $B(t) = \exp[\int_0^t r(s)\, ds]$ is the cash account at time t. The discounted price was a martingale under the same measure Q for all maturity dates U.

Let us now change the numeraire from $B(t)$ to $P(t, T)$. Thus we consider

$$Y(t, U) = \frac{P(t, U)}{P(t, T)}.$$

Then

$$dY(t, U) = \frac{1}{P(t, T)} \, dP(t, U) + P(t, U) \, d\left(\frac{1}{P(t, T)}\right)$$
$$+ dP(t, U) \, d\left(\frac{1}{P(t, T)}\right)$$
$$\Rightarrow \quad dY(t, U) = Y(t, U)(S(t, U) - S(t, T)) \, d\tilde{W}(t)$$
$$+ Y(t, U)S^2(t, T) \, dt - Y(t, U)S(t, U)S(t, T) \, dt.$$

Suppose that we define a new process $\hat{W}(t)$ with

$$\hat{W}(0) = 0 \quad \text{and} \quad d\hat{W}(t) = d\tilde{W}(t) - S(t, T) \, dt.$$

Then

$$dY(t, U) = Y(t, U)(S(t, U) - S(t, T))(d\hat{W}(t) + S(t, T) \, dt)$$
$$- Y(t, U)S(t, T)(S(t, U) - S(t, T)) \, dt$$
$$= Y(t, U)(S(t, U) - S(t, T)) \, d\hat{W}(t). \tag{7.1}$$

Thus, it appears that we may be able to find a suitable measure under which the $Y(t, U)$ for all $U > t$ are martingales.

7.2 Change of Measure

Define $\gamma(t) = -S(t, T)$. By Girsanov's theorem (see Appendix A) there exists a measure P_T which is equivalent to Q (and hence equivalent to the real-world measure P) such that

$$\hat{W}(t) = \tilde{W}(t) + \int_0^t \gamma(s) \, ds$$

is a Brownian motion under P_T. This transformation requires the technical (Novikov) condition $E_Q[\exp(\int_0^U \frac{1}{2}\gamma(t)^2 \, dt)] < \infty$ to hold.

P_T is called a *forward measure*. Note that this forward measure depends upon the forward maturity date T but not upon the maturity date U of the bond under consideration. Thus, the $Y(t, U)$ for $U > T$ are all martingales under the measure P_T.

7.3 Derivative Payments

Let X be a derivative payment contingent upon $\mathcal{F}_T$ payable at time T.

Define

$$D(t) = E_{P_T}\left[\frac{X}{P(T, T)} \,\middle|\, \mathcal{F}_t\right] = E_{P_T}[X \mid \mathcal{F}_t].$$

Hence $D(t)$ is a martingale under P_T. But $Y(t, U)$ is also a martingale under P_T. Therefore, by the Martingale Representation Theorem, there exists a previsible process $\phi(t)$ such that

$$dD(t) = \phi(t)\,dY(t, U).$$

7.4 A Replicating Strategy

Let $\psi(t) = D(t) - \phi(t)Y(t, U)$.

Consider the investment strategy which holds, at time t, $\psi(t)$ units of the numeraire $P(t, T)$ and $\phi(t)$ units of the U-bond. (Previously, when we were considering dynamics under the risk-neutral measure Q, the replicating strategy held different numbers of units of *cash* plus the U-bond.) The value of this portfolio at time t is $V(t) = P(t, T)D(t)$.

Now

$$
\begin{aligned}
dP(t, T) &= P(t, T)(r(t)\,dt + S(t, T)\,d\tilde{W}(t)) \\
&= P(t, T)(r(t)\,dt + S(t, T)(d\hat{W}(t) + S(t, T)\,dt)) \\
&= P(t, T)((r(t) + S^2(t, T))\,dt + S(t, T)\,d\hat{W}(t)), \\
dD(t) &= \phi(t)\,dY(t, U) \\
&= \phi(t)Y(t, U)(S(t, U) - S(t, T))\,d\hat{W}(t) \qquad \text{(from equation (7.1)).}
\end{aligned}
$$

Thus,

$$
\begin{aligned}
dV(t) &= D(t)\,dP(t, T) + P(t, T)\,dD(t) + dP(t, T)\,dD(t) \\
&= P(t, T)D(t)((r(t) + S^2(t, T))\,dt + S(t, T)\,d\hat{W}(t)) \\
&\quad + \phi(t)P(t, T)Y(t, U)(S(t, U) - S(t, T))\,d\hat{W}(t) \\
&\quad + \phi(t)P(t, T)Y(t, U)[(r(t) + S^2(t, T))\,dt + S(t, T)\,d\hat{W}(t)] \\
&\qquad \times [(S(t, U) - S(t, T))\,d\hat{W}(t)] \\
&= V(t)((r(t) + S^2(t, T))\,dt + S(t, T)\,d\hat{W}(t)) \\
&\quad + \phi(t)P(t, U)(S(t, U) - S(t, T))\,d\hat{W}(t) \\
&\quad + \phi(t)P(t, U)S(t, T)(S(t, U) - S(t, T))\,dt.
\end{aligned}
$$

The instantaneous investment gain is

$$
\begin{aligned}
\psi(t)\,&dP(t, T) + \phi(t)\,dP(t, U) \\
&= (D(t) - \phi(t)Y(t, U))P(t, T)((r(t) + S^2(t, T))\,dt + S(t, T)\,d\hat{W}(t)) \\
&\quad + \phi(t)P(t, U)(r(t)\,dt + S(t, U)(d\hat{W}(t) + S(t, T)\,dt))
\end{aligned}
$$

$$= P(t, T)D(t)[(r(t) + S^2(t, T))\, dt + S(t, T)\, d\hat{W}(t)]$$
$$+ \phi(t)P(t, T)Y(t, U)[-r(t) - S^2(t, T) + r(t) + S(t, U)S(t, T)]\, dt$$
$$+ \phi(t)P(t, T)Y(t, U)[-S(t, T) + S(t, U)]\, d\hat{W}(t)$$
$$= dV(t).$$

Hence $(\psi(t), \phi(t))$ is a self-financing strategy. Furthermore, $V(T) = X$, so that the strategy is also a replicating strategy.

This development allows us to state, now, the following theorem.

Theorem 7.1. *The price at time t for the derivative X payable at time T is*

$$V(t) = P(t, T)E_{P_T}[X \mid \mathcal{F}_t].$$

This *price* can be seen to be independent of the choice of the hedging asset $P(t, U)$, although the *hedging strategy* itself does depend upon the choice of U.

7.5 Evaluation of a Derivative Price

Suppose that the derivative payment X at T is a function of $P(T, U)$. Under the equivalent martingale measure Q we have

$$V(t) = E_Q\left[\exp\left(-\int_t^T r(s)\, ds\right) X(P(T, U)) \;\middle|\; \mathcal{F}_t\right].$$

Evaluation of this expectation requires a knowledge of the joint distribution of $P(T, U)$ and $\exp(-\int_t^T r(s)\, ds)$ under Q. Even under as simple a model as the Vasicek model, this is a difficult task to carry out (see, for example, the proof of Theorem 4.7 and the exercises in the same chapter on the Vasicek model).

On the other hand, under the forward-measure approach we have

$$V(t) = P(t, T)E_{P_T}[X(P(T, U)) \mid \mathcal{F}_t].$$

It is only necessary for us now to establish the distribution of $P(T, U)$ under P_T. It is very likely that this second expectation will be easier to calculate.

Lemma 7.2. $E_{P_T}[P(T, U) \mid \mathcal{F}_t]$ *is the forward price at time t for delivery of $P(T, U)$ at time T.*

Proof. Since $Y(t, U)$ is a martingale under P_T we have

$$E_{P_T}[Y(T, U) \mid \mathcal{F}_t] = Y(t, U),$$
$$\Rightarrow \quad E_{P_T}\left[\frac{P(T, U)}{P(T, T)} \;\middle|\; \mathcal{F}_t\right] = E_{P_T}[P(T, U) \mid \mathcal{F}_t] = \frac{P(t, U)}{P(t, T)},$$

which is the forward price.   □

Lemma 7.3. *Suppose that the zero-coupon-bond volatility term structure, $S(v, V)$, is deterministic. Then, under P_T, $P(T, U)$ is lognormal with*

$$\operatorname{Var}_{P_T}[\log P(T, U) \mid \mathcal{F}_t] = b^2 = \int_t^T (S(v, U) - S(v, T))^2 \, dv$$

and

$$E_{P_T}[\log P(T, U) \mid \mathcal{F}_t] = a = \log \frac{P(t, U)}{P(t, T)} - \tfrac{1}{2}b^2.$$

Proof. Define $L(t, U) = \log Y(t, U)$. Then

$$
\begin{aligned}
dL(t, U) &= \frac{1}{Y(t, U)} \, dY(t, U) - \frac{1}{2Y(t, U)^2} (dY(t, U))^2 \\
&= (S(t, U) - S(t, T)) \, d\hat{W}(t) - \tfrac{1}{2}(S(t, U) - S(t, T))^2 \, dt \\
\Rightarrow \quad L(T, U) &= L(t, U) - \frac{1}{2} \int_t^T (S(v, U) - S(v, T))^2 \, dv \\
&\quad + \int_t^T (S(v, U) - S(v, T)) \, d\hat{W}(v).
\end{aligned}
$$

Since $S(v, V)$ is deterministic this implies that $L(T, U)$, given $\mathcal{F}_t$, is Normally distributed under P_T with mean a and variance b^2 as defined in the statement of the lemma. Furthermore, $L(T, U) = \log P(T, U)$ so the proof is complete. □

Under the Vasicek model,

$$
\begin{aligned}
S(v, U) &= \frac{\sigma}{\alpha}(1 - e^{-\alpha(U-v)}) \\
\Rightarrow \quad b^2 &= \frac{\sigma^2}{\alpha^2} \int_t^T (e^{-\alpha(T-v)} - e^{-\alpha(U-v)})^2 \, dv \\
&= \frac{\sigma^2}{2\alpha^3}(1 - e^{-\alpha(U-T)})^2 (1 - e^{-2\alpha(T-t)}).
\end{aligned}
$$

However, the development that follows requires only that $S(v, U)$ is a deterministic function, rather than specifically the form given under the Vasicek model. The Vasicek model is mentioned here as a follow-up to the formulae given in Section 4.5.

Now we make use of the following lemma.

Lemma 7.4. *Suppose that a random variable Y has a lognormal distribution with parameters a and b^2; that is, $\log Y \sim N(a, b^2)$. Then for any constant $K > 0$:*

$$E[(Y - K)_+] = E[Y]\Phi(h) - K\Phi(h - b),$$

where

$$h = \frac{a + b^2 - \log K}{b} = \frac{\log(E[Y]/K) + \tfrac{1}{2}b^2}{b}$$

and $\Phi(x)$ is the cumulative distribution function of the standard Normal distribution.

We can apply this lemma to the problem of how to price a call option that matures at T with a strike price of K and which is written on $P(t, U)$. We have

$$E_{P_T}[(P(T, U) - K)_+ \mid \mathcal{F}_t] = E_{P_T}[P(T, U) \mid \mathcal{F}_t]\Phi(h) - K\Phi(h - b)$$
$$= \frac{P(t, U)}{P(t, T)}\Phi(h) - K\Phi(h - b) \quad \text{(by Lemma 7.2)},$$

where

$$h = \frac{1}{b}\left(\log \frac{P(t, U)}{P(t, T)} - \log K + \tfrac{1}{2}b^2\right)$$
$$= \frac{1}{b}\log \frac{P(t, U)}{KP(t, T)} + \tfrac{1}{2}b$$

and

$$b^2 = \int_t^T (S(v, U) - S(v, T))^2 \, dv.$$

Hence the price at time t of the call option is

$$V(t) = P(t, T)E_{P_T}[(P(T, U) - K)_+ \mid \mathcal{F}_t]$$
$$= P(t, U)\Phi(h) - KP(t, T)\Phi(h - b).$$

Note the similarity between this and the Black–Scholes formula for equity call options. This similarity only holds where bond prices are lognormally distributed.

7.6 Equity Options with Stochastic Interest

The forward measure can also be used to price equity options when the risk-free rate of interest, $r(t)$, is stochastic. Although equity-linked options are really beyond the scope of this book it is, nevertheless, interesting to include one simple example.

Suppose then that we have the usual one-factor model for bond prices and interest rates

$$dP(t, T) = P(t, T)[r(t) \, dt + S(t, T) \, d\tilde{W}_1(t)],$$

where $\tilde{W}_1(t)$ is a Brownian motion under the risk-neutral measure Q. Now consider an equity index $R(t)$ which includes full reinvestment of dividends (that is, $R(t)$ is a tradable asset). Under Q suppose that $R(t)$ has the SDE

$$dR(t) = R(t)[r(t) \, dt + \sigma_1 \, d\tilde{W}_1(t) + \sigma_2 \, d\tilde{W}_2(t)],$$

where $\tilde{W}_2(t)$ is a second Brownian motion under Q that is independent of $\tilde{W}_1(t)$. (Remember that all tradable assets must have expected growth rate $r(t)$ under Q.) Let $\mathcal{F}_t$ be the sigma-algebra generated by $\{(\tilde{W}_1(s), \tilde{W}_2(s)) : 0 \leqslant s \leqslant t\}$.

Remark 7.5. Let $\gamma_1(t)$ and $\gamma_2(t)$ be the market prices of risk for $\tilde{W}_1(t)$ and $\tilde{W}_2(t)$ which transfer us from Q to the real-world measure P. Then

$$dR(t) = R(t)[\mu(t) + \sigma_1 W_1(t) + \sigma_2 W_2(t)],$$

where $\mu(t) = r(t) + \sigma_1\gamma_1(t) + \sigma_2\gamma_2(t)$ and $(W_1(t), W_2(t))$ is a standard two-dimensional Brownian motion under P. We require only that $\gamma_1(t)$ and $\gamma_2(t)$ are previsible and satisfy the Novikov condition. It follows that the dynamics of $R(t)$ under P can allow for mean reversion of returns if that is a desired characteristic.

Now let P_T be the measure equivalent to Q under which the prices of all tradable assets under the extended market (bonds plus equity) discounted by $P(t, T)$ are martingales.

As before for the one-factor model, this requires $d\hat{W}_1(t) = d\tilde{W}_1(t) - S(t, T)\,dt$ with $\hat{W}_1(t)$ a Brownian motion under P_T. Now consider the discounted price process $\hat{R}(t) = R(t)/P(t, T)$. It is straightforward to show that

$$d\hat{R}(t) = \hat{R}(t)[-S(t, T)(\sigma_1 - S(t, T))\,dt + (\sigma_1 - S(t, T))\,d\tilde{W}_1(t) + \sigma_2\,d\tilde{W}_2(t)].$$

Hence,

$$d\hat{R}(t) = \hat{R}(t)[(\sigma_1 - S(t, T))\,d\hat{W}_1(t) + \sigma_2\,d\hat{W}_2(t)],$$

where $d\hat{W}_1(t)$ has been defined already as

$$d\tilde{W}_1(t) - S(t, T)\,dt \quad \text{and} \quad d\hat{W}_2(t) = d\tilde{W}_2(t).$$

Thus, the market price of risk transferring from Q to P_T is *zero* for the second source of risk, $\tilde{W}_2(t)$.

Lemma 7.6. *If the zero-coupon-bond volatilities $S(v, V)$ are deterministic, then $\hat{R}(\tau)$ is lognormal with*

$$R(T) = \hat{R}(T)$$

$$\Rightarrow \quad E_{P_T}[R(T)|\mathcal{F}_t] = \frac{R(t)}{P(t, T)},$$

$$b^2 = \operatorname{Var}_{P_T}[\log R(T) \mid \mathcal{F}_t] = \int_t^T (\sigma_1 - S(v, T))^2\,dv + \sigma_2^2(T - t),$$

$$E_{P_T}[\log R(T) \mid \mathcal{F}_t] = -\tfrac{1}{2}b^2 + \log\frac{R(t)}{P(t, T)} = a, \quad say.$$

Consider a general derivative which pays $f(R(T))$ at time T. Under the usual Q-measure, the price at time t of this derivative is

$$V(t) = E_Q\left[\exp\left(-\int_t^T r(s)\,ds\right)f(R(T)) \;\middle|\; \mathcal{F}_t\right],$$

which may be complex to derive. We now demonstrate how the forward-measure approach can be used to price the derivative.

Let $D(t) = E_{P_T}[f(R(T)) \mid \mathcal{F}_t]$. This is a martingale under P_T.

Recall that $\hat{P}(t, U) = P(t, U)/P(t, T)$ and $\hat{R}(t) = R(t)/P(t, T)$ are both also martingales under P_T which are not perfectly correlated. By the multi-dimensional version of the Martingale Representation Theorem (Theorem A.10), there exist previsible processes $\phi_P(t)$ and $\phi_R(t)$ such that

$$dD(t) = \phi_P(t)\, d\hat{P}(t, U) + \phi_R(t)\, d\hat{R}(t).$$

Define $\psi(t) = D(t) - \phi_P(t)\hat{P}(t, U) - \phi_R(t)\hat{R}(t)$. Consider the portfolio process which holds at time t,

$$\left.\begin{aligned} &\phi_P(t) \text{ units of } P(t, U), \\ &\phi_R(t) \text{ units of } R(t), \\ &\psi(t) \text{ units of } P(t, T). \end{aligned}\right\} \tag{7.2}$$

Then it is straightforward to show that this portfolio is self-financing.

The value at time t of this portfolio is

$$V(t) = P(t, T)D(t) = P(t, T)E_{P_T}[f(R(T)) \mid \mathcal{F}_t].$$

It follows that the portfolio is also replicating (taking $t = T$) and that $V(t)$ is the unique, no-arbitrage price at time t.

Theorem 7.7. *Consider the European call option $f(R(T)) = (R(T) - K)_+$. The price for this option at time t is*

$$V(t) = P(t, T)[F(t)\Phi(h) - K\Phi(h - b)],$$

where

$$h = \frac{\log(F(t)/K) + \frac{1}{2}b^2}{b},$$

$$b^2 = \int_t^T (\sigma_1 - S(v, T))^2\, dv + \sigma_2^2(T - t),$$

$$F(t) = \frac{R(t)}{P(t, T)} \quad \text{is the forward price for } R(T).$$

Proof. This is a straightforward application of Lemma 7.4. □

7.7 Exercises

Exercise 7.1. Recall that Theorem 7.1 states that for a derivative X payable at time T the value at time $t < T$ is $V(t) = P(t, T)E_{P_T}[X \mid \mathcal{F}_t]$.

Now suppose that $\tau > T$. Prove that

$$V(t) = P(t, \tau) E_{P_\tau}\left[\left.\frac{X}{P(T, \tau)}\right| \mathcal{F}_t\right]$$

by constructing a suitable self-financing and replicating strategy.

Exercise 7.2. Show that the portfolio in equation (7.2) is self-financing.

Exercise 7.3. Consider the general one-factor model developed in Section 4.2. Under Q, suppose that the risk-free rate of interest has SDE

$$dr(t) = a(t, r(t))\, dt + b(t, r(t))\, d\tilde{W}(t),$$

where $\tilde{W}(t)$ is a standard Brownian motion under the risk-neutral measure Q. Derive the equivalent SDE for $r(t)$ under the forward measure P_T.

Exercise 7.4. An interest rate cap is a derivative that guarantees that the rate of interest on a loan at any given time will be the lesser of the prevailing rate, $r(t)$, and the cap rate, r_c. Thus, over the outstanding term of the loan from t to T the effective payoff at s on the derivative relative to an uncapped loan is $\max\{r(s) - r_c, 0\}\, ds$.

This is effectively a collection of interest rate call options where the payoff at s is $\max\{r(s) - r_c, 0\}$.

(a) Write down an expression for the value at time t of such an interest rate call option, using the equivalent martingale measure Q.

(b) Suppose that under Q the risk-free rate $r(t)$ follows the Vasicek model:

$$dr(t) = \alpha(\mu - r(t))\, dt + \sigma\, d\tilde{W}(t).$$

   (i) Find the forward measure P_T under which

$$Y(t, U) = P(t, U).P(0, T)/P(t, T)$$

   is a martingale and find the change of measure drift $\gamma(s)$.

   (ii) Show that under P_T, $r(s)$ given $\mathcal{F}_t$ has a Normal distribution with

$$E_{P_T}[r(s) \mid \mathcal{F}_t] = e^{-\alpha(s-t)}r(t) + \left(\mu - \frac{\sigma^2}{\alpha^2}\right)(1 - e^{-\alpha(s-t)})$$
$$+ \frac{\sigma^2}{2\alpha^2}e^{-\alpha(T-s)}(1 - e^{-2\alpha(s-t)})$$

and

$$\operatorname{Var}_{P_T}[r(s) \mid \mathcal{F}_t] = \frac{\sigma^2}{2\alpha}(1 - e^{-2\alpha(s-t)}).$$

(iii) Hence find an expression for

$$E_{P_T}[(r(T) - r_c)_+ \mid \mathcal{F}_t].$$

(iv) Finally, find an expression for the price of this option.

Exercise 7.5. Investigate the relationship between the equity option pricing formula in Theorem 7.7 and the 'traditional' Black–Scholes formula (which assumes $S(u, V) = 0$ for all u, V).

(a) Keep σ_1 and σ_2 constant and vary the magnitude of the function $S(v, T)$. By how much does the option price change?

(b) Take the form of $S(v, T)$ as given. Vary σ_1 and σ_2 in such a way that $\sigma_1^2 + \sigma_2^2$ is held constant (that is, keep equity volatility constant but vary the correlation between bond and equity price changes). How sensitive is the option price to this correlation?

(c) How do these sensitivities change with respect to the term to maturity, $T - t$?

8

Positive Interest

8.1 Introduction

In this chapter we adapt the concept of forward measures introduced in Chapter 7 to describe a more recent approach to the development of new term-structure models.

It is clear that the work of Heath, Jarrow and Morton (1992) provided quantitative analysts with a very useful framework within which they could develop specific term-structure models. However, the framework does not lend itself naturally to the development of models which keep interest rates positive. In addition, models developed within the general framework are often non-Markov in structure (that is, their dynamics cannot be summarized by a finite number of state variables) unless we are careful about the choice of volatility structures.

More recently, a new approach has been developed by Flesaker and Hughston (1996), Rogers (1997) and Rutkowski (1997). They developed a general framework which does ensure that interest rates remain positive. Furthermore, the framework lends itself more easily to the development of Markov interest rate models.[1] An important historical stepping stone towards the Flesaker–Hughston, Rutkowski and Rogers frameworks was provided by Constantinides (1992) and we comment on this later.

8.2 Mathematical Development

The following theorems were developed in different forms separately by Rogers (1997) and Rutkowski (1997). Their development was more general but also more straightforward than that of Flesaker and Hughston (1996), which we describe in the next section. Let $(\Omega, \mathcal{F}, \hat{P})$ be a probability triple and let $\hat{W}(t)$ be a standard Brownian motion in $\mathbb{R}^d$ under $\hat{P}$. Let $\mathcal{F}_t$ be the sigma-algebra generated by $\{\hat{W}(s) : 0 \leqslant s \leqslant t\}$.

---

[1] For a more rigorous development than is presented here the reader is referred to the original papers by Rutkowski (1997) and Rogers (1997).

Let $A(t)$ be a strictly positive diffusion process in $\mathbb{R}$ which is adapted to $\mathcal{F}_t$. Thus, there exist previsible processes $\mu_A(t)$ (a scalar) and $\sigma_A(t)$ (a $d \times 1$ vector) such that

$$dA(t) = A(t)[\mu_A(t)\,dt + \sigma_A(t)'\,d\hat{W}(t)].$$

Now define

$$P(t, T) = \frac{E_{\hat{P}}[A(T) \mid \mathcal{F}_t]}{A(t)}. \tag{8.1}$$

We now show that $P(t, T)$ can be interpreted as the price at time t of a zero-coupon bond which matures at time T and that the proposed framework gives rise to arbitrage-free dynamics for the term structure.

Because of these properties, the probability measure $\hat{P}$ is sometimes referred to as the *pricing measure*. This measure is different from, but equivalent to (assuming the usual conditions hold), the risk-neutral measure Q and the real-world measure P.[2]

Theorem 8.1. *The term-structure model described by equation (8.1) is arbitrage free.*

Proof. Since $A(t)$ is a strictly positive diffusion process we can note that $D(t, T) = E_{\hat{P}}[A(T) \mid \mathcal{F}_t]$ is a strictly positive martingale under $\hat{P}$. It follows that we can write $dD(t, T) = D(t, T)\sigma_D(t, T)'\,d\hat{W}(t)$ for some function σ_D. Thus,

$$\begin{aligned}
dP(t, T) &= d\left(\frac{D(t, T)}{A(t)}\right) \\
&= dD(t, T)\frac{1}{A(t)} + D(t, T)\,d\left(\frac{1}{A(t)}\right) + dD(t, T)\,d\left(\frac{1}{A(t)}\right) \\
&= \frac{D(t, T)}{A(t)}[\sigma_D(t, T)'\,d\hat{W}(t) + (-\mu_A(t) + |\sigma_A(t)|^2)\,dt \\
&\qquad\qquad - \sigma_A(t)'\,d\hat{W}(t) - \sigma_A(t)'\sigma_D(t, T)\,dt] \\
&= P(t, T)[r(t)\,dt + S(t, T)'\,d\tilde{W}(t)], \tag{8.2}
\end{aligned}$$

where

$$\begin{aligned}
r(t) &= -\mu_A(t), \\
S(t, T) &= \sigma_D(t, T) - \sigma_A(t), \\
d\tilde{W}(t) &= d\hat{W}(t) - \sigma_A(t)\,dt.
\end{aligned}$$

With appropriate conditions on $\sigma_A(t)$ the Girsanov Theorem (see Appendix A) tells us that there exists a measure Q equivalent to $\hat{P}$ under which $\tilde{W}(t)$ is a standard Brownian motion in $\mathbb{R}^d$.

---

[2]While Q is the unique risk-neutral measure if the market is complete, there are many so-called pricing measures, of which Q and $\hat{P}$ are but two.

Thus, we have a form for $dP(t, T)$ which is consistent with the arbitrage-free model developed under the Heath–Jarrow–Morton framework (in particular, compare equation (8.2) with equation (5.4)).

Now note that the change of measure from $\hat{P}$ to Q is the same for all maturity dates T. It follows that Q is the unique equivalent measure under which the prices of all bonds have equal expected growth rate $r(t)$.

Next, define

$$\left.\begin{aligned} B(t) &= \exp\left[\int_0^t r(u)\,du\right], \\ \xi(t) &= \exp\left[-\int_0^t \sigma_A(u)'\,d\hat{W}(u) + \frac{1}{2}\int_0^t |\sigma_A(u)|^2\,du\right]. \end{aligned}\right\} \tag{8.3}$$

The change of measure from $\hat{P}$ to Q implied by equation (8.2) tells us that $\xi(t)^{-1}$ represents the Radon–Nikodým derivative over the interval $0 \leqslant u \leqslant t$. Thus, the measure Q is equivalent to $\hat{P}$ over the interval $t \leqslant u \leqslant T$ with Radon–Nikodým derivative

$$\frac{dQ}{d\hat{P}}(t, T) = \frac{\xi(t)}{\xi(T)}. \tag{8.4}$$

We also have[3] $A(t) = B(t)^{-1}\xi(t)^{-1}$. It follows that

$$\begin{aligned} P(t, T) &= A(t)^{-1} E_{\hat{P}}[A(T) \mid \mathcal{F}_t] \\ &= B(t) E_{\hat{P}}\left[\frac{\xi(t)}{B(T)\xi(T)} \,\middle|\, \mathcal{F}_t\right] \\ &= B(t) E_Q[B(T)^{-1} \mid \mathcal{F}_t] \qquad \text{(by equation (8.4)).} \end{aligned}$$

It follows that all discounted price processes $P(t, T)/B(t)$ are martingales under Q. If we interpret $B(t)$ as the cash account in the usual way, we recall the Fundamental Theorem of Asset Pricing (Theorem 2.2), which tells us that the model is arbitrage free. $\qquad\square$

Theorem 8.2. *Suppose that the strictly positive diffusion $A(t)$ is also a supermartingale under $\hat{P}$. Then $r(t)$ remains positive almost surely.*

Proof. If $A(t)$ is a supermartingale, then $\mu_A(t)$ must be less than or equal to zero for all t and $\mathcal{F}_t$ almost surely. It follows that $r(t) = -\mu_A(t) \geqslant 0$ almost surely.

Alternatively, since $A(t)$ is a supermartingale, $E_{\hat{P}}[A(T) \mid \mathcal{F}_t]$ must be a decreasing function of T. Hence $\partial P(t, T)/\partial T \leqslant 0$ for all T. This implies that the forward rates $f(t, T)$ are positive for all T and, in particular, $r(t) = \lim_{T \to t} f(t, T)$ is positive. $\qquad\square$

---

[3]For example, consider $d(\log A(t))$.

The dependence of $A(t)$ on a d-dimensional Brownian motion $\hat{W}(t)$ is made more explicit by Rogers (1997). Let $f : \mathbb{R}^d \to \mathbb{R}^+$ be some strictly positive, twice-differentiable function. Let $X(t)$ be some diffusion process in $\mathbb{R}^d$ so that there exist previsible processes $\mu_X(t)$ (a $d \times 1$ vector) and $\sigma_X(t)$ (a $d \times d$ matrix) with

$$dX(t) = \mu_X(t)\, dt + \sigma_X(t)\, d\hat{W}(t).$$

We then define

$$A(t) = \frac{e^{-\alpha t} f(X(t))}{f(X(0))}, \tag{8.5}$$

for some positive constant α. This gives rise to a multitude of new models with any number of factors, some of which are described in Section 8.5. This formulation is often referred to as the *potential approach*.

It is important to note that $A(t)$ as defined in equation (8.5) is not always a supermartingale. It is therefore necessary to check whether or not the drift of $A(t)$ is always negative. Fortunately, there are many candidate models which satisfy the supermartingale condition for positive interest. In some cases there will be a unique value for α for which the minimum value attainable by $r(t)$ is exactly zero (see Section 8.5 below, Examples 3 and 4).

In other models where $X(t)$ is a stationary diffusion under $\hat{P}$, we can interpret α as the constant, limiting spot rate; that is, $l(t) = \lim_{T \to \infty} R(t, T) = \alpha$ for all t.

8.3   The Flesaker and Hughston Approach

The following special case of these results was introduced by Flesaker and Hughston (1996).

Corollary 8.3. *Suppose that $M(t, T)$ is a strictly positive family of martingales under $\hat{P}$ adapted to $\mathcal{F}_t$. Let*

$$P(t, T) = \frac{\int_T^\infty \phi(s) M(t, s)\, ds}{\int_t^\infty \phi(s) M(t, s)\, ds} \tag{8.6}$$

for some deterministic, strictly positive function $\phi(s)$. Then $P(t, T)$ gives rise to an arbitrage-free model for the term structure of interest rates with positive interest.

Proof. Let $A(t) = \int_t^\infty \phi(s) M(t, s)\, ds$. Then

$$E_{\hat{P}}[A(T) \mid \mathcal{F}_t] = E_{\hat{P}}\left[\int_T^\infty \phi(s) M(T, s)\, ds \,\bigg|\, \mathcal{F}_t \right]$$

$$= \int_T^\infty \phi(s) M(t, s)\, ds < A(t),$$

since $\phi(s)$ and $M(t, s)$ are strictly positive. Also, since $\phi(s)$ and $M(t, s)$ are strictly positive, so is $A(t)$. Hence, $A(t)$ is a strictly positive supermartingale. This satisfies the requirements for Theorems 8.1 and 8.2. It follows that the model is arbitrage free.   □

8.3.1 Calibration Using the Flesaker and Hughston Approach

The intention of Flesaker and Hughston (1996) was that the models developed under their framework should be of the no-arbitrage variety; that is, with initial observed prices matching initial theoretical prices.

Let us assume (without loss of generality) that $M(0, T) = 1$ for all T and that initial bond prices $P_{obs}(0, T)$ are known for all $T > 0$.

If we define

$$\phi(s) = -\left.\frac{\partial P_{obs}(0, T)}{\partial T}\right|_{T=s},$$

it is straightforward to show that with $P(0, T)$ defined by equation (8.6) we have $P(0, T) = P_{obs}(0, T)$ for all T.

8.3.2 Forward Rates

It is straightforward to show that the forward-rate curve is

$$f(t, T) = \frac{\phi(T)M(t, T)}{\int_T^\infty \phi(s)M(t, s)\, ds}.$$

It follows that

$$r(t) = \frac{\phi(t)M(t, t)}{\int_t^\infty \phi(s)M(t, s)\, ds}.$$

8.4 Derivative Pricing

Next we consider a derivative which pays $V(T)$ at time T, where $V(T)$ is an $\mathcal{F}_T$-measurable function. Let the price at time $t < T$ of this derivative be $V(t)$. The standard valuation formula for this contract is $V(t) = B(t)E_Q[V(T)/B(T) \mid \mathcal{F}_t]$. Now recall that $B(u) = A(u)^{-1}\xi(u)^{-1}$. Thus we have

$$V(t) = \frac{1}{A(t)}E_Q\left[\frac{V(T)A(T)\xi(T)}{\xi(t)} \;\middle|\; \mathcal{F}_t\right]$$
$$= \frac{1}{A(t)}E_{\hat{P}}[V(T)A(T) \mid \mathcal{F}_t]. \tag{8.7}$$

Remark 8.4. The general approach of Rutkowski (1997) and Rogers (1997) was preceded by a paper by Constantinides (1992).

Constantinides proposed a formula for interest rate derivative pricing similar to equation (8.7),

$$V(t) = \frac{E_P[V(T)M(T) \mid \mathcal{F}_t]}{M(t)},$$

where $M(t)$ is some diffusion process and P is the real-world measure. The key advance by Flesaker and Hughston, Rogers and Rutkowski was to jump from the use of P to the more general pricing measure $\hat{P}$. The use of a general $\hat{P}$ allows the easier development of new arbitrage-free, positive interest models.

Additionally, Constantinides (1992) requires examination of a specific model (similar to Examples 2 and 3 below) to find conditions for positive interest. This is in contrast to the general supermartingale condition on $A(t)$ (Theorem 8.2).

Constantinides refers to $M(t)$ as the *pricing kernel* or *state-price density process*. If we define $\zeta(t, T)$ as the Radon–Nikodým derivative $d\hat{P}/dP$, then we have the simple relationship $M(T) = A(T)\zeta(0, T)$.

8.5   Examples

Example 1: Rational lognormal model.   Both Flesaker and Hughston (1996) and Rutkowski (1997) discuss the model generated by the process $A(t) = f(t) + g(t)M(t)$, where $M(t)$ is a strictly positive martingale under $\hat{P}$ with $M(0) = 1$ and $f(t)$ and $g(t)$ are strictly positive, decreasing, deterministic functions. This leads to the pricing formula

$$P(t, T) = \frac{f(T) + g(T)M(t)}{f(t) + g(t)M(t)}.$$

In particular, suppose that we have a one-factor model ($d = 1$) with

$$dM(t) = \sigma_M(t)M(t)\,d\hat{W}(t),$$

where $\sigma_M(t)$ is some strictly positive, deterministic function. Then

$$M(t) = \exp\left[\int_0^t \sigma_M(u)\,d\hat{W}(u) - \frac{1}{2}\int_0^t \sigma_M(u)^2\,du\right]$$

$$\Rightarrow \quad \log M(t) \sim N\left(-\frac{1}{2}\int_t^T \sigma_M(u)^2\,du, \int_t^T \sigma_M(u)^2\,du\right),$$

$$r(t) = -\frac{f'(t) + g'(t)M(t)}{f(t) + g(t)M(t)} \geqslant 0,$$

since $f(t)$ and $g(t)$ are decreasing functions. It is this lognormality of $M(t)$ and the form of $P(t, T)$ which leads to the name *rational lognormal* model.

The principal advantage of this model is that it leads to simple formulae for the prices of certain derivative contracts.

As an example, consider a caplet. For a given term to maturity $\tau > 0$ we define the LIBOR rate at time T as

$$L(T) = (P(T, T + \tau)^{-1} - 1)/\tau$$

or, that is,

$$P(T, T + \tau) = (1 + \tau L(T))^{-1}.$$

The caplet pays $V(T + \tau) = \tau(L(T) - K)_+$ at time $T + \tau$. The amount of this payment is, of course, known at time T and has value at time T

$$V(T) = \tau P(T, T + \tau)(L(T) - K)_+ = (1 - (1 + \tau K)P(T, T + \tau))_+.$$

For the given form for $A(t)$ we can express the price at time t for this contract as

$$V(t) = \frac{E_{\hat{P}}[(\alpha + \beta M(T))_+ \mid \mathcal{F}_t]}{f(t) + g(t)M(t)}, \tag{8.8}$$

where

$$\alpha = f(T) - (1 + \tau K)f(T + \tau) \quad \text{and} \quad \beta = g(T) - (1 + \tau K)g(T + \tau).$$

Since $M(T)$ is lognormal we can exploit Lemma 7.4 to derive a Black–Scholes-type valuation formula.

There are significant limitations to this model.

- Bond prices (as noted by Flesaker and Hughston) have strict upper and lower bounds determined by the functions $f(t)$ and $g(t)$; that is,

$$\frac{f(T)}{f(t)} \geqslant P(t, T) \geqslant \frac{g(T)}{g(t)}$$

  (or vice versa). This means that the risk-free rate $r(t)$ is also bounded, that is, by $-f'(t)/f(t)$ and $-g'(t)/g(t)$. This limits the usefulness of the model to short-maturity contracts which are close to being at-the-money.

- Closer inspection of equation (8.8) indicates that, depending upon the values of τ and K and the forms of $f(t)$ and $g(t)$, the caplet may be redundant. In particular, the function

$$\alpha + \beta M(T) = f(T) - (1 + \tau K)f(T + \tau) + \{g(T) - (1 + \tau K)g(T + \tau)\}M(T)$$

  might be entirely positive over the range $0 < M(T) < \infty$, entirely negative, positive then negative, or negative then positive. The first two render the model useless (that is, the caplet is redundant) and are related to strike prices of bonds being outside the upper or lower bounds mentioned previously. Only the latter two outcomes give rise to Black–Scholes-type valuation formulae.

Example 2: Gaussian model. Suppose that $dX(t) = -\theta X(t)\, dt + d\hat{W}(t)$ and $f(x) = e^{\sigma x}$ for $\theta > 0$ and $\sigma > 0$. Since $X(t)$ is an Ornstein–Uhlenbeck process under $\hat{P}$, $X(T)$ given $X(t)$ has a Normal distribution with mean $X(t)\exp(-\theta(T-t))$ and variance $[1 - \exp(-2\theta(T - t))]/2\theta$.

Now apply equation (8.5); that is,

$$A(t) = \frac{e^{-\alpha t}f(X(t))}{f(X(0))} \tag{8.9}$$

$$\Rightarrow \quad P(t, T) = \exp\left[-\alpha(T - t) - \sigma X(t)(1 - e^{-\theta(T-t)}) + \frac{\sigma^2}{4\theta}(1 - e^{-2\theta(T-t)})\right].$$

Now

$$f(t, T) = -\frac{\partial}{\partial T} \log P(t, T)$$

$$= \alpha + \theta \sigma X(t) e^{-\theta(T-t)} - \tfrac{1}{2}\sigma^2 e^{-2\theta(T-t)}$$

$$\Rightarrow \quad r(t) = \alpha - \tfrac{1}{2}\sigma^2 + \theta \sigma X(t), \tag{8.10}$$

$$dr(t) = -\theta(r(t) - \hat{\mu})\,dt + \hat{\sigma}\,d\hat{W}(t),$$

where

$$\hat{\mu} = \alpha - \tfrac{1}{2}\sigma^2 \quad \text{and} \quad \hat{\sigma} = \sigma\theta.$$

The market price of risk transferring us from $\hat{P}$ to Q was shown earlier to be equal to the volatility of $A(t)$, which, here, is $\sigma_A(t) = \sigma$.[4] Thus we have $dr(t) = -\theta(r(t) - \hat{\mu})\,dt + \hat{\sigma}(d\hat{W}(t) + \sigma\,dt)$, where $\tilde{W}(t)$ is a standard Brownian motion under the risk-neutral measure Q. This means that we have

$$dr(t) = -\theta(r(t) - \tilde{\mu})\,dt + \hat{\sigma}\,d\tilde{W}(t),$$

where

$$\tilde{\mu} = \alpha + \tfrac{1}{2}\sigma^2 = \alpha + \frac{\hat{\sigma}^2}{2\theta^2}.$$

We recognize this as the Vasicek model. We can recall that the Vasicek model admits the possibility of negative interest rates and we can note that with the present formulation $A(t)$ does not always have the required negative drift.

A more general, multifactor, Gaussian model is described by Rogers (1997).

Example 3: Squared Gaussian model (see also Rogers (1997) and Ahn, Dittmar and Gallant (2002)). Let $X(t)$ be a d-dimensional diffusion $(d > 1)$ with

$$dX(t) = -BX(t)\,dt + C\,d\hat{W}(t),$$

where B is the diagonal $d \times d$ matrix $\text{diag}(\beta_1, \ldots, \beta_d)$ and C is a $d \times d$ matrix of full rank. Furthermore, define $R = (\rho_{ij})_{i,j=1}^d = CC'$. We require C to be defined such that $\rho_{ii} = 1$ for all i; that is, R can be regarded as the instantaneous correlation matrix. It follows that, under $\hat{P}$, $X(T)$ is multivariate normal with

$$(\mu_{t,T})_i = E_{\hat{P}}[X_i(T) \mid \mathcal{F}_t] = e^{-\beta_i(T-t)} X_i(t),$$

$$(V_{t,T})_{ii} = \text{Var}_{\hat{P}}[X_i(T) \mid \mathcal{F}_t] = \frac{(1 - e^{-2\beta_i(T-t)})}{2\beta_i},$$

$$(V_{t,T})_{ij} = \text{Cov}_{\hat{P}}[X_i(T), X_j(T) \mid \mathcal{F}_t] = \frac{(1 - e^{-(\beta_i+\beta_j)(T-t)})}{\beta_i + \beta_j}\rho_{ij},$$

---

[4]The market price of risk can, of course, be derived directly by considering the stochastic differential equation for $P(t, T)$.

where

$$\rho_{ij} = \sum_k c_{ik} c_{jk}.$$

Now take $f(x) = \exp(\frac{1}{2}(x-c)'\theta(x-c))$ for some constant vector c and symmetric, positive-definite matrix θ.

Suppose $t = 0$. After some tedious algebra we find that if we define $A(t) = e^{-\alpha t} f(X(t))/f(X(0))$, then

$$A(t) = \exp[-\alpha t + \tfrac{1}{2}(X(t) - c)'\theta(X(t) - c) - \tfrac{1}{2}(X(0) - c)'\theta(X(0) - c)]$$

and

$$P(t, T) = e^{-\alpha(T-t)} \det(I - \theta V_{t,T})^{-1/2}$$
$$\times \exp[\tfrac{1}{2}\mu'_{t,T}(I - \theta V_{t,T})^{-1}\theta \mu_{t,T} - \tfrac{1}{2}\mu'_{t,t}\theta\mu_{t,t}].$$

Define $Y(t) = \log A(t)$. Since $dA(t) = A(t)[\mu_A(t) + \sigma_A(t)' \, d\hat{W}(t)]$ we have $dY(t) = \mu_A(t) \, dt - \frac{1}{2}\sigma_A(t)'\sigma_A(t) \, dt + \sigma_A(t)' \, d\hat{W}(t)$. We can then show (again with some tedious algebra) that

$$dY(t) = -\alpha + (X(t) - c)'\theta(-BX(t) \, dt + C \, d\hat{W}(t)) + \tfrac{1}{2} \mathrm{tr}(\theta CC')$$
$$\Rightarrow \quad \sigma_A(t) = C'\theta(X(t) - c)$$
$$\mu_A(t) = -\alpha - (X(t) - c)'\theta BX(t) + \tfrac{1}{2} \mathrm{tr}(\theta CC')$$
$$+ \tfrac{1}{2}(X(t) - c)'\theta CC'\theta(X(t) - c).$$

(Recall that $\sigma_A(t)$ gives the vector of market prices of risk.) Rearranging this and recalling that $r(t) = -\mu_A(t)$ we find that

$$r(t) = \tfrac{1}{2}(X(t) - S^{-1}v)'S(X(t) - S^{-1}v) - \tfrac{1}{2}v'S^{-1}v + k,$$

where $S = \theta B + B\theta - \theta CC'\theta$, $v = (B - \theta CC')\theta c$ and $k = \alpha - \frac{1}{2} \mathrm{tr}(\theta CC') - \frac{1}{2}c'\theta CC'\theta c$. Ideally, the minimum value attainable by $r(t)$ is exactly 0, that is, if $k - \frac{1}{2}v'S^{-1}v = 0$. It follows that

$$\alpha = \tfrac{1}{2} \mathrm{tr}(\theta CC') + \tfrac{1}{2}c'\theta CC'\theta c + \tfrac{1}{2}v'S^{-1}v.$$

Recall that, since $X(t)$ is stationary and time homogeneous, we have

$$\lim_{T \to \infty} R(t, T) = \alpha \quad \text{for all } t.$$

This interpretation for α imposes loose constraints on the other parameters in the model. There are clear similarities with the Cox–Ingersoll–Ross model. In particular, if the number of factors, d, is at least 2, then the risk-free rate of interest will never hit 0. Unlike the CIR model (where it is sufficient to know the value of $\sum_{i=1}^{d} X_i(t)^2$), this model is truly multifactor since the dynamics of $r(t)$ do depend upon full knowledge of each of $X_1(t), \ldots, X_d(t)$.

Example 4: Hyperbolic Gaussian model (see Rogers 1997). Consider the one-factor model $(d = 1)$ $dX(t) = -\beta X(t)\,dt + d\hat{W}(t)$ and take $f(x) = \cosh \gamma x = \frac{1}{2}(e^{\gamma x} + e^{-\gamma x})$ in equation (8.5). We have previously considered the case $f(x) = e^{\gamma x}$, which gives us the Vasicek model and consequently the possibility of negative interest rates. The use of $\cosh(\gamma x)$ results in positive interest. For this model, given $X(0)$, we have $\mu_{0,T} = X(0)e^{-\beta T}$ and $V_{0,T} = (1 - e^{-2\beta T})/2\beta$. Then

$$E_{\hat{P}}[f(X(T)) \mid X(0) = x] = \exp\left[\frac{1}{2}\gamma^2 \frac{(1 - e^{-2\beta T})}{2\beta}\right] \cosh(\gamma x e^{-\beta T})$$

$$\Rightarrow \quad P(0, T) = \exp\left[-\alpha T + \frac{1}{2}\gamma^2 \frac{(1 - e^{-2\beta T})}{2\beta}\right] \frac{\cosh(\gamma x e^{-\beta T})}{\cosh(\gamma x)}.$$

For general t and $X(t)$ this results in

$$r(t) = \alpha - \frac{\gamma^2}{4\beta} + \beta\gamma X(t)\tanh(\gamma X(t)).$$

This is minimized when $X(t) = 0$. If we wish to allow $r(t)$ to get arbitrarily close to 0, then we require $\alpha = \gamma^2/4\beta$. The form of dependence of $r(t)$ on $X(t)$ means that $r(t)$ acts like a linear Gaussian process for large $r(t)$ and like a squared-Gaussian process near 0.

A slightly more general example is considered by Rogers (1997).

Example 5: Integrated Gaussian model. Here we summarize the family of multifactor models developed by Cairns (1999, 2004).

Consider the Flesaker and Hughston (1996) framework (equation (8.6)). Suppose that $\hat{W}_1(t), \ldots, \hat{W}_d(t)$ are d correlated Brownian motions under $\hat{P}$ with

$$d\langle \hat{W}_i, \hat{W}_j \rangle(t) = \rho_{ij}\,dt \quad \text{for all } i, j.$$

Suppose that we have $M(0, s) = 1$ for all s and

$$dM(t, s) = M(t, s)\sum_{i=1}^{d} \sigma_i e^{-\alpha_i(s-t)}\,d\hat{W}_i(t)$$

$\Rightarrow$

$$M(t, s) = \exp\left[\sum_i \sigma_i \hat{X}_i(t)e^{-\alpha_i t} - \frac{1}{2}\sum_{i,j} \sigma_i \sigma_j \rho_{ij} e^{-(\alpha_i + \alpha_j)(s-t)} \frac{1 - e^{-(\alpha_i + \alpha_j)t}}{\alpha_i + \alpha_j}\right],$$

where

$$\hat{X}_i(t) = \int_0^t e^{-\alpha_i(t-u)}\,d\hat{W}_i(u) \quad \text{for } i = 1, \ldots, d. \tag{8.11}$$

From the form of equation (8.11) we recognize $\hat{X}_i(t)$ as a standard Ornstein–Uhlenbeck process under $\hat{P}$.

Now take

$$\phi(s) = \phi \exp\left[-\beta s + \sum_i \sigma_i e^{-\alpha_i s} x_i - \frac{1}{2} \sum_{i,j} \sigma_i \sigma_j \rho_{ij} e^{-(\alpha_i + \alpha_j)s}\right].$$

Then

$$P(t, T) = \frac{\int_T^\infty \phi(s) M(t, s)\, ds}{\int_t^\infty \phi(s) M(t, s)\, ds} = \frac{\int_{T-t}^\infty H(u, X(t))\, du}{\int_0^\infty H(u, X(t))\, du}, \tag{8.12}$$

where

$$H(u, y) = \exp\left[-\beta u + \sum_i \sigma_i y_i e^{-\alpha_i u} - \frac{1}{2} \sum_{i,j} \frac{\sigma_i \sigma_j \rho_{ij}}{(\alpha_i + \alpha_j)} e^{-(\alpha_i + \alpha_j)u}\right],$$

$$X_i(t) = x_i e^{-\alpha_i t} + \hat{X}_i(t).$$

From the form of equation (8.12) we see that the model is Markov and time homogeneous.

Clearly, there is no closed-form, analytical expression for bond prices. Nevertheless, bond prices are straightforward to evaluate; we are only ever required to calculate numerically (for example, using Simpson's rule) a one-dimensional integral, regardless of the number of factors in the model.

For pricing a European-style derivative which pays $V(T) = V(T, X(T))$ at time T we find that

$$V(t) = e^{-\beta(T-t)} \frac{E_{\hat{P}}[V(T, X(T)) \int_0^\infty H(u, X(T))\, du \mid X(t)]}{\int_0^\infty H(u, X(t))\, du}.$$

Numerically, any difficulty in evaluating a derivative price lies in the calculation of $V(T, x)$ and $\int_0^\infty H(u, x)\, du$. The probabilistic aspects of derivative-price evaluation are simple since $X(T) \mid X(t)$ has a multivariate Normal distribution.

This family of models shares an important characteristic with the lognormal models of Black, Derman and Toy (1990) and Black and Karasinski (1991). If we refer back to Figure 1.3, we see that the yield curve in Japan has a flattened 'S' shape. More mathematically, this suggests that as $r(t)$ tends to zero the slopes of the various yield curves (forward, spot and par) also tend to zero at very small terms to maturity. This characteristic is present in the Cairns (2004) model but is not achievable in models that are asymptotically squared Gaussian as $r(t)$ tends to zero (see, for instance, Examples 3 and 4 above, or the Cox–Ingersoll–Ross model). For the same reason, the present model permits yields of all types and terms to maturity to be arbitrarily close to zero (again, unlike squared-Gaussian models). This characteristic is critical for derivative contracts that move into the money when medium- or long-term interest rates drop below a specific value.

Sample par-yield curves (coupons payable annually) are plotted in Figure 8.1 for a two-factor version of the model.

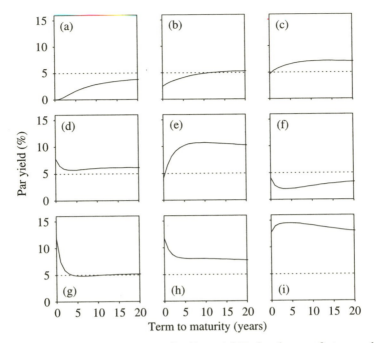

Figure 8.1. Sample par-yield curves $(\log(1 + \text{yield}))$ for the two-factor model with $\beta = 0.05$, $\alpha = (1, 0.1)$, $\sigma = (1, 1)$ and $\rho_{12} = -0.5$. The horizontal dotted line gives the limiting value for spot rates and forward rates. (a) $X = (-5, 1)$, (b) $X = (0, 2)$, (c) $X = (0, 3)$, (d) $X = (1, 2.3)$, (e) $X = (-1, 4.5)$, (f) $X = (2, 0)$, (g) $X = (2, 1.5)$, (h) $X = (1, 3)$, (i) $X = (0, 5)$.

8.6 Exercises

Exercise 8.1. Consider the Gaussian model with $A(t)$ defined by equation (8.9) (Section 8.5, Example 2). If $dA(t) = A(t)(\mu_A(t)\,dt + \sigma_A(t)\,d\hat{W}(t))$, verify that $\mu_A(t) = -r(t)$, where $r(t)$ is defined in equation (8.10) and that $\sigma_A(t) = \sigma$.

Exercise 8.2. Consider the integral Gaussian model (Section 8.5, Example 5). Show that

(a) $H(T, x) \sim e^{-\beta T}$ as $T \to \infty$, for all x;

(b) $\int_T^\infty H(u, x)\,du \sim \beta^{-1} e^{-\beta T}$ as $T \to \infty$, for all x;

(c) $f(t, T) \to \beta$ as $T \to \infty$, for all x.

Exercise 8.3. Consider the rational lognormal model defined in Section 8.5, Example 1. Find the zero-coupon bond volatility $S(t, T)$ under $\hat{P}$.

9

Market Models

Market models came to prominence in the late 1990s with the offer to practitioners of distinct advantages over earlier approaches such as the Heath–Jarrow–Morton framework or the no-arbitrage models described in Chapter 5. In particular, the approach shifted from a concentration on unobservable, instantaneous rates of interest such as the risk-free rate, $r(t)$, or forward rates, $f(t, T)$, to genuinely quoted market rates of interest such as LIBOR (see, for example, Brace and Musiela 1994). This was combined with the key modelling assumption that *relevant* market rates of interest are lognormal, resulting in analytical formulae for some commonly traded derivatives.[1] Both aspects create an environment which makes calibration of a model relatively straightforward compared with models arising from alternative frameworks. This particular combination was developed by Miltersen, Sandmann and Sondermann (1997) with formulae derived for zero-coupon-bond options, caps and floors. This was improved on subsequently by Brace, Gatarek and Musiela (1997), Jamshidian (1997) and Musiela and Rutkowski (1997) in ways which we will now describe.

9.1 Market Rates of Interest

We define two types of market interest rate. First we define LIBOR (London Interbank Offer Rate) and forward LIBOR. LIBOR defines an annualized, simple rate of interest that will be delivered at the end of a defined period. In particular, the τ-LIBOR, $L(T, T, T + \tau)$, means that an investment of 1 at time T will grow to $1 + \tau L(T, T, T + \tau)$ at time $T + \tau$. LIBOR is always quoted as an annual rate of interest even though it typically applies over shorter periods.

Mathematically, the *forward* LIBOR rate is defined as

$$L(t, T, T + \tau) = \frac{1}{\tau} \left[\frac{P(t, T)}{P(t, T + \tau)} - 1 \right].$$

---

[1] Recall (Section 4.9) that the use of a lognormal model for the short rate was first developed by Black, Derman and Toy (1990) and Black and Karasinski (1991).

This means that we can make a contract at time t under which we pay 1 at time T and receive back $1 + \tau L(t, T, T + \tau)$ at time $T + \tau$. The interest period τ is often referred to as the *tenor*. Typically, $\tau = 3$ months, 6 months or 1 year and we talk about 3-month LIBOR, etc.

The second common type of market interest rates is the swap and the forward swap rates. Under a swap contract (recall Exercise 3.2(c)) there are swap payment dates $T_1, T_2, \ldots, T_M$, where $T_k = T_0 + k\tau$. At time T_k the contract states that we pay a fixed rate of τK and receive in return the floating τ-LIBOR, $\tau L(T_{k-1}, T_{k-1}, T_k)$. At time T_0 when a contract is initiated, the market swap rate, K, is that at which the value of the arrangement is zero. It follows then that this is

$$K_\tau(T_0, T_0, T_M) = \frac{1 - P(T_0, T_M)}{\tau \sum_{k=1}^{M} P(T_0, T_k)}.$$

A contract which is struck at an earlier time t but with the same payment dates $T_1, \ldots, T_M$ is called a forward swap contract. The fair swap rate is that at which the contract has zero value; that is,

$$K_\tau(t, T_0, T_M) = \frac{P(t, T_0) - P(t, T_M)}{\tau \sum_{k=1}^{M} P(t, T_k)}.$$

The party which pays the fixed rate is usually referred to as *the payer* and the party which pays the floating rate is usually referred to as *the receiver* (of the fixed rate).

Before we derive the key results we will restate and rephrase two useful lemmas presented earlier in this book.

Lemma 9.1. *Suppose* $\log X \sim N(\mu, \sigma^2)$. *Then for any* $x > 0$,

$$E[(X - x)_+] = E(X)\Phi(d_1) - x\Phi(d_2),$$

where $d_1 = (\mu + \sigma^2 - \log x)/\sigma$ *and* $d_2 = d_1 - \sigma$.

Lemma 9.2. *Let* $X(t)$ *represent the price of a tradable asset, with* $X(t)$ *strictly positive. Suppose that we are using* $X(t)$ *as the numeraire. Then there exists a unique martingale measure* P_X *equivalent to the risk-neutral measure* Q *under which the prices of all tradable assets discounted by* $X(t)$ *are martingales. Furthermore, let* $V(S)$ *be some derivative payoff at time* S. *Then the price for this derivative at some earlier time* t *is*

$$V(t) = X(t)E_{P_X}\left[\frac{V(S)}{X(S)} \,\middle|\, \mathcal{F}_t\right].$$

9.2 LIBOR Market Models: the BGM Approach

For the time being, we will concentrate on a single maturity date T and tenor τ. For notational simplicity we write $L(t) \equiv L(t, T, T + \tau)$.

It is a requirement of our model that $L(t)$ remains strictly positive. Thus we assume that

$$dL(t) = L(t)(\mu(t)\,dt + v(t)'\,d\tilde{W}(t)),$$

where $\tilde{W}(t)$ is some M-dimensional Brownian motion under Q, $\mu(t)$ is a suitable previsible process, and $v(t) \equiv v(t, T, T+\tau)$ is assumed to be a deterministic, $M \times 1$ vector[2] with $v(t) \equiv 0$ for $T \leqslant t \leqslant T + \tau$. Now

$$L(t) = \frac{1}{\tau}\left(\frac{P(t, T)}{P(t, T+\tau)} - 1\right)$$

$$\Rightarrow \quad L(t)P(t, T+\tau) = \frac{1}{\tau}[P(t, T) - P(t, T+\tau)].$$

It follows that $V(t) = L(t)P(t, T+\tau)$ can be seen to be a tradable asset (that is, it is a self-financing portfolio). Let us also write $X(t) = P(t, T+\tau)$. Then $L(t) = V(t)/X(t)$ and we are now in a position to call on Lemma 9.2 to note that there exists a measure P_X under which $V(t)/X(t) = L(t)$ is a martingale (as are the prices of *all* tradable assets discounted by $P(t, T+\tau)$). Since $X(t) = P(t, T+\tau)$, the measure P_X is usually referred to as the forward measure $P_{T+\tau}$ (as in Chapter 7). If we assume further that $L(t)$ remains strictly positive (which follows if $P(t, s)$ is always strictly decreasing in s), then we can write

$$dL(t) = L(t)v(t)'\,d\hat{W}(t),$$

where $\hat{W}(t)$ is an M-dimensional Brownian motion under $P_{T+\tau}$. Finally, since $v(t)$ is deterministic, we find that $L(T)$ is lognormal under $P_{T+\tau}$ with

$$\mathrm{Var}_{P_{T+\tau}}[\log L(T) \mid \mathcal{F}_t] = \int_t^T v(s)'v(s)\,ds = \int_t^T |v(s)|^2\,ds,$$

$$E_{P_{T+\tau}}[\log L(T) \mid \mathcal{F}_t] = \log L(t) - \frac{1}{2}\int_t^T |v(s)|^2\,ds.$$

A *caplet* pays $V(T+\tau) = (L(T)-c)_+$ at time $T+\tau$. For $T \leqslant t \leqslant T+\tau$ we have $V(t) = P(t, T+\tau)(L(T) - c)_+$. Now, using the numeraire $X(t) = P(t, T+\tau)$, we employ Lemma 9.2 to indicate that for $t < T$ we have

$$V(t) = P(t, T+\tau)E_{P_{T+\tau}}\left[\frac{V(T+\tau)}{P(T+\tau, T+\tau)} \,\middle|\, \mathcal{F}_t\right]$$

$$= P(t, T+\tau)E_{P_{T+\tau}}[(L(T) - c)_+ \mid \mathcal{F}_t].$$

Finally, by Lemma 9.1, we have, for $t < T$,

$$V(t) = P(t, T+\tau)[L(t)\Phi(d_1) - c\Phi(d_2)], \qquad (9.1)$$

---

[2]The assumption that $v(t)$ is deterministic is central to the lognormality of $L(T)$ under $P_{T+\tau}$.

where

$$d_1 = \frac{\log(L(t)/c) + \frac{1}{2}\sigma_v^2}{\sigma_v}, \qquad d_2 = d_1 - \sigma_v, \qquad \sigma_v^2 = \int_t^T |v(s)|^2 \, ds.$$

A *cap* is a collection of caplets that pays $(L(T_{k-1}, T_{k-1}, T_k) - c)_+$ at time T_k for $k = 1, \ldots, M$.

Theorem 9.3. *The total value of this cap is*

$$V(t) = \sum_{k=1}^M P(t, T_k)[L(t, T_{k-1}, T_k)\Phi(d_{1k}) - c\Phi(d_{2k})],$$

where

$$d_{1k} = \frac{\log(L(t, T_{k-1}, T_k)/c) + \frac{1}{2}\sigma_{vk}^2}{\sigma_{vk}}, \qquad d_{2k} = d_{1k} - \sigma_{vk},$$

$$\sigma_{vk}^2 = \int_t^{T_{k-1}} |v(s, T_{k-1}, T_k)|^2 \, ds.$$

Brace, Gatarek and Musiela (1997) (BGM) give rigorous proofs covering the various technical aspects of this approach (see also Hunt and Kennedy 2000).

In Section 9.3 we will discuss how to price more complex derivatives in a way which is consistent with the LIBOR market model.

Brace, Gatarek and Musiela also consider how to extend the LIBOR market model to find approximate prices for swaption contracts (although a more satisfactory method is described in Section 9.4).

Consider a *payer swaption*[3] that matures at time T_0 with exercise swap rate K. This entitles the holder to enter into a swap contract at rate K, with net payments of $\tau(L(T_{k-1}, T_{k-1}, T_k) - K)$ at times T_k for $k = 1, \ldots, M$. This can be shown to have a value at time T_0 of

$$V(T_0) = \tau(K_\tau(T_0, T_0, T_M) - K)_+ \sum_{k=1}^M P(T_0, T_k). \qquad (9.2)$$

Let A be the event that the swaption is exercised at T_0 and I_A be the corresponding indicator random variable, that is, $I_A = 1$ if the option is exercised and $I_A = 0$ otherwise. Also, define $c_k = K\tau$ for $k = 1, \ldots, M - 1$ and $c_M = 1 + K\tau$. Then, for $t < T_0$, $C(t) = \sum_{k=1}^M c_k P(t, T_k)$ is the price of a coupon bond paying a coupon of $K\tau$ at times $T_1, \ldots, T_M$. We can then see that

$$V(T_0) = (1 - C(T_0))_+ = (1 - C(T_0))I_A,$$

since $P(T_0, T_M) + \tau K_\tau(T_0, T_0, T_M) \sum_{k=1}^M P(T_0, T_k) = 1$.

---

[3]The name *payer swaption* derives from the fact that the option is with the *payer* of the fixed rate K. A *receiver swaption* is one where the option lies with the *receiver* of the fixed rate.

At earlier times $t < T_0$ we have

$$V(t) = E_Q \left[\frac{B(t)}{B(T_0)} I_A \;\middle|\; \mathcal{F}_t \right] - E_Q \left[\sum_{k=1}^M c_k \frac{B(t)}{B(T_k)} I_A \;\middle|\; \mathcal{F}_t \right]$$

$$= P(t, T_0) E_{P_{T_0}} [I_A \mid \mathcal{F}_t] - \sum_{k=1}^M c_k P(t, T_k) E_{P_{T_k}} [I_A \mid \mathcal{F}_t], \qquad (9.3)$$

where $B(t)$ is the usual cash account at time t with $dB(t) = r(t)B(t)\,dt$. In the multifactor HJM framework prices evolve in the following way

$$dP(t, T) = P(t, T)[r(t)\,dt + S(t, T)'\,d\tilde{W}(t)],$$

where $\tilde{W}(t)$ is an M-dimensional Brownian motion under Q.

Assumption 9.4. $S(t, T) \equiv 0$ for all $0 \leqslant T - t < \tau$ (*in particular, for* $T \in \{T_0 + k\tau : k \in \mathbb{Z}\}$).

This implies that

$$B(T_k) = B(T_{k-1})/P(T_{k-1}, T_k) \qquad (9.4)$$

for all k and that $r(t)$ is deterministic between the T_k. For convenience, in what follows we will use the simplified notation:

$$L_k(t) \equiv L(t, T_{k-1}, T_k) \quad \text{and} \quad v_k(t) \equiv v(t, T_{k-1}, T_k).$$

Consider the dynamics of the LIBOR processes $L_k(t)$:

$$dL_k(t) = L_k(t)(\mu_k(t)\,dt + v_k(t)'\,d\tilde{W}(t)).$$

But

$$L_k(t) = \frac{1}{\tau} \left[\frac{P(t, T_{k-1})}{P(t, T_k)} - 1 \right]$$

$$\Rightarrow \quad dL_k(t) = \frac{P(t, T_{k-1})}{\tau P(t, T_k)} (S(t, T_{k-1}) - S(t, T_k))'(-S(t, T_k)\,dt + d\tilde{W}(t))$$

$$\Rightarrow \quad L_k(t)v_k(t) = \frac{P(t, T_{k-1})}{\tau P(t, T_k)} (S(t, T_{k-1}) - S(t, T_k))$$

$$= \frac{1 + \tau L_k(t)}{\tau} (S(t, T_{k-1}) - S(t, T_k))$$

$$\Rightarrow \quad S(t, T_{k-1}) - S(t, T_k) = \frac{\tau L_k(t)}{1 + \tau L_k(t)} v_k(t). \qquad (9.5)$$

In the mathematical development which follows we switch from time to time between the risk-neutral measure Q and the forward measures P_{T_k}. In Section 5.3.3

we showed that under the forward measure P_{T_i} the process $W_{T_i}(t) = \tilde{W}(t) - \int_0^t S(u, T_i)\, du$ was a Brownian motion. In the present context it follows that

$$dW_{T_i}(t) + S(t, T_i)\, dt = dW_{T_j}(t) + S(t, T_j)\, dt$$

$$\Rightarrow \quad dW_{T_i}(t) = dW_{T_j}(t) + \sum_{k=1}^{i} \frac{\tau L_k(t)}{1 + \tau L_k(t)} v_k(t)\, dt - \sum_{k=1}^{j} \frac{\tau L_k(t)}{1 + \tau L_k(t)} v_k(t)\, dt,$$

$$(9.6)$$

by equation (9.5).

Let us return now to develop further equation (9.3) for the price of the swaption. Within this equation we have

$$P(t, T_k) E_{P_{T_k}}[I_A \mid \mathcal{F}_t]$$

$$= E_Q\left[\frac{I_A B(t)}{B(T_k)} \,\Big|\, \mathcal{F}_t \right]$$

$$= E_Q\left[\frac{B(t)}{B(T_{k+1})} \frac{1}{P(T_k, T_{k+1})} I_A \,\Big|\, \mathcal{F}_t \right] \qquad \text{(by equation (9.4))}$$

$$= E_Q\left[\frac{B(t)}{B(T_{k+1})} (1 + \tau L_{k+1}(T_k)) I_A \,\Big|\, \mathcal{F}_t \right]$$

$$= P(t, T_{k+1}) E_{P_{T_{k+1}}}[I_A \mid \mathcal{F}_t] + \tau P(t, T_{k+1}) E_{P_{T_{k+1}}}[L_{k+1}(T_k) I_A \mid \mathcal{F}_t].$$

Consequently,

$$P(t, T_0) E_{P_{T_0}}[I_A \mid \mathcal{F}_t]$$

$$= P(t, T_M) E_{P_{T_M}}[I_A \mid \mathcal{F}_t] + \tau \sum_{k=1}^{M} P(t, T_k) E_{P_{T_k}}[L_k(T_{k-1}) I_A \mid \mathcal{F}_t].$$

Hence

$$V(t) = P(t, T_0) E_{P_{T_0}}[I_A \mid \mathcal{F}_t] - \sum_{k=1}^{M} c_k P(t, T_k) E_{P_{T_k}}[I_A \mid \mathcal{F}_t]$$

$$= P(t, T_M) E_{P_{T_M}}[I_A \mid \mathcal{F}_t] + \tau \sum_{k=1}^{M} P(t, T_k) E_{P_{T_k}}[L_k(T_{k-1}) I_A \mid \mathcal{F}_t]$$

$$- \tau K \sum_{k=1}^{M} P(t, T_k) E_{P_{T_k}}[I_A \mid \mathcal{F}_t] - P(t, T_M) E_{P_{T_M}}[I_A \mid \mathcal{F}_t]$$

$$= \tau \sum_{k=1}^{M} P(t, T_k) E_{P_{T_k}}[(L_k(T_{k-1}) - K) I_A \mid \mathcal{F}_t]. \qquad (9.7)$$

We return next to the event A. This depends upon the price of the coupon bond

$$C(T_0) = \sum_{k=1}^{M} c_k P(T_0, T_k) = \sum_{k=1}^{M} c_k \prod_{i=1}^{k} (1 + \tau L_i(T_0))^{-1}, \qquad (9.8)$$

with $A = \{C(T_0) \leqslant 1\}$. Now we can exploit the fact that $L_k(s)$ is a martingale under P_{T_k} to get

$$L_i(T_0) = L_i(t) \exp\left[-\frac{1}{2} \int_t^{T_0} |v_i(s)|^2 + \int_t^{T_0} v_i(s)' \, dW_{T_i}(s) \right]. \qquad (9.9)$$

Let

$$X_i = \int_t^{T_0} v_i(s)' \, dW_{T_i}(s)$$

$$= \int_t^{T_0} v_i(s)' \, dW_{T_j}(s) + \int_t^{T_0} \sum_{k=1}^{i} \frac{\tau L_k(s)}{1 + \tau L_k(s)} v_k(s)' v_i(s) \, ds$$

$$- \int_t^{T_0} \sum_{k=1}^{j} \frac{\tau L_k(s)}{1 + \tau L_k(s)} v_k(s)' v_i(s) \, ds. \qquad (9.10)$$

Approximation 1.

$$\int_t^{T_0} \frac{\tau L_k(s)}{1 + \tau L_k(s)} v_k(s)' v_i(s) \, ds \approx \frac{\tau L_k(t)}{1 + \tau L_k(t)} \Delta_{ki},$$

where $\Delta_{ki} = \int_t^{T_0} v_k(s)' v_i(s) \, ds$. This approximation relies on $L_k(s) \approx L_k(t)$ for $t \leqslant s \leqslant T_0$. It follows that an approximation for X_i under P_{T_j} is

$$X_i^j = \int_t^{T_0} v_i(s)' \, dW_{T_j}(s) + \sum_{k=1}^{i} \frac{\tau L_k(t)}{1 + \tau L_k(t)} \Delta_{ki} - \sum_{k=1}^{j} \frac{\tau L_k(t)}{1 + \tau L_k(t)} \Delta_{ki}$$

$$\Rightarrow \quad X^j \sim \text{MVN}(\mu^j, \Delta),$$

where

$$\Delta = (\Delta_{ki}) \quad \text{and} \quad \mu_i^j = \sum_{k=1}^{i} \frac{\tau L_k(t)}{1 + \tau L_k(t)} \Delta_{ki} - \sum_{k=1}^{j} \frac{\tau L_k(t)}{1 + \tau L_k(t)} \Delta_{ki}.$$

Furthermore, Brace, Gatarek and Musiela (1997) report that numerical tests suggest that the principal eigenvalue of the matrix Δ is typically much larger in magnitude than the remaining eigenvalues. This suggests the following.

Approximation 2A.

$$\Delta_{ki} \approx \Gamma_k \Gamma_i \quad \text{for some constants } \Gamma_1, \ldots, \Gamma_M.$$

Next, define

$$d_0 = 0,$$

$$d_i = \sum_{k=1}^{i} \frac{\tau L_k(t)}{1 + \tau L_k(t)} \Gamma_k \quad \text{for } i = 1, \ldots, M$$

$$\Rightarrow \quad \mu_i^j = \Gamma_i(d_i - d_j).$$

We now combine equations (9.8), (9.9) and (9.10) and then apply the subsequent approximations:

$$
\begin{aligned}
C(T_0) &= \sum_{k=1}^{M} c_k \prod_{i=1}^{k} \{1 + \tau L_i(t) \exp(X_i - \tfrac{1}{2}\Delta_{ii})\}^{-1} \\
&\approx \sum_{k=1}^{M} c_k \prod_{i=1}^{k} \{1 + \tau L_i(t) \exp(X_i^j - \tfrac{1}{2}\Delta_{ii})\}^{-1} \\
&\approx \sum_{k=1}^{M} c_k \prod_{i=1}^{k} \{1 + \tau L_i(t) \exp(\Gamma_i(z_i^j + d_i - d_j) - \tfrac{1}{2}\Delta_{ii})\}^{-1} \\
&= g_j(z_1^j, \ldots, z_M^j),
\end{aligned}
$$

where the z_i^j are identically distributed, $N(0, 1)$, random variables under P_{T_j}. Since $\Delta \approx \Gamma\Gamma'$, the z_i^j are approximately perfectly correlated.

Approximation 2B. Hence $g_j(z_1^j, \ldots, z_M^j) \approx \tilde{g}_j(z)$ for some scalar $z \sim N(0, 1)$ under P_{T_j}.

Proposition 9.5. *The event A corresponds approximately to the event that $\tilde{g}_j(z) \leqslant 1$.*

Note that

$$\tilde{g}_j'(z) < 0 \quad \text{for all } z, \qquad \lim_{z \to -\infty} \tilde{g}_j(z) = 1 + M\tau K, \qquad \lim_{z \to +\infty} \tilde{g}_j(z) = 0.$$

These three observations indicate that for each $j = 0, \ldots, M$ there exists $\tilde{z}_j$ such that $\tilde{g}_j(\tilde{z}_j) = 1$ and that if $\tilde{z}_0$ solves $\tilde{g}_0(z) = 1$, then, for each $j = 1, \ldots, M$, $\tilde{z}_j = \tilde{z}_0 + d_j$.

We return now to equation (9.7). This includes terms of the form $E_{P_{T_k}}[I_A \mid \mathcal{F}_t]$ and $E_{P_{T_k}}[L_k(T_{k-1})I_A \mid \mathcal{F}_t]$ for $k = 1, \ldots, M$, for which we must derive approximate expressions for:

$$E_{P_{T_k}}[I_A \mid \mathcal{F}_t] \approx \Pr_{P_{T_k}}(\tilde{g}_j(Z) \leqslant 1) \quad \text{where } Z \sim N(0, 1) \text{ under } P_{T_k}$$
$$= \Pr_{P_{T_k}}(Z \geqslant \tilde{z}_0 + d_k)$$
$$= \Phi(-\tilde{z}_0 - d_k),$$

$$E_{P_{T_k}}[L_k(T_{k-1})I_A \mid \mathcal{F}_t] = E_{P_{T_k}}\left[I_A L_k(t) \exp\left(\int_t^{T_0} v_k(s)\, dW_{T_k}(s) - \tfrac{1}{2}\Delta_{kk}\right) \,\middle|\, \mathcal{F}_t\right]$$
$$= L_k(t) E_{P_{T_k}}\left[\frac{d\hat{P}_{T_k}}{dP_{T_k}} I_A \,\middle|\, \mathcal{F}_t\right],$$

where

$$\frac{d\hat{P}_{T_k}}{dP_{T_k}} \equiv \exp\left(\int_t^{T_0} v_k(s)\, dW_{T_k}(s) - \tfrac{1}{2}\Delta_{kk}\right).$$

This implies that

$$E_{P_{T_k}}[L_k(T_{k-1})I_A \mid \mathcal{F}_t] = L_k(t)\Pr_{\hat{P}_{T_k}}(Z > \tilde{z}_0 + d_k),$$

where, again, $Z \sim N(0, 1)$ under P_{T_k}. In order to evaluate this final expression we must establish the distribution of Z under $\hat{P}_{T_k}$. Under $\hat{P}_{T_k}$,

$$\hat{W}_{T_k}(s) = W_{T_k}(s) - \int_t^s v_k(u)\, du$$

is a Brownian motion by the Girsanov Theorem (Theorem A.12). Also, for any i

$$X_i^k \approx \Gamma_i(Z + d_i - d_k)$$
$$\Rightarrow \quad \Gamma_i Z \approx \int_t^{T_0} v_i(s)'\, dW_{T_k}(s)$$
$$= \int_t^{T_0} v_i(s)'\, d\hat{W}_{T_k}(s) + \int_t^{T_0} v_i(s)'v_k(s)\, ds$$
$$\approx \Gamma_i(\hat{Z} + \Gamma_k),$$

where $\hat{Z} \sim N(0, 1)$ under $\hat{P}_{T_k}$. Hence,

$$E_{P_{T_k}}[L_k(T_{k-1})I_A \mid \mathcal{F}_t] \approx L_k(t)\Pr_{\hat{P}_{T_k}}(\hat{Z} + \Gamma_k > \tilde{z}_0 + d_k)$$
$$= \Phi(-\tilde{z}_0 - d_k + \Gamma_k).$$

Proposition 9.6. *The price at time t of the swaption is given by*

$$V(t) \approx \tau \sum_{k=1}^M P(t, T_k)[L_k(t)\Phi(-\tilde{z}_0 - d_k + \Gamma_k) - K\Phi(-\tilde{z}_0 - d_k)],$$

where $\tilde{z}_0$ is the solution to $\tilde{g}_0(z) = 1$ and

$$\tilde{g}_0(z) = \sum_{k=1}^{M} c_k \prod_{i=1}^{k} \{1 + \tau L_i(t) \exp(\Gamma_i(z + d_i) - \tfrac{1}{2}\Delta_{ii})\}^{-1}.$$

Hunt and Kennedy (2000) discuss briefly the application of LIBOR models to more complex contracts including limit caps and barrier swaptions. In such cases they suggest that simulation is a necessary part of the pricing exercise. In both cases the contracts are sufficiently close to the simpler cap and swaption contracts to require careful calibration and evaluation of derivative prices.

9.3   Simulation of LIBOR Market Models

In the previous section we introduced the LIBOR market model. We developed a simple and exact formula for pricing caps and gave an approximate and more complex formula for pricing swaption contracts.

It is useful to be able to price other derivative contracts in a way which is consistent with the pricing of caps; this can be done via simulation.

Recall the dynamics under Q of the various LIBOR rates given in the equations leading up to (9.5):

$$
\begin{aligned}
\mathrm{d}L_k(t) &\equiv \mathrm{d}L(t, T_{k-1}, T_k) \\
&= \frac{P(t, T_{k-1})}{\tau P(t, T_k)} (S(t, T_{k-1}) - S(t, T_k))'(-S(t, T_k)\,\mathrm{d}t + \mathrm{d}\tilde{W}(t)) \\
&= L_k(t)v_k(t)'(-S(t, T_k)\,\mathrm{d}t + \mathrm{d}\tilde{W}(t)) \qquad\qquad (9.11) \\
\Rightarrow\quad S(t, T_{k-1}) - S(t, T_k) &= \frac{\tau L_k(t)}{1 + \tau L_k(t)} v_k(t). \qquad\qquad (9.12)
\end{aligned}
$$

Now suppose that we have a derivative payment at T_1 and that the payoff on the derivative can be fully determined by the behaviour of the $L_k(t)$ for $k = 1, \ldots, M$ up to and including time T_1. Based on Chapter 7, the price at time $t < T_1$ is then the product of $P(t, T_1)$ and the expected value under P_{T_1} of the payoff at T_1. So we need to be able to simulate each of the $L_k(t)$ under P_{T_1}. What we actually have so far is a model for each of the $L_k(t)$ under corresponding measures P_{T_k}, so what we need to be able to do now is translate the dynamics of $L_k(t)$ under P_{T_k} into dynamics under P_{T_1}.

From (9.11) we can deduce first that, for each k,

$$W_{T_k}(t) = \tilde{W}(t) - \int_0^t S(u, T_k)\,\mathrm{d}u$$

is a standard Brownian motion under P_{T_k}. Second, for $i < j$, we have

$$dW_{T_j}(t) = dW_{T_i}(t) + (S(t, T_j) - S(t, T_i))\,dt$$

$$= dW_{T_i}(t) + \sum_{k=i+1}^{j} (S(t, T_k) - S(t, T_{k-1}))\,dt$$

$$= dW_{T_i}(t) - \sum_{k=i+1}^{j} \frac{\tau L_k(t)}{1 + \tau L_k(t)} v_k(t)\,dt \qquad \text{(by (9.12)).}$$

It follows that simulating under P_{T_1} we have

$$dL_1(t) = L_1(t)v_1(t)'\,dW_{T_1}(t)$$

and

$$dL_j(t) = L_j(t)v_j(t)'\left(dW_{T_1}(t) - \sum_{k=2}^{j} \frac{\tau L_k(t)}{1 + \tau L_k(t)} v_k(t)\,dt\right)$$

for $j = 2, \ldots, M$.

9.4 Swap Market Models

It can be seen from the preceding discussion that the Brace, Gatarek and Musiela (1997) method for pricing swaptions is rather cumbersome and contains a number of awkward approximations and assumptions. Jamshidian (1997) and Musiela and Rutkowski (1997) developed a more direct approach that does not suffer from these complications.

For pricing the swaption described in Section 9.2 we use a different numeraire from before; namely,

$$X(t) = \tau \sum_{k=1}^{M} P(t, T_k),$$

with associated forward measure P_X. Under P_X the prices of all tradable assets discounted by $X(t)$ are martingales. Now the forward swap rate $K_\tau(t, T_0, T_M)$ equals $(P(t, T_0) - P(t, T_M))/X(t)$ and $(P(t, T_0) - P(t, T_M))$ equals the value of a tradable asset. It follows that $K_\tau(t, T_0, T_M)$ is a martingale under P_X. For notational convenience we will write $K(t) = K_\tau(t, T_0, T_M)$. Suppose that

$$dK(t) = K(t)\gamma(t)\,dW_X(t),$$

where $W_X(t)$ is a Brownian motion under P_X and $\gamma(t)$ is deterministic.

Theorem 9.7. *The value $V(t)$ at time t of the T_0 maturity swaption with exercise rate K is*

$$V(t) = \left(\tau \sum_{k=1}^{M} P(t, T_k)\right)[K(t)\Phi(d_1) - K\Phi(d_2)],$$

where

$$d_1 = \frac{\log(K(t)/K) + \frac{1}{2}S_K^2}{S_K}, \qquad d_2 = d_1 - S_K, \qquad S_K^2 = \int_t^{T_0} |\gamma(s)|^2 \, ds.$$
$$(9.13)$$

Proof. The SDE for $K(t)$ gives us the solution

$$K(T_0) = K(t) \exp\left[-\frac{1}{2}\int_t^{T_0} |\gamma(s)|^2 \, ds + \int_t^{T_0} \gamma(s) \, dW_X(s)\right].$$

Since $\gamma(s)$ is deterministic it follows that, under P_X, $K(T_0)$ is lognormal with

$$\mathrm{Var}_{P_X}[\log K(T_0) \mid \mathcal{F}_t] = \int_t^{T_0} |\gamma(s)|^2 \, ds = S_K^2,$$

$$E_{P_X}[\log K(T_0) \mid \mathcal{F}_t] = \log K(t) - \frac{1}{2}S_K^2.$$

We have

$$V(T_0) = \tau \sum_{k=1}^{M} P(T_0, T_k)(K(T_0) - K)_+ = X(T_0)(K(T_0) - K)_+$$

$$\Rightarrow \quad V(t) = X(t)E_{P_X}\left[\frac{V(T_0)}{X(T_0)} \,\middle|\, \mathcal{F}_t\right] \text{ by Lemma 9.2}$$

$$= \left(\tau \sum_{k=1}^{M} P(t, T_k)\right)E_{P_X}[(K(T_0) - K)_+ \mid \mathcal{F}_t]$$

$$= \left(\tau \sum_{k=1}^{M} P(t, T_k)\right)[K(t)\Phi(d_1) - K\Phi(d_2)],$$

using the lognormality of $K(T_0)$ in combination with Lemma 9.1. $\qquad\square$

Definition 9.8. The implied swaption volatility for the swaption defined in Theorem 9.7 is $\bar{\sigma}_K$, where

$$\bar{\sigma}_K^2 = \frac{S_K^2}{T_0 - t}$$

and S_K^2 is defined in equation (9.13). The implied swaption volatility represents the average volatility of the swap rate $K(u)$ between t and T_0. Usually the exercise rate K is the current at-the-money rate $K(t)$.

For further discussion and numerical examples, see Brigo and Mercurio (2001) and Pelsser (2000).

Clearly, this method offers an intellectual advantage over the Brace–Gatarek–Musiela (BGM) swaption pricing formula. This is in the sense that it provides us with an exact pricing formula consistent with the basic assumption that swap rates are lognormal. On the other hand, the BGM approach offers consistency between LIBOR and swap rates, whereas the swap market model treats each class in isolation.

Additionally, we have to be continually aware (and this statement applies to all models described in this book) that just because we are able to derive analytical formulae for derivative prices does not mean that we have a good model. For example, the lognormal assumption for LIBOR may not be very accurate.

9.5 Exercises

Exercise 9.1. Explain why

(a) the assumption that 3-month LIBORs are lognormal is not consistent with the assumption that 6-month LIBORs are lognormal;

(b) the assumption that 6-month LIBORs are lognormal is not consistent with the assumption that swap rates with 6-monthly payment dates are lognormal.

Exercise 9.2. Investigate numerically the accuracy of Approximations 1 and 2 in Section 9.2.

Exercise 9.3. Recall how the HJM dynamics for forward rates $f(t, T)$ ties in with the Hull and White model (Section 5.2.2), which has a mean reverting risk-free rate. Show how the same principles can be used to generate a discrete-time model for 6-month LIBOR which is mean reverting.

Exercise 9.4. Consider the caplet valuation formula (9.1). Describe one way of hedging to replicate the payoff on this contract.

Exercise 9.5. Suppose that $V(t)$ is the price of a tradable asset and that $V(t) > 0$ for all t almost surely with SDE $dV(t) = V(t)[r(t) dt + S_V(t)' d\tilde{W}(t)]$, $S_V(t)$ is a previsible $n \times 1$ vector volatility process and $\tilde{W}(t)$ is a standard n-dimensional Brownian motion under the risk-neutral measure Q.

Suppose also that $X(t)$ is another tradable asset.

Find the change of measure which will make the discounted price process $Z(t) = X(t)/V(t)$ a martingale. Show that this change of measure is the same for all tradable assets $X(t)$.

Exercise 9.6. Table 9.1 gives the forward 12-month LIBOR rates at time $t = 0$ for various maturity dates and for the prices of at-the-money (ATM) caplets. In the table, $L(t, T_k, T_{k+1}) = P(t, T_k)/P(t, T_{k+1}) - 1$, where $T_{k+1} = T_k + 1$ for all k.

For a cap rate of c the caplet price given is for 1000 units; that is, the price is in respect of a derivative payoff of $1000(L(T_k, T_k, T_{k+1}) - c)_+$ at time T_{k+1}.

Table 9.1.

Forward LIBOR	Rate (%)	ATM caplet price
$L(0, 0, 1)$	4.17	0
$L(0, 1, 2)$	5.49	3.7534
$L(0, 2, 3)$	5.95	5.1526
$L(0, 3, 4)$	6.02	5.7395
$L(0, 4, 5)$	5.91	5.8987
$L(0, 5, 6)$	5.75	5.8505
$L(0, 6, 7)$	5.60	5.7116
$L(0, 7, 8)$	5.46	5.5396
$L(0, 8, 9)$	5.34	5.3609
$L(0, 9, 10)$	5.25	5.1863
$L(0, 10, 11)$	5.18	5.0196
$L(0, 11, 12)$	5.13	4.8610
$L(0, 12, 13)$	5.08	4.7099
$L(0, 13, 14)$	5.05	4.5651
$L(0, 14, 15)$	5.03	4.4257
$L(0, 15, 16)$	5.01	4.2907
$L(0, 16, 17)$	5.00	4.1597
$L(0, 17, 18)$	4.99	4.0321
$L(0, 18, 19)$	4.99	3.9076
$L(0, 19, 20)$	4.99	3.7858

(a) Calculate the implied LIBOR volatilities for these data.

(b) Calculate the price of a 10-year interest rate cap where the cap rate is 6%, per 1000 nominal.

(c) Given these prices, calculate the swap-rate curve of the same date assuming annual payment dates.

(d) Calculate the forward swap rate for a swap contract commencing at time 10 and ceasing at time 20.

(e) For each T_k, the forward LIBOR rate $L(t, T_{k-1}, T_k)$ is lognormally distributed under the forward measure P_{T_k}. However, under the same measure P_{T_k}, the other forward LIBOR rates are not lognormally distributed.

Assume that the instantaneous *correlation* between the changes in different LIBOR rates $L(t, T_k, T_{k+1})$ and $L(t, T_j, T_{j+1})$ is $\rho^{|k-j|}$, for some $0 < \rho < 1$.

Show how to shift between the different forward measures P_{T_k} and $P_{T_{k+1}}$ and write down the SDE for different LIBOR rates under the same measure P_{T_k}.

Simulate under P_{10} the forward LIBOR rates $L(10, T_k, T_{k+1})$ for $k = 10, \ldots,$ 19.

Hence, simulate the 10-year swap rate at time 10; that is

$$K(10, 10, 20) = \frac{1 - P(10, 20)}{\sum_{j=11}^{20} P(10, j)}.$$

Investigate how close this is to lognormality and test your conclusion for sensitivity to changes in ρ.

On the assumption that $K(10, 10, 20)$ is lognormal, estimate the variance of $\log K(10, 10, 20)$. Use this to feed into the Jamshidian swap-rate market model to calculate the price at time 0 of an ATM swaption on $K(10, 10, 20)$. Compare this with the price you get based on your simulation results for $K(10, 10, 20)$ under the measure P_{10}.

Exercise 9.7. Give a brief description of the swap market model, giving some numerical examples.

Compare and contrast this with the LIBOR market model.

Describe how the LIBOR market model can be used to calculate approximate prices for swaption contracts.

Exercise 9.8. Derive a replicating strategy for caplets and swaptions based on the LIBOR and swap market models respectively.

Carry out a simulation exercise to see how this performs with discrete-time rebalancing.

10

Numerical Methods

We have seen in previous chapters that analytical pricing formulae exist for bonds and for European options for specific models (for example, Vasicek, Cox–Ingersoll–Ross, Hull–White). In this chapter we discuss how to price bonds and derivatives when analytical formulae do not exist.

We will concentrate in the earlier parts of this chapter on one-factor models for the term structure, leaving comment on multifactor models until later on.

10.1 Choice of Measure

Before we discuss specific numerical techniques there is a more basic issue to discuss: the choice of measure.

The natural or most obvious place to start is to price a derivative under the risk-neutral measure Q; that is,

$$V(t) = E_Q\left[\exp\left(-\int_t^T r(s)\,\mathrm{d}s\right)V(T) \,\bigg|\, \mathcal{F}_t\right].$$

This approach is the most widely used in practice and is the basis of the trinomial lattice and finite-difference methods described later in this chapter. In probabilistic terms, this requires derivation of the joint distribution of $\int_t^T r(s)\,\mathrm{d}s$ and the derivative payoff $V(T)$ at the maturity or exercise date, T.

However, there is an alternative that we have seen earlier which offers specific advantages for European-type derivatives.

Specifically, we can use the forward measure, P_T, under which $V(t)/P(t,T)$ is a martingale. We saw in Chapter 7 that

$$V(t) = P(t,T)E_{P_T}[V(T) \mid \mathcal{F}_t].$$

In particular, we saw how this simplified the derivation of analytical expressions for option prices for the Vasicek model. In more general circumstances the original problem splits into two parts: first, establish the zero-coupon bond price $P(t,T)$; second, establish analytically or numerically the distribution of $V(T)$ under P_T. In effect, this means that the pricing problem under certain circumstances is reduced from two dimensions to one.

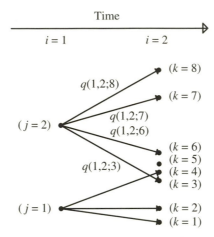

Figure 10.1. Example of a general lattice model with multiple branching. The branching probabilities $q(i, j; k)$ represent the probability under Q that the process moves to state $(i + 1, k)$ at time t_{i+1} given it is in state (i, j) at time t_i.

10.2 Lattice Methods

Earlier in this book we introduced the Ho and Lee binomial model (Section 3.2). At that point the motivation was to present a simple framework which demonstrated the no-arbitrage approach to bond pricing. However, we can now consider the binomial model as a candidate for approximation of continuous-time models. Indeed we note that the continuous-time analogue of the Ho and Lee binomial model was presented in Section 5.2.1.

The binomial model uses dynamic hedging to replicate any derivative payoff. We can note the following points in favour of this model.

- The binomial structure in the model means that the market is complete.

- The dynamic hedging strategy for replication of a derivative payoff is a simple by-product of the numerical procedure.

- The no-arbitrage price of a derivative is uniquely determined.

- The risk-neutral measure, Q, plays no part in the pricing of a derivative, except as a convenient computational tool.

The major disadvantage of this model is that there is insufficient structure in the binomial lattice to allow us to approximate accurately and simultaneously both the drift (for example, mean-reverting) and volatility of many continuous-time diffusion models.

In contrast, the trinomial (and more general) lattice models we will shortly discuss have the following characteristics.

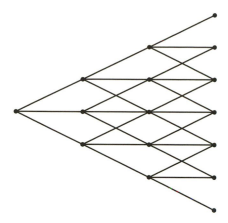

Figure 10.2. Example of a regular trinomial lattice. Over each time step the risk-free rate can either stay at the same level, go up by one step or go down by one step.

- We require explicit specification of the risk-neutral measure, Q.

- Appropriate specification of the branching probabilities allows us to model both the drift and volatility of the continuous-time model.

- Dynamic hedging strategies are not by-products of the basic procedure.

These characteristics appear to make more general lattice models less attractive. However, the ability to match both drift and volatility makes the use of such an approach more effective than the binomial model for pricing derivatives.

10.2.1 The General Lattice Model

In the general case we assume that we sample the process at times $t_0 < t_1 < \cdots < t_m$. At time t_i there are n_i nodes with node j ($j = 1, 2, \ldots, n_i$) at time t_i referred to as state (i, j). Each state has associated with it the continuously compounding risk-free rate of interest $r(i, j)$ which applies between times t_i and t_{i+1}. We will also refer later to $r(t)$ or $r(t_i)$ as the random value representing the risk-free rate of interest at time t_i. We define the

$$q(i, j; k) = \Pr_Q[\text{state } (i + 1, k) \text{ at time } t_{i+1} \mid \text{state } (i, j) \text{ at time } t_i]$$

to be the transition probabilities for this lattice (see Figure 10.1).

It is common to introduce a certain amount of regularity into the lattice so that $t_i = t_0 + i \Delta t$ (where Δt is the length of the time step) with similar regular spacing in either the $r(i, j)$ or some transformation of the risk-free rates. Additionally, the model is often time homogeneous, in which case the transition probabilities $q(i, j; k)$ and the risk-free rate of interest are independent of the time index, i.

10.2.2 Derivative Pricing

Consider now, under this general lattice model, a derivative contract with maturity date $T = t_m$. The derivative is known to pay $V(m, k)$ at time T if the process finishes in state (m, k). For a European-style contract we calculate prices working backwards one time-step at a time; that is, at $t_{m-1}, t_{m-2}, \ldots, t_1, t_0 = 0$. In particular, given the prices at time t_i in each of the states $(i, 1), \ldots, (i, n_i)$, we calculate prices at time t_{i-1} as follows. For each $j = 1, \ldots, n_{i-1}$,

$$V(i - 1, j) = \exp[-r(i - 1, j)(t_i - t_{i-1})] \sum_{k=1}^{n_i} q(i - 1, j; k)V(i, k).$$

For an American-style contract which pays $U(i - 1, j)$ upon early exercise in state $(i - 1, j)$ at time t_{i-1} we have

$$
V(i - 1, j)
$$
$$
= \max\left\{ \exp[-r(i - 1, j)(t_i - t_{i-1})] \sum_{k=1}^{n_i} q(i - 1, j; k)V(i, k), U(i - 1, j) \right\}.
$$

10.2.3 The Hull and White Trinomial Lattice Model

We will now develop in detail the regular trinomial lattice described by Hull and White (1994a). Consider first the time homogeneous case. The basic branching pattern for the risk-free rate of interest, $r(t)$, is illustrated in Figure 10.2. Over each time interval, Δt, the risk-free rate can either stay the same, increase by a fixed amount, Δr, or decrease by the same amount, Δr. Thus, given $r(i \Delta t)$ at time $i \Delta t$,

$$
r((i + 1)\Delta t) = \begin{cases} r(i\Delta t) + \Delta r, \\ r(i\Delta t), \\ r(i\Delta t) - \Delta r. \end{cases}
$$

We can note, trivially, the recombining nature of the tree; namely, that over two time steps the combinations *up–down, level-level, down–up* arrive at the same point (as, for example, do *up-level* and *level-up*).

In labelling the states at different times, it is convenient from the computational point of view that a level move from state (i, j) corresponds to a move into state $(i + 1, j)$, an up move to $(i + 1, j + 1)$, and a down move to $(i + 1, j - 1)$. If we start from state $(0, 0)$, then, after n time steps, we will have $2n + 1$ states: $(n, -n), (n, -n + 1), \ldots, (n, n - 1), (n, n)$. (This labelling of states is in contrast to the general model in which we limited ourselves to states $(i, 1), \ldots, (i, n_i)$ at time t_i.)

It is possible (and indeed likely as we let Δt tend to zero) that either

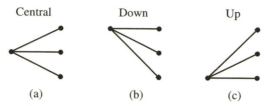

Figure 10.3. Hull and White branching patterns: downward branching applies at the uppermost points in the lattice; upward branching applies at the lowest points in the lattice; central branching applies at all points in between.

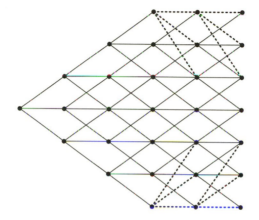

Figure 10.4. Example of a Hull and White trinomial lattice. Once the branching has reached its full extent at the end of the run-in period, downward and upward branching applies at the uppermost and lowest nodes in the lattice (dotted lines).

- the ultimate spread of the regular lattice (Figure 10.2) will be far wider than the process $r(t)$ is ever likely to reach; or

- it will be impossible at the extremes to find positive branching probabilities which allow us to match the drift and volatility of the underlying continuous-time model.

This led Hull and White to impose a limit on the extent of the lattice. At the maximum value for $r(t)$, central branching (Figure 10.3(a)) switches to downward branching (Figure 10.3(b)) with $r(t + \Delta t)$ taking the value $r(t)$, $r(t) - \Delta r$ and $r(t) - 2\Delta r$. (Downward branching still has three branches to allow us to match the drift and the volatility of the underlying continuous-time process.) Similarly, at the minimum value for $r(t)$ we switch from central branching to upward branching (Figure 10.3(c)). The net result is a restricted lattice of the type illustrated in Figure 10.4.

As has been mentioned above, the branching probabilities are set in such a way as to match (approximately) the drift and volatility of the underlying continuous-time process. We will make use of the following notation:

$$q_{uu} = \Pr_Q[\text{state } (i+1, j+2) \text{ at time } (i+1)\Delta t \mid \text{state } (i, j) \text{ at time } i\Delta t],$$

$$q_u = \Pr_Q[(i+1, j+1) \mid (i, j)], \qquad\qquad q_c = \Pr_Q[(i+1, j) \mid (i, j)],$$

$$q_d = \Pr_Q[(i+1, j-1) \mid (i, j)], \qquad\qquad q_{dd} = \Pr_Q[(i+1, j-2) \mid (i, j)].$$

Our moment-matching requirements are then based on the SDE for $r(t)$, $dr(t) = \mu_r(t, r(t)) \, dt + \sigma_r(t, r(t)) \, d\tilde{W}(t)$, as follows:

$$E_Q[r(t + \Delta t) - r(t) \mid r(t)] \approx \mu_r(t, r(t))\Delta t$$
$$= q_{uu} \times 2\Delta r + q_u \times \Delta r + q_c \times 0 + q_d \times (-\Delta r) + q_{dd} \times (-2\Delta r), \tag{10.1}$$

$$E_Q[(r(t + \Delta t) - r(t))^2 \mid r(t)] \approx \sigma_r(t, r(t))^2 \Delta t + \mu_r(t, r(t))^2 \Delta t^2$$
$$= q_{uu} \times 4\Delta r^2 + q_u \times \Delta r^2 + q_d \times \Delta r^2 + q_{dd} \times 4\Delta r^2, \tag{10.2}$$

$$q_{uu} + q_u + q_c + q_d + q_{dd} = 1. \tag{10.3}$$

Central branching ($q_{uu} = q_{dd} = 0$). We now solve equations (10.1)–(10.3) to get

$$q_u = \frac{1}{2}\left[\frac{\sigma_r(t, r(t))^2 \Delta t}{\Delta r^2} + \frac{\mu_r(t, r(t))\Delta t}{\Delta r} + \frac{\mu_r(t, r(t))^2 \Delta t^2}{\Delta r^2} \right],$$

$$q_d = \frac{1}{2}\left[\frac{\sigma_r(t, r(t))^2 \Delta t}{\Delta r^2} - \frac{\mu_r(t, r(t))\Delta t}{\Delta r} + \frac{\mu_r(t, r(t))^2 \Delta t^2}{\Delta r^2} \right]$$

$$\Rightarrow \quad q_c = 1 - q_u - q_d = 1 - \frac{\sigma(t, r(t))^2 \Delta t}{\Delta r^2} - \frac{\mu_r(t, r(t))^2 \Delta t^2}{\Delta r^2}.$$

For convenience we define

$$k_r = k_r(t, r(t)) = \frac{\Delta r^2}{\sigma_r(t, r(t))^2 \Delta t} \quad \text{and} \quad \theta_r = \theta_r(t, r(t)) = \frac{\mu_r(t, r(t))\Delta t}{\Delta r}$$

$$\Rightarrow \quad q_u = \frac{1}{2k_r} + \frac{\theta_r}{2} + \frac{\theta_r^2}{2}, \qquad q_c = 1 - \frac{1}{k_r} - \theta_r^2, \qquad q_d = \frac{1}{2k_r} - \frac{\theta_r}{2} + \frac{\theta_r^2}{2}. \tag{10.4}$$

Down branching ($q_{uu} = q_u = 0$). Again we solve equations (10.1)–(10.3) to get

$$q_c = 1 + \frac{1}{2k_r} + \frac{3\theta_r}{2} + \frac{\theta_r^2}{2}, \qquad q_d = -\frac{1}{k_r} - 2\theta_r - \theta_r^2, \qquad q_{dd} = \frac{1}{2k_r} + \frac{\theta_r}{2} + \frac{\theta_r^2}{2}. \tag{10.5}$$

Up branching ($q_{dd} = q_d = 0$).

$$q_{uu} = \frac{1}{2k_r} - \frac{\theta_r}{2} + \frac{\theta_r^2}{2}, \qquad q_u = -\frac{1}{k_r} + 2\theta_r - \theta_r^2, \qquad q_c = 1 + \frac{1}{2k_r} - \frac{3\theta_r}{2} + \frac{\theta_r^2}{2}.$$
(10.6)

This scheme works effectively, in general, for smooth drift and volatility functions. However, where the volatility depends upon $r(t)$, it is computationally convenient to transform the process into one with constant (and, without loss of generality, unit) volatility. We have $dr(t) = \mu_r(t, r(t)) \, dt + \sigma_r(t, r(t)) \, d\tilde{W}(t)$, where $\tilde{W}(t)$ is a Brownian motion under the risk-neutral measure Q. Now consider $X(t) = f(t, r(t))$. We assume that $f(t, r)$ is strictly monotonic over the domain of r, meaning that $r(t) = f^{-1}(t, X(t))$ is well defined. By Itô's formula we have

$$dX(t) = \mu_X(t, X(t)) \, dt + \sigma_X(t, X(t)) \, d\tilde{W}(t)$$
$$= \left(\frac{\partial f}{\partial t} + \mu_r(t, r(t)) \frac{\partial f}{\partial r} + \tfrac{1}{2} \sigma_r(t, r(t))^2 \frac{\partial^2 f}{\partial r^2} \right) dt + \sigma_r(t, r(t)) \frac{\partial f}{\partial r} \, d\tilde{W}(t).$$

If we require $X(t)$ to have unit volatility ($\sigma_X(t, X(t)) \equiv 1$), then we require

$$\frac{\partial f}{\partial r} = \frac{1}{\sigma_r(t, r(t))}.$$
(10.7)

For example, in the case of the Cox–Ingersoll–Ross model we have $\sigma_r(t, r(t)) = \sigma \sqrt{r(t)}$, which implies that $f(t, r) = 2\sigma^{-1}\sqrt{r} + c(t)$ for some deterministic function $c(t)$.

We will now consider in more detail the trinomial lattice for the transformed process $X(t)$.

The drift of $X(t)$. We now have

$$\mu_X(t, X(t)) = \frac{\partial f}{\partial t}(t, r(t)) + \mu_r(t, r(t)) \frac{\partial f}{\partial r}(t, r(t)) + \tfrac{1}{2} \sigma_r(t, r(t))^2 \frac{\partial^2 f}{\partial r^2}(t, r(t))$$
$$= \frac{\partial f}{\partial t}(t, r(t)) + \frac{\mu_r(t, r(t))}{\sigma_r(t, r(t))} - \frac{1}{2} \frac{\partial \sigma_r}{\partial r}(t, r(t))$$

as a consequence of equation (10.7).

The factor k. Recall that we defined $k_r(t, r(t)) = \Delta r^2 / \sigma_r(t, r(t))^2 \Delta t$. Under the transformed process we have

$$k_X(t, X(t)) = \Delta x^2 / \sigma_X(t, X(t))^2 \Delta t = \Delta x^2 / \Delta t,$$

since $\sigma_X(t, x) \equiv 1$. Thus, $k_X(t, X(t)) = k$ is constant. We now combine this with the drift of $X(t)$ to define

$$\theta_X(t, X(t)) = \theta = \mu_X(t, X(t)) \frac{\Delta t}{\Delta x} = \mu_X(t, X(t)) \sqrt{\frac{\Delta t}{k}}.$$

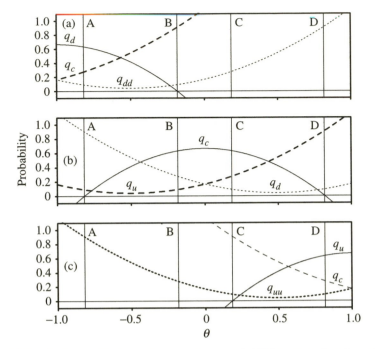

Figure 10.5. Hull and White model branching probabilities as a function of $\theta =$ $\mu_X(t, X(t))\sqrt{\Delta t/k}$: $k = 3$, $\Delta x = \sqrt{3\Delta t}$: (a) down branching; (b) central branching; (c) up branching. The switch from central branching (b) to up branching (c) must occur between thresholds C and D. The switch from central branching (b) to down branching (a) must occur between thresholds A and B. For a mean-reverting model high values of θ correspond to low values of $X(t)$ and $r(t)$.

Formulae for the branching probabilities for the transformed process $X(t)$ take the same functional form in k and θ as specified in equations (10.4)–(10.6).

Hull and White suggest that we take $k = 3$ (that is, $\Delta x = \sqrt{3\Delta t}$) indicating that this is a good choice for k on grounds of numerical efficiency. We clarify this as follows.

- Central, up and down branching probabilities as a function of θ are plotted in Figure 10.5.
- $\theta = 0$ corresponds to zero drift in the process $X(t)$ with a degree of symmetry in the central branching probabilities around $\theta = 0$.
- We require $k > 1$ to ensure that $q_c > 0$ for some values of θ with central branching (particularly the maximum at $\theta = 0$).
- If $k < 4$, then the central branching probabilities q_d and q_u are positive for all θ.
- Similar requirements exist on k for the down and up branching probabilities.

- Switching from central to up branching must take place, in terms of θ between the thresholds at $1 - \sqrt{(k-1)/k}$ (barrier C in Figure 10.5) and $\sqrt{(k-1)/k}$ (barrier D). These requirements ensure that the branching probabilities are positive at all nodes in the lattice. Similarly, switching from central to down branching must take place between $\theta = -1 + \sqrt{(k-1)/k}$ and $\theta = -\sqrt{(k-1)/k}$ (barriers B and A).

  We also require that $1 - \sqrt{(k-1)/k} < \sqrt{(k-1)/k}$ for these conditions to be feasible (that is, to ensure that barrier C lies strictly to the left of barrier D). This simplifies to the requirement that $k > 4/3$.

- In summary, k should be taken somewhere in the range $4/3$ to 4. As noted above, Hull and White suggest $k = 3$. The barriers are then at -0.816, -0.184, $+0.184$ and $+0.816$.

  It is possible to switch between central and up or down branching at any point between these barriers. However, Hull and White suggest that the switch takes place at the earliest opportunity; that is, at $\theta = \pm(1 - \sqrt{(k-1)/k})$. This minimizes the size of the lattice for given values of Δt and k without, in general, losing substantial accuracy in subsequent calculations (since the probabilities of ever attaining the upper or lower barriers become negligible as Δt gets small).

Further comment on the transformation of $r(t)$. We noted above that it is convenient to transform the process $r(t)$ into one with unit volatility: $X(t) = f(t, r(t))$. We note that we get the same effect whenever we add a deterministic (and continuously differentiable) function of time $c(t)$; that is, $\tilde{X}(t) = f(t, r(t))$ still has unit volatility.

Hull and White (1994a) suggest that for the Hull and White (1990) model (see Section 5.2.2),

$$dr(t) = \alpha(\mu(t) - r(t)) \, dt + \sigma \, d\tilde{W}(t),$$

one should use the transformation

$$X(t) = \frac{r(t) - \epsilon(t)}{\sigma},$$

where

$$\epsilon(t) = e^{-\alpha t} r(0) + \alpha \int_0^t e^{-\alpha(t-s)} \mu(s) \, ds$$

$$\Rightarrow \quad dX(t) = -\alpha X(t) \, dt + d\tilde{W}(t).$$

It follows that $X(0) = 0$ and $\epsilon(t) = E_Q[r(t) \mid r(0)]$.

A specific advantage of this transformation is that $X(t)$ is a time-homogeneous process.

Derivative pricing under the trinomial lattice.   As with the general lattice model,
our procedure for calculating derivative prices follows a backwards recursion. In
the trinomial tree we have

$$q(i, j; k) = \begin{cases} q_{uu} = q_{uu}(i, j) & \text{if } k = j + 2, \\ q_u & k = j + 1, \\ q_c & k = j, \\ q_d & k = j - 1, \\ q_{dd} & k = j - 2, \\ 0 & \text{otherwise,} \end{cases}$$

where the probabilities are specified by the appropriate type of branching. Furthermore, each point (i, j) in the lattice has associated with it a risk-free rate of interest
$r(i, j)$ which is determined by the value of the transformed process, $X(i, j)$, at the
same lattice point.

For a European-style contract with maturity date $T = m \Delta t$, we start with derivative payoffs of $V(m, j)$ for $j = n_l, \ldots, n_u$. (Note that the number of nodes at each
time can be regarded as constant.) As we work backwards, if the $V(i, j)$ are known
for $j = n_l, \ldots, n_u$, we then have

$$V(i - 1, j) = \begin{cases} \exp(-r(i - 1, j)\Delta t) \displaystyle\sum_{k=j-2}^{j} q(i - 1, j; k) V(i, k) \\ \qquad\qquad\qquad \text{if } j = n_u \text{ (down branching),} \\[1em] \exp(-r(i - 1, j)\Delta t) \displaystyle\sum_{k=j-1}^{j+1} q(i - 1, j; k) V(i, k) \\ \qquad\qquad\qquad \text{if } n_l < j < n_u \text{ (central branching),} \\[1em] \exp(-r(i - 1, j)\Delta t) \displaystyle\sum_{k=j}^{j+2} q(i - 1, j; k) V(i, k) \\ \qquad\qquad\qquad \text{if } j = n_l \text{ (up branching).} \end{cases}$$

Corresponding formulae for American-style derivatives can be written down easily.

We will consider numerical examples for the trinomial lattice later in this chapter
when we compare the procedure with other numerical schemes.

10.3   Finite-Difference Methods

We can use lattices in a different way. In particular, we return to the partial differential equation that we derived (equation (4.10)) earlier in the book. In general, the
martingale approach is regarded as the appropriate starting point for pricing, rather

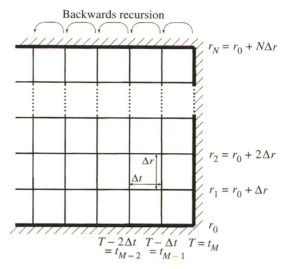

Figure 10.6. Lattice framework for numerical solution of PDEs.
Thick lines round the borders indicate where boundary conditions need to be specified.

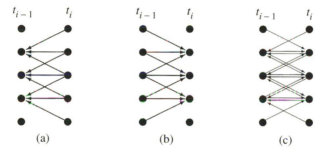

Figure 10.7. Possible finite-difference schemes:
(a) explicit; (b) implicit; (c) Crank–Nicolson.

than the use of PDEs. However, PDEs can be used as an additional pricing methodology, largely because there is an extensive literature on the numerical solution of PDEs. Consider the one-factor model for $r(t)$:

$$dr(t) = \mu_r(t, r(t)) \, dt + \sigma_r(t, r(t)) \, d\tilde{W}(t),$$

where $\tilde{W}(t)$ is a standard Brownian motion under the risk-neutral measure Q. For a general European-style derivative which pays $V(T) = \psi(r(T))$ at time T we get the PDE equivalent to (4.10):

$$\frac{\partial V}{\partial t} + \frac{\partial V}{\partial r} \mu_r(t, r) + \frac{1}{2} \frac{\partial^2 V}{\partial r^2} \sigma_r(t, r)^2 = rV. \tag{10.8}$$

We solve this PDE numerically by using a lattice approximation, starting at the maturity date for the derivative and working backwards, recursively through the lattice (Figure 10.6).

Numerical solution of this PDE (equation (10.8)) is somewhat simplified if we transform $r(t)$ into an alternative diffusion $X(t) = f(r(t))$ that has unit (or constant) volatility, as we did for the trinomial lattice. This assists the determination of convergence and stability of the numerical procedures. Where there is an explicit time dependence in $\sigma_r(t, r(t))$, the transformation may need to be of the form $X(t) = f(t, r(t))$. Thus, we choose $f(t, r)$ such that

$$\frac{\partial f}{\partial r}(t, r) = \frac{1}{\sigma_r(t, r)}.$$

Equivalently, we write $r(t) = r(t, X(t)) = f^{-1}(t, X(t))$, where

$$\frac{\partial r}{\partial x}(t, x) = \sigma_r(t, r(t, x)).$$

Then equation (10.8) becomes

$$\frac{\partial V}{\partial t} + \frac{\partial V}{\partial x}\theta_x(t, x) + \frac{1}{2}\frac{\partial^2 V}{\partial x^2} = r(t, x)V, \tag{10.9}$$

where

$$\theta_x(t, x) = \left(\frac{\partial r}{\partial x}(t, x)\right)^{-1}\left[\mu_r(t, r(t, x)) - \frac{1}{2}\frac{\partial^2 r}{\partial x^2}(t, x) - \frac{\partial r}{\partial t}(t, x)\right]$$

$$= \frac{\mu_r(t, r(t, x))}{\sigma_r(t, r(t, x))} - \frac{1}{2}\frac{\partial \sigma_r}{\partial r}(t, r(t, x)) - \frac{1}{\sigma_r(t, r(t, x))}\frac{\partial r}{\partial t}(t, x), \tag{10.10}$$

with boundary condition $V(T, x) = \psi(r(T, x))$. Note that for this choice of function $X(t) = f(t, r(t))$ the SDE for $X(t)$ is $dX(t) = \theta_x(t, X(t)) + d\tilde{W}(t)$. Sometimes it is easier to calculate $\theta_x(t, x)$ directly using Itô's formula rather than trying to work out equation (10.10).[1]

We now define the lattice which will be used as an approximation to the continuous (t, x) space. Let Δt be the time step and Δx be the step size for the state variable x (Figure 10.6, replacing r with x). The discretized state space runs from x_0 to x_N, where $x_i = x_0 + i\Delta x$. The time line runs from t_0 to $t_M = T$ with $t_j = t_0 + j\Delta t$.

For obvious computational reasons, the state space needs to be finite. This means that we not only need the exact boundary condition at T, but also some form of boundary conditions (either exact or approximate) along the upper and lower boundaries at x_0 and x_N.

---

[1] Note that $\theta_x(t, x)$ in (10.9) is defined differently from $\theta_r(t, r)$ in (10.4).

Within the lattice we will use the following notation:

$$V(i, j) \equiv V(t_i, x_j), \qquad \frac{\partial V}{\partial t}(i, j) \equiv \frac{\partial V}{\partial t}(t_i, x_j), \quad \text{etc.,}$$

$$\theta(i, j) \equiv \theta_x(t_i, x_j), \qquad r(i, j) \equiv r(t_i, x_j).$$

For points that are not on the boundary we use the following approximations for the partial derivatives:

$$\frac{\partial V}{\partial t}(i, j) = \frac{V(i + 1, j) - V(i, j)}{\Delta t},$$

or

$$\frac{\partial V}{\partial t}(i + 1, j) = \frac{V(i + 1, j) - V(i, j)}{\Delta t},$$

$$\frac{\partial V}{\partial x}(i, j) = \frac{V(i, j + 1) - V(i, j - 1)}{2\Delta x},$$

$$\frac{\partial^2 V}{\partial x^2}(i, j) = \frac{V(i, j + 1) - 2V(i, j) + V(i, j - 1)}{(\Delta x)^2}.$$

Boundary conditions can be of various forms; for example, on the boundaries x_0, x_N, $V(x, t)$ may be specified explicitly, or $\partial V/\partial x$ may be specified explicitly implying that, for instance, we take

$$V(i, N) = V(i, N - 1) + \frac{\partial V}{\partial x}(i, N) \times \Delta x.$$

Alternatively, Vetzal (1998) suggests that if the volatility of $X(t)$ is constant and $\theta_x(t, x) \to -\infty$ as x tends to its true upper limit, and $\theta_x(t, x) \to +\infty$ as x tends to its true lower limit, then traditional forms of boundary condition are unimportant (provided x_0 and x_N are sufficiently extreme). Consequently, Vetzal proposes the use of alternative approximations for the partial derivatives of V at the boundaries; that is,

$$\frac{\partial V}{\partial x}(i, 0) = \frac{-V(i, 2) + 4V(i, 1) - 3V(i, 0)}{2\Delta x},$$

$$\frac{\partial^2 V}{\partial x^2}(i, 0) = \frac{V(i, 2) - 2V(i, 1) + V(i, 0)}{(\Delta x)^2},$$

$$\frac{\partial V}{\partial x}(i, N) = \frac{3V(i, N) - 4V(i, N - 1) + V(i, N - 2)}{2\Delta x},$$

$$\frac{\partial^2 V}{\partial x^2}(i, N) = \frac{V(i, N) - 2V(i, N - 1) + V(i, N - 2)}{(\Delta x)^2}.$$

We apply these approximations in three different ways to approximate the solution to the PDE. The methods are all described as finite-difference methods and the three approaches are illustrated in Figure 10.7. Each approach results in a set of $N + 1$

linear equations with $N + 1$ unknowns, $V(i, 0), \ldots, V(i, N)$, defined in terms of $V(i+1, 0), \ldots, V(i+1, N)$. In the case of the explicit finite-difference scheme the linear equations immediately give us the new values at time t_i. In the implicit and Crank–Nicolson[2] schemes the linear equations are (nearly) tridiagonal in nature, which requires a certain amount of extra computational effort compared with the explicit scheme. An advantage of the implicit and Crank–Nicolson schemes is, however, that they are unconditionally stable, whereas the explicit scheme may not be.

10.3.1   Explicit Finite Differences

We start with the PDE (equation (10.9)) and then introduce the approximations for the partial derivatives in a suitable way. The approximations in the present case give us

$$\frac{V(i+1, j) - V(i, j)}{\Delta t} + \theta(i, j)\frac{(V(i+1, j+1) - V(i+1, j-1))}{2\Delta x}$$
$$+ \frac{1}{2}\frac{(V(i+1, j+1) - 2V(i+1, j) + V(i+1, j-1))}{(\Delta x)^2} = r(i, j)V(i, j).$$

(Alternatives to this scheme replace $r(i, j)$ by $r(i+1, j)$ or $\theta(i, j)$ by $\theta(i+1, j)$.)

Hence we express $V(i, j)$ explicitly in terms of the $V(i+1, k)$:

$$(1 + r(i, j)\Delta t)V(i, j)$$
$$= V(i+1, j+1)p_u(i, j) + V(i+1, j)p_m(i, j) + V(i+1, j-1)p_d(i, j),$$
$$(10.11)$$

where

$$p_u(i, j) = \frac{\theta(i, j)\Delta t}{2\Delta x} + \frac{\Delta t}{2(\Delta x)^2},$$

$$p_m(i, j) = 1 - \frac{\Delta t}{(\Delta x)^2},$$

$$p_d(i, j) = -\frac{\theta(i, j)\Delta t}{2\Delta x} + \frac{\Delta t}{2(\Delta x)^2}.$$

Similarly, using the Vetzal approximations at the boundaries we find that

$$(1 + r(i, 0)\Delta t)V(i, 0)$$
$$= V(i+1, 2)p_{uu}(i, 0) + V(i+1, 1)p_u(i, 0) + V(i+1, 0)p_m(i, 0), \quad (10.12)$$

---

[2]Some textbooks use the alternative spelling 'Nicholson'. A possible cause of this is the original paper, Crank and Nicolson (1947). The paper itself has the (presumably correct) spelling 'Nicolson', whereas the journal contents use the (incorrect) spelling 'Nicholson'! Crank also refers to 'Nicolson' in his book.

where

$$p_{uu}(i, 0) = -\frac{\theta(i, 0)\Delta t}{2\Delta x} + \frac{\Delta t}{2(\Delta x)^2},$$

$$p_u(i, 0) = \frac{2\theta(i, 0)\Delta t}{\Delta x} - \frac{\Delta t}{(\Delta x)^2},$$

$$p_m(i, 0) = 1 - \frac{3\theta(i, 0)\Delta t}{2\Delta x} + \frac{\Delta t}{2(\Delta x)^2}$$

and

$$(1 + r(i, N)\Delta t)V(i, N)$$
$$= V(i+1, N)p_m(i, N) + V(i+1, N-1)p_d(i, N)$$
$$+ V(i+1, N-2)p_{dd}(i, N), \quad (10.13)$$

where

$$p_m(i, N) = 1 + \frac{3\theta(i, N)\Delta t}{2\Delta x} + \frac{\Delta t}{2(\Delta x)^2},$$

$$p_d(i, N) = -\frac{2\theta(i, N)\Delta t}{\Delta x} - \frac{\Delta t}{(\Delta x)^2},$$

$$p_{dd}(i, N) = \frac{\theta(i, N)\Delta t}{2\Delta x} + \frac{\Delta t}{2(\Delta x)^2}.$$

This scheme is, in fact, quite similar to the trinomial lattice described earlier (see equations (10.4)–(10.6)). The differences are in the approximation in the method of discounting, the range of the lattice and the second-order elements in the probabilities. Recall also that $\theta_x(t, x)$ in (10.9) is defined differently from $\theta_r(t, r)$ in (10.4).

Stability of the scheme is a key issue. By stability we mean: are errors introduced as a result of discretization magnified or damped as we move backwards through the lattice? If they become magnified with each step back, then the differences between the approximation and the true solution become unbounded as we step backwards in time. Such a scheme is described as unstable. (For a graphical illustration of instability, see Smith (1985, Fig. 2.3).)

The explicit finite-difference scheme can be shown to be *unstable* (given $X(t)$ has unit volatility) if

$$\frac{\Delta t}{(\Delta x)^2} > 1.$$

As an example, suppose we progressively refine the lattice by letting $\Delta t \to 0$ and, at each stage, keeping $\Delta x = \sqrt{\Delta t/2}$. Then for each Δt we have an unstable scheme ($\Delta t/(\Delta x)^2 = 2 > 1$), so for a given (t, x) we cannot expect the approximation to converge to the true $V(t, x)$ as $\Delta t \to 0$. In contrast, if $\Delta x = \sqrt{2\Delta t}$, the scheme is stable and should converge as $\Delta t \to 0$.

10.3.2 Implicit Finite Differences

The basic idea here is very similar to the explicit finite-difference scheme, except that we reverse the indices used in the approximations (see Figure 10.7: the arrows indicate how the nodes influence each other).

For the implicit scheme we make a different approximation to equation (10.9). Thus, for $j = 1, \ldots, N - 1$ we have

$$\frac{V(i+1, j) - V(i, j)}{\Delta t} + \theta(i+1, j)\frac{(V(i, j+1) - V(i, j-1))}{2\Delta x}$$
$$+ \frac{1}{2}\frac{(V(i, j+1) - 2V(i, j) + V(i, j-1))}{(\Delta x)^2} = r(i+1, j)V(i+1, j).$$

Hence

$$(1 - r(i+1, j)\Delta t)V(i+1, j)$$
$$= V(i, j+1)q_u(i, j) + V(i, j)q_m(i, j) + V(i, j-1)q_d(i, j), \quad (10.14)$$

where

$$q_u(i, j) = -\frac{\theta(i+1, j)\Delta t}{2\Delta x} - \frac{\Delta t}{2(\Delta x)^2},$$

$$q_m(i, j) = 1 + \frac{\Delta t}{(\Delta x)^2},$$

$$q_d(i, j) = \frac{\theta(i+1, j)\Delta t}{2\Delta x} - \frac{\Delta t}{2(\Delta x)^2}.$$

For the boundaries using the Vetzal approximations we find that

$$(1 - r(i+1, 0)\Delta t)V(i+1, 0)$$
$$= V(i, 2)q_{uu}(i, 0) + V(i, 1)q_u(i, 0) + V(i, 0)q_m(i, 0), \quad (10.15)$$

where

$$q_{uu}(i, 0) = \frac{\theta(i+1, 0)\Delta t}{2\Delta x} - \frac{\Delta t}{2(\Delta x)^2},$$

$$q_u(i, 0) = -\frac{2\theta(i+1, 0)\Delta t}{\Delta x} + \frac{\Delta t}{(\Delta x)^2},$$

$$q_m(i, 0) = 1 + \frac{3\theta(i+1, 0)\Delta t}{2\Delta x} - \frac{\Delta t}{2(\Delta x)^2}$$

and

$$(1 - r(i+1, N)\Delta t)V(i+1, N)$$
$$= V(i, N)q_m(i, N) + V(i, N-1)q_d(i, N) + V(i, N-2)q_{dd}(i, N), \quad (10.16)$$

where

$$q_m(i, N) = 1 - \frac{3\theta(i+1, N)\Delta t}{2\Delta x} - \frac{\Delta t}{2(\Delta x)^2},$$

$$q_d(i, N) = \frac{2\theta(i+1, N)\Delta t}{\Delta x} + \frac{\Delta t}{(\Delta x)^2},$$

$$q_{dd}(i, N) = -\frac{\theta(i+1, N)\Delta t}{2\Delta x} - \frac{\Delta t}{2(\Delta x)^2}.$$

It follows that the values for $V(i, 0), \ldots, V(i, N)$ are implicit in equations (10.14)–(10.16), in contrast to the explicit nature of equations (10.11)–(10.13).

To progress, we write the equations in matrix form, $\Omega(i)V(i) = \lambda(i)$ (which we will abbreviate to $\Omega V = \lambda$ where appropriate), where

$$\lambda(i)_j = (1 - r(i+1, j)\Delta t)V(i+1, j), \quad j = 0, \ldots, N,$$

$$V(i)_j = V(i, j), \quad j = 0, \ldots, N,$$

$$\Omega(i) = [\omega_{j,k}]_{j,k=0}^N.$$

The matrix $\Omega(i)$ is nearly tridiagonal, that is,

$$\omega_{N,N-2} = q_{dd}(i, N),$$

$$\omega_{j,j-1} = q_d(i, j) \quad \text{for } j = 1, \ldots, N \text{ only,}$$

$$\omega_{j,j} = q_m(i, j) \quad \text{for } j = 0, \ldots, N,$$

$$\omega_{j,j+1} = q_u(i, j) \quad \text{for } j = 0, \ldots, N-1 \text{ only,}$$

$$\omega_{0,2} = q_{uu}(i, 0),$$

$$\omega_{j,k} = 0 \quad \text{otherwise.}$$

10.3.3 Solution of a Nearly Tridiagonal System of Equations

To solve $\Omega V = \lambda$ we first tridiagonalize the problem. Thus, we aim to transform Ω into a fully tridiagonal matrix $\tilde{\Omega} = [\tilde{\omega}_{j,k}]_{j,k=0}^N$ and λ into a new vector associated with $\tilde{\Omega}, \tilde{\lambda} = [\tilde{\lambda}_j]_{j=0}^N$. $\tilde{\Omega}$ is calculated as follows. Rows 1 to $N-1$ are left unaltered: that is,

$$\tilde{\omega}_{j,k} = \omega_{j,k} \quad \text{for } j = 1, \ldots, N-1, \ k = 0, \ldots, N.$$

Row 0 of the new matrix is calculated by taking row 0 of Ω and subtracting $\omega_{0,2}/\omega_{1,2}$ times row 1 of Ω. Thus,

$$\tilde{\omega}_{0,k} = \omega_{0,k} - \frac{\omega_{0,2}}{\omega_{1,2}}\omega_{1,k} \quad \text{for } k = 0, \ldots, N,$$

which implies that

$$\tilde{\omega}_{0,2} = 0 \quad \text{and, trivially,} \quad \tilde{\omega}_{0,k} = 0 \quad \text{for } k = 3, \ldots, N.$$

Similarly, row N of the new matrix is found by taking row N of Ω and subtracting $\omega_{N,N-2}/\omega_{N-1,N-2}$ times row $N-1$ of Ω: that is,

$$\tilde{\omega}_{N,k} = \omega_{N,k} - \frac{\omega_{N,N-2}}{\omega_{N-1,N-2}}\omega_{N-1,k} \quad \text{for } k = 0, \ldots, N,$$

which implies that $\tilde{\omega}_{N,k} = 0$ for $k = 0, \ldots, N-2$. The vector $\tilde{\lambda}$ is calculated in a similar way: thus,

$$\tilde{\lambda}_j = \lambda_j \quad \text{for } j = 1, \ldots, N-1,$$

$$\tilde{\lambda}_0 = \lambda_0 - \frac{\omega_{0,2}}{\omega_{1,2}}\lambda_1, \qquad \tilde{\lambda}_N = \lambda_N - \frac{\omega_{N,N-2}}{\omega_{N-1,N-2}}\lambda_{N-1}.$$

We now have to solve the tridiagonal problem $\tilde{\Omega} V = \tilde{\lambda}$.

We define a further matrix $\bar{\Omega}$ and vector $\bar{\lambda}$ as follows:

$$\bar{\omega}_{0k} = \tilde{\omega}_{0k} \quad \text{for } k = 0, \ldots, N, \qquad \bar{\lambda}_0 = \tilde{\lambda}_0.$$

For $j = 1, \ldots, N$ we use a recursive scheme. Let $\tilde{\phi}_j = \tilde{\omega}_{j,j-1}/\bar{\omega}_{j-1,j-1}$. Row j of $\bar{\Omega}$ is then defined as row j of $\tilde{\Omega}$ minus $\tilde{\phi}_j$ times row $j-1$ of $\bar{\Omega}$: that is,

$$\bar{\omega}_{j,k} = \tilde{\omega}_{j,k} - \tilde{\phi}_j\bar{\omega}_{j-1,k} \quad \text{for } k = 0, \ldots, N, \qquad \bar{\lambda}_j = \tilde{\lambda}_j - \tilde{\phi}_j\bar{\lambda}_{j-1}.$$

(Essentially, we are knocking out the lower off-diagonal elements one by one.)

We now have the problem in the form $\bar{\Omega} V = \bar{\lambda}$, where $\bar{\Omega}$ is upper triangular (indeed $\bar{\omega}_{j,k} \neq 0$ only when $k = j$ or $k = j+1$). Such a system of equations is straightforward to solve recursively. Thus,

$$V(i, N) = \frac{\bar{\lambda}_N}{\bar{\omega}_{N,N}},$$

$$V(i, j) = \frac{\bar{\lambda}_j - \bar{\omega}_{j,j+1}V(i, j+1)}{\bar{\omega}_{j,j}} \quad \text{for } j = N-1, N-2, \ldots, 0.$$

It is clear that the implicit scheme is computationally more expensive than the explicit scheme, which does not involve the solution of a tridiagonal system of equations. However, the implicit scheme offers the advantage of being unconditionally stable. This allows us a greater range of relationships between Δx and Δt. For example, for the same size of lattice (number of nodes) it may be that the accuracy of the implicit scheme can be improved by reducing Δx and increasing Δt. Such a change may not be possible under the explicit scheme as it may result in $\Delta t/\Delta x^2 > 1$, which is unstable. We will illustrate this point later.

10.3.4 Orders of Convergence

It is appropriate to consider how accurate the proposed schemes are relative to the true solution. One approach to this is to consider a sequence of discrete schemes with index $l = 1, 2, \ldots$ and step sizes $\Delta t(l) \to 0$ and $\Delta x(l) \to 0$.

Errors for both the explicit and implicit finite-difference schemes can be shown to be $O(\Delta t) + O(\Delta x^2)$ as $l \to \infty$ ($\Delta t(l) \to 0$ and $\Delta x(l) \to 0$). It is appropriate, therefore, that we set $\Delta t(l) = \gamma \Delta x(l)^2$ for some γ. In the case of the explicit scheme we require $\gamma \leqslant 1$ for stability.

The final finite-difference scheme we will discuss now, the Crank–Nicolson method, can be shown to have errors which are of order $O(\Delta t^2) + O(\Delta x^2)$. As a consequence, it is more appropriate, in order to maximize the speed of convergence, to take $\Delta t(l) = \gamma \Delta x(l)$.

10.3.5 The Crank–Nicolson Method

Let α be some constant between 0 and 1. Then we use the following approximation to the PDE (equation (10.9)). For $j = 1, \ldots, N - 1$,

$$\frac{V(i+1, j) - V(i, j)}{\Delta t} + (1 - \alpha)\theta(i, j)\frac{(V(i+1, j+1) - V(i+1, j-1))}{2\Delta x}$$

$$+ \alpha\theta(i+1, j)\frac{(V(i, j+1) - V(i, j-1))}{2\Delta x}$$

$$+ \frac{(1-\alpha)}{2}\frac{(V(i+1, j+1) - 2V(i+1, j) + V(i+1, j-1))}{(\Delta x)^2}$$

$$+ \frac{\alpha}{2}\frac{(V(i, j+1) - 2V(i, j) + V(i, j-1))}{(\Delta x)^2}$$

$$= (1 - \alpha)r(i, j)V(i, j) + \alpha r(i+1, j)V(i+1, j).$$

Hence

$$\alpha\{V(i, j+1)q_u(i, j) + V(i, j)q_m(i, j) + V(i, j-1)q_d(i, j)\}$$
$$+ (1 - \alpha)(1 + r(i, j)\Delta t)V(i, j)$$
$$= (1 - \alpha)\{V(i+1, j+1)p_u(i, j)$$
$$+ V(i+1, j)p_m(i, j) + V(i+1, j-1)p_d(i, j)\}$$
$$+ \alpha(1 - r(i+1, j)\Delta t)V(i+1, j). \tag{10.17}$$

Similarly, using the Vetzal approximations, we get

$$\alpha\{V(i, 2)q_{uu}(i, 0) + V(i, 1)q_u(i, 0) + V(i, 0)q_m(i, 0)\}$$
$$+ (1 - \alpha)(1 + r(i, 0)\Delta t)V(i, 0)$$
$$= (1 - \alpha)\{V(i+1, 2)p_{uu}(i, 0)$$
$$+ V(i+1, 1)p_u(i, 0) + V(i+1, 0)p_m(i, 0)\}$$
$$+ \alpha(1 - r(i+1, 0)\Delta t)V(i+1, 0) \quad (10.18)$$

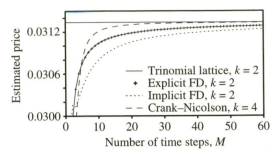

Figure 10.8. Comparison of price approximations for different numerical methods, with different numbers of time steps. $\Delta t = 1/M$. Trinomial, explicit and implicit schemes use $\Delta x = \sqrt{k \Delta t}$. Crank–Nicolson uses $\Delta x = k \Delta t$. (Vasicek model: $\mu = 0.05$, $\alpha = 0.1$, $\sigma = 0.02$, $r(0) = 0.05$. Call option on $P(T, T + 10)$, $T = 1$, $K = 0.627\,268$.)

and

$$\alpha\{V(i, N)q_m(i, N) + V(i, N - 1)q_d(i, N) + V(i, N - 2)q_{dd}(i, N)\}$$
$$+ (1 - \alpha)(1 + r(i, N)\Delta t)V(i, N)$$
$$= (1 - \alpha)\{V(i + 1, N)p_m(i, N)$$
$$+ V(i + 1, N - 1)p_d(i, N) + V(i + 1, N - 2)p_{dd}(i, N)\}$$
$$+ \alpha(1 - r(i + 1, N)\Delta t)V(i + 1, N). \tag{10.19}$$

We can readily see that $\alpha = 0$ gives us the explicit finite-difference scheme, while $\alpha = 1$ gives us the implicit scheme. The Crank–Nicolson method uses $\alpha = \frac{1}{2}$.

Whatever the value of α, equations (10.17)–(10.19) define a nearly tridiagonal system of equations that we can solve in exactly the same way as the implicit scheme.

In the numerical examples below, we will see that the Crank–Nicolson method is very effective for a European call option. In other situations this may not be the case; in particular, if the option payoff is a discontinuous function of the state variable (for example, a binary option), then the method becomes much less effective. Such discontinuities can, however, be smoothed out by applying, for example, the implicit scheme for the first few time steps backwards before switching to the Crank–Nicolson method.

10.4 Numerical Examples

We will now consider some numerical examples which will illustrate various aspects of the trinomial lattice and finite-difference methods.

As a means of checking the accuracy of the proposed methods, we will illustrate their application in an example where the exact price is known. Thus, we will look at the pricing of a call option on a zero-coupon bond using the Vasicek model $dr(t) = -\alpha(r(t) - \mu)\,dt + \sigma\,d\tilde{W}(t)$. The parameter values used in this example

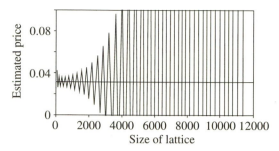

Figure 10.9. Example of instability of explicit finite-difference method when $\Delta x = \sqrt{k\Delta t}$ for $0.9 = k < 1$. Size of lattice is number of Δt time steps $\times$ number of Δx steps.

are $\mu = 0.05$, $\alpha = 0.1$ and $\sigma = 0.02$ with $r(0) = 0.05$ in both examples. The two examples we will consider are European call options with $P(t, 11)$ as the underlying, $T = 1$ as the exercise date, and strike price $K = 0.627\,268$ (which corresponds to $r(T) = 0.05$) or $K = 0.55$. These strike prices give prices for the call option at time 0 of $0.031\,338$ and $0.082\,620$ respectively, using the usual call option formula (see Theorem 4.7).

For ease of computation we transform $r(t)$ to

$$X(t) = (r(t) - \mu - e^{-\alpha t}(r(0) - \mu))/\sigma$$

so that $X(0) = 0$ and $dX(t) = -\alpha X(t)\,dt + d\tilde{W}(t)$.

In the first example (Figures 10.8–10.11) the use of $K = 0.627\,268$ means that the strike price naturally falls exactly at one of the nodes in the lattice. In the second example (Figures 10.12 and 10.13) we will see what happens when this is not the case.

We are concerned with how accurate the various methods are, both relative to each other and to the coarseness of the lattice. With regard to lattice size, we divide the time interval $[0, 1]$ into M time steps of length $\Delta t = 1/M$. The state variable x theoretically ranges from $-\infty$ to $+\infty$. We restrict this to the range $[x_0, x_N]$.

Δx. In the trinomial lattice, explicit finite-difference and implicit finite-difference methods we assume that $\Delta x = \sqrt{k\Delta t}$ for various values of k. In the Crank–Nicolson method we assume that $\Delta x = k\Delta t$.

$[x_0, x_N]$. In the explicit and implicit finite-difference methods and the Crank–Nicolson method we choose x_0 and x_N (where N depends upon the location of x_N and Δx) to correspond approximately to ± 5 standard deviations of the stationary distribution of $X(t)$ (that is, $\pm 5/\sqrt{2\alpha}$). This is a subjective pair of limits. On the other hand, the trinomial lattice method has prescribed rules which mean that the interval $[x_0, x_N]$ expands at rate $O(\Delta t^{-1/2})$. In the computations that follow, therefore, the sizes of the lattices grow in line with $M^{3/2}$ for the explicit and implicit schemes and in line with M^2 for the trinomial and Crank–Nicolson schemes.

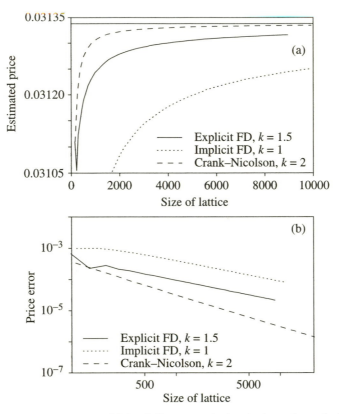

Figure 10.10.   (a) Comparison of finite-difference methods, plotting estimated price against size of lattice (number of Δt time steps $\times$ number of Δx steps). (b) Price errors versus lattice size, L, both on a log scale. Linearity indicates price error $= O(L^{-2/3})$ for explicit and implicit schemes and $= O(L^{-1})$ for Crank–Nicolson. (Vasicek model: $\mu = 0.05$, $\alpha = 0.1$, $\sigma = 0.02$, $r(0) = 0.05$. Call option on $P(T, T + 10)$, $T = 1$, $K = 0.627\,268$.)

In the present case we choose N to be even and set $x_{N/2} = X(0)$ and $x_{N/2+k} = X(0) + k\Delta x$ for $-N/2 \leqslant k \leqslant N/2$. This means that the derivative price we seek corresponds to $V(i = 0, j = N/2)$ without the need for interpolation.

Example 1: $K = 0.627\,268$.   In Figure 10.8 we see how the four methods compare with one another for specified values of k. The explicit scheme and the trinomial lattice are almost (but not exactly) identical in the use of probabilities and the price approximations are indistinguishable in the diagram. (For this reason we do not consider the trinomial lattice method any further.) Crank–Nicolson can be seen to converge much more quickly (at rate M^{-2}, while the others converge at rate M^{-1}). The implicit scheme appears to be less good, although in Example 2 the reverse is true.

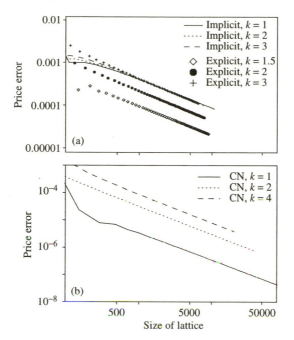

Figure 10.11. Price errors versus lattice size for different relationships between Δx and Δt. (a) $\Delta x = \sqrt{k\Delta t}$; (b) $\Delta x = k\Delta t$. (Vasicek model: $\mu = 0.05$, $\alpha = 0.1$, $\sigma = 0.02$, $r(0) = 0.05$. Call option on $P(T, T + 10)$, $T = 1$, $K = 0.627\,268$.)

In Figure 10.9 we show what happens if we use the explicit finite-difference method with $k = 0.9$. As was mentioned earlier, we require $k \geqslant 1$ for stability. This figure clearly demonstrates the instability that arises when k does not satisfy this requirement.

From the computational point of view, the size of the lattice (that is, $L = (N + 1) \times (M + 1)$) is more important than the number of time steps. For example, the Crank–Nicolson method typically uses a smaller value of Δx and a larger value of Δt than the explicit or implicit schemes. Figure 10.10(a) shows how the price approximations for the various methods vary with lattice size for given values of k. We can observe that the Crank–Nicolson scheme is still the best. In Figure 10.10(b) we plot both price errors and lattice size on a log scale. We can see clear asymptotic linearity which is consistent with L^{-1} convergence for the Crank–Nicolson scheme and $L^{-2/3}$ convergence for the explicit and implicit schemes.

This would suggest that the Crank–Nicolson method is clearly the best choice. However, we have to be cautious as the Crank–Nicolson method requires costly (in time) solution of the tridiagonal system of equations. Thus, for the same size of lattice, approximate prices can be computed several times faster using the explicit scheme.

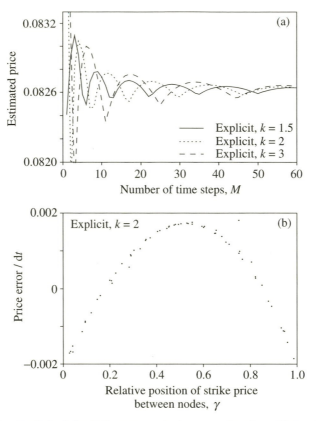

Figure 10.12. Explicit finite-difference method: example of oscillation in the price error. (a) The problem exists for all values of k. (b) Strike price lies between nodes at $P(T, T + 10, x_i + \Delta x)$ and $P(T, T + 10, x_i)$. Points plotted for each $n = 1, \ldots, 60$, $\Delta t = 1/n$, $\Delta x = \sqrt{k\Delta t}$: price error/$\Delta t$ versus relative position of K between nodes $\gamma = (K - P(T, T + 10, x_i + \Delta x))/(P(T, T + 10, x_i) - P(T, T + 10, x_i + \Delta x))$. Rescaling of price error demonstrates $O(\Delta t)$ convergence. (Vasicek model: $\mu = 0.05$, $\alpha = 0.1$, $\sigma = 0.02$, $r(0) = 0.05$. Call option on $P(T, T + 10)$, $T = 1$, $K = 0.55$.)

In Figure 10.11 we show how the price errors depend upon the value of k used. We can see that, in this example at least, the accuracy of the implicit scheme does not depend significantly upon k. Both the explicit and Crank–Nicolson schemes on the other hand do depend upon k significantly. However, in Example 2 we find (Figure 10.13) that $k = 2$ is best for the Crank–Nicolson method. So the patterns we see in one example may not be repeated elsewhere, making firm conclusions about the best methods difficult.

Example 2: $K = 0.55$. This example demonstrates the consequences of setting up lattices in which we are not able to control the position of the strike price between

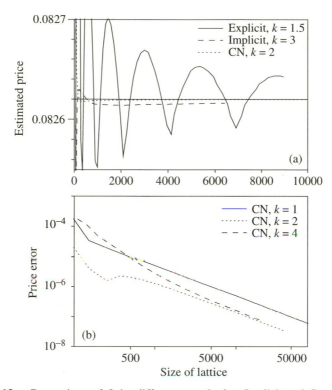

Figure 10.13. Comparison of finite-difference methods. Implicit and Crank–Nicolson methods do not oscillate, and are much more accurate. (Vasicek model: $\mu = 0.05$, $\alpha = 0.1$, $\sigma = 0.02$, $r(0) = 0.05$. Call option on $P(T, T + 10)$, $T = 1$, $K = 0.55$.)

nodes. In particular, when we consider the explicit finite-difference method (Figure 10.12(a)) we find that the estimated price, as a function of the number of time steps, oscillates around the true price.

There is a logical explanation for this behaviour: it is driven by the position of the strike price relative to the nearest nodes above and below. Let i be the index at which we have $P(T, T + 10, x_i) \geqslant K > P(T, T + 10, x_{i+1})$. We measure how close K is to $P(T, T + 10, x_i)$ relative to the gap between the two nodes: that is, $\gamma = (K - P(T, T + 10, x_{i+1}))/(P(T, T + 10, x_i) - P(T, T + 10, x_{i+1}))$. We find that when γ is close to 0 or 1 then we underprice the option, and when γ is near $1/2$ we overprice the option. The latter is, perhaps, easier to explain. The explicit method is, in some sense, having to do lots of interpolation. Since K lies between x_i and x_{i+1} the shape of the call option payoff means that interpolation will overvalue, slightly, outcomes which finish close to K. Hence the option is overpriced. The form of this mispricing is illustrated in Figure 10.12(b). Recall, that price errors are $O(\Delta t)$. Thus, it was natural in the investigation of these oscillations to divide price errors by Δt. In the figure (which plots price error/Δt versus γ) we can see that this has a most striking effect: all of the points lie close to a single almost-parabolic curve. The fact that these points all lie so close to this curve confirms that we have $O(\Delta t)$ convergence.

In Figure 10.13(a) we show how the three finite-difference methods compare in terms of accuracy. We can note that the implicit and Crank–Nicolson schemes both have smooth convergence rather than oscillatory. Furthermore, both give rather smaller errors. However, one might note that the average price within a cycle on the explicit scheme is rather closer to the true price. In Figure 10.13(b) we illustrate how the price error depends on lattice size and on k for the Crank–Nicolson method. In contrast to Example 1, $k = 1$ gives the poorest results.

10.5   Simulation Methods

As we have discussed above with the lattice methods, we are typically faced with a derivative valuation problem where we cannot evaluate analytically the expected value

$$V(t) = E_Q\left[\exp\left(-\int_t^T r(u)\,\mathrm{d}u\right)V(T)\,\bigg|\,\mathcal{F}_t\right],$$

even assuming we are able to evaluate the derivative payoff $V(T)$ accurately.

The lattice methods described in the previous sections assumed that there was one dimension in addition to time. When the number of dimensions increases (for example, because we wish to use a two- or three-factor interest rate model rather than a single-factor model) lattice-based computing times increase very significantly. Simulation or Monte Carlo methods (first applied to derivative pricing by Boyle (1977)) offer a popular and effective alternative to lattice methods under these circumstances.

Consider, then, the random variable $X = \exp(-\int_t^T r(u)\,\mathrm{d}u)V(T)$. We will consider in this section how we can simulate the value of X, or related quantities, and then take suitable averages to estimate its expected value $V(t) = E_Q[X]$.

10.5.1   Basic Monte Carlo

We take as an example a general, one-factor model for $r(t)$ with SDE

$$\mathrm{d}r(t) = a(t, r(t))\,\mathrm{d}t + b(t, r(t))\,\mathrm{d}\tilde{W}(t),$$

where $\tilde{W}(t)$ is a standard Brownian motion under Q. We will assume, without loss of generality, that we wish to evaluate the price at time 0 of a derivative payable at time T.

The following 'algorithm' assumes that at time T the derivative payoff $V(T)$ can be evaluated accurately (alternatively, $V(T)$ can be evaluated by continuing the simulation beyond T).

Step 1. Discretize the period $[0, T]$ into n intervals of equal length $\Delta t = T/n$ and define $t_j = j\Delta t$.

Step 2. For simulation i ($i = 1, \ldots, N$), simulate n i.i.d. standard normal random variables $\epsilon_j^{(i)} \sim N(0, 1)$ for $j = 1, \ldots, n$.

Step 3. Calculate the simulated path i of $r(t)$ recursively as follows: $r^{(i)}(0) = r(0)$ is given; for $j = 1, \ldots, n$ let

$$r^{(i)}(t_j) = r^{(i)}(t_{j-1}) + a(t_{j-1}, r^{(i)}(t_{j-1}))\Delta t + b(t_{j-1}, r^{(i)}(t_{j-1}))\sqrt{\Delta t}\epsilon_j^{(i)}. \quad (10.20)$$

Step 4. Evaluate $V^{(i)}(T) \equiv V(T, r^{(i)}(t_n))$ for simulation i.

Step 5. Evaluate the random discounted value of the derivative payoff

$$X_i = \exp\left[-\sum_{j=0}^{n-1} r^{(i)}(t_j)\Delta t\right] V^{(i)}(T). \quad (10.21)$$

Step 6. Finally, calculate the mean value

$$\bar{X} = \sum_{i=1}^{N} X_i. \quad (10.22)$$

This is our estimate of the derivative price $V(0)$.

This procedure gives rise to two types of error: simulation error and discretization error. First, the number of sample paths, N, is, of course, finite. This means that $\bar{X}$ is subject to a certain amount of randomness. Second, the discretization of the time period $[0, T]$ results in some errors:

- approximation of the integral $\int_0^T r^{(i)}(u) \, du$ by the sum $\sum_{j=0}^{n-1} r^{(i)}(t_j)\Delta t$;
- the recursive calculation of the $r^{(i)}(t_j)$ in equation (10.20) means that, in general, the simulated distribution of the $r(t_j)$ is not exactly the same as its true distribution.

The first type of error can be reduced by increasing the number of simulations, N. However, the existence of the second error type means that we should not attempt to do enormous numbers of simulations without also making some attempts to reduce the step size Δt. Without these attempts being made, we will converge to the wrong value.

For a given step size, Δt, the second type of error can be reduced to a limited extent. In particular, with certain models such as the Vasicek and Black–Karasinski models we can modify equation (10.20) to simulate exactly the $r(t_j)$.

Another way to reduce the discretization error is to use the forward-measure approach:

$$V(0) = P(0, T)E_{P_T}[V(T) \mid \mathcal{F}_0], \quad (10.23)$$

where P_T is the forward measure under which the prices of all tradable assets divided by $P(t, T)$ are martingales (Chapter 7). The effectiveness of using this method to reduce the second type of error relies on our ability to compute $P(0, T)$ accurately. For example, in the case of a no-arbitrage model, $P(0, T)$ is part of the input into the model. The problem then shifts to establishing how the drift $a(t, r(t))$ differs between Q and P_T (see, for example, Exercise 7.3).

We will concentrate now on the first type of error: simulation error. Under the basic Monte Carlo simulation we can estimate the standard error as follows.

- Calculate the sample variance

$$S^2 = \frac{1}{N-1} \sum_{i=1}^{N} (X_i - \bar{X})^2.$$

- The estimated standard error of the estimator $\bar{X}$ is then

$$\mathrm{SE}(\bar{X}) = \frac{S}{\sqrt{N}}.$$

It is often the case that S is not especially small relative to $\bar{X}$. If this is combined with the $1/\sqrt{N}$ convergence rate, we can see that for a required degree of accuracy N may have to be very large. This problem has led to the development of alternative methods to basic Monte Carlo, which can reduce the standard error (that is, variance-reduction techniques). Most of these techniques involve methods which affect the underlying variance of the X_i (that is, S^2 above) rather than the underlying $1/\sqrt{N}$ convergence. Nevertheless, in some cases S^2 can be reduced by as much as a factor of 100, resulting in real computational gains. Notable exceptions to the $O(1/\sqrt{N})$ convergence rule are Latin hypercube sampling and quasi-Monte Carlo, which usually demonstrate $O(1/N)$ convergence.

We should add a small cautionary note here. An alternative method may, for example, reduce S by a factor of 2. Sometimes the additional complexity in the method will increase computing times by, say, a factor of 2 for each simulated path. So, although we require fewer simulations, we may find that there is no reduction in times to achieve the same degree of accuracy.

Basic Monte Carlo simulation of multifactor models is almost as straightforward as for one-factor models. For example, suppose $r(t) = f(X(t))$, where $X(t)$ is some m-dimensional diffusion with SDE

$$dX(t) = a(t, X(t)) \, dt + B(t, X(t)) \, d\tilde{W}(t),$$

$a(t, X(t))$ is an $m \times 1$ vector process, $B(t, X(t))$ is an $m \times m$ volatility matrix and $\tilde{W}(t)$ is a standard m-dimensional Brownian motion under Q. We simulate n

i.i.d. vectors $\epsilon_j^{(i)}$, where the individual elements of each vector $\epsilon_j^{(i)}$ are themselves
i.i.d. $\sim N(0, 1)$. Given $X^{(i)}(0) = X(0)$ we calculate recursively

$$X^{(i)}(t_j) = X^{(i)}(t_{j-1}) + a(t_{j-1}, X^{(i)}(t_{j-1}))\Delta t + B(t_{j-1}, X^{(i)}(t_{j-1}))\epsilon_j^{(i)}\sqrt{\Delta t}.$$

We will now give some brief descriptions of the principal variance-reduction
techniques. The various methods can all work very effectively in certain cases, but
they are also sometimes complex to implement. Here we will present the main ideas
only, referring the reader to relevant papers and books where possible.

10.5.2 Stratified and Latin Hypercube Sampling

Consider a simplified example where we can simulate X directly. Basic Monte
Carlo simulation is equivalent to simulation of a uniform random variable on the
interval $[0, 1]$ (that is, $U_i \sim U[0, 1]$) and then let $X_i = F_X^{-1}(U_i)$, where $F_X(x)$ is
the cumulative distribution function of X.[3]

In this same example we can replace the simulated values $U_1, \ldots, U_N$ by deterministic values

$$U_i = \frac{i - 0.5}{N} \quad \text{for } i = 1, \ldots, N.$$

This approach is called *stratified sampling* and achieves a more uniform spread
of values from the interval $[0, 1]$ than Monte Carlo simulation. In particular, it
generally results in more accurate estimates of functionals of X, such as its mean
value. It is important to note that, since the U_i are not random in any way, we
cannot calculate a standard error for the estimated mean $\bar{X}$ (every time we repeat
the calculation we will get the same value!).[4] However, we can note that as we
increase N the estimated value of $\bar{X}$ converges to its true value at the rate $1/N$.
In its own right, this is a significant improvement over basic Monte Carlo. In this
one-dimensional case the approach is very similar to numerical integration.

When we are simulating in d dimensions rather than 1, it ceases to be feasible
to sample each point in a hypercube lattice with N^d nodes. One solution to this
problem is to use *Latin hypercube sampling*.

Under Latin hypercube sampling, the simulation procedure for a one-factor interest rate model is very similar to the five-step procedure for basic Monte Carlo.
However, the method for simulation of the $\epsilon_j^{(i)}$ is rather different. So, we modify
the original Step 2 as follows.

---

[3]For example, if $X \sim N(\mu, \sigma^2)$, then $F_X(x) = \Phi((x - \mu)/\sigma)$ and $F_X^{-1}(u) = \mu + \sigma\Phi^{-1}(u)$.

[4]However, the sampling scheme can be randomized by taking $U_i = (i - 1 + \xi_i)/N$, where the ξ_i are
i.i.d. $\sim U[0, 1]$.

Step 2a. For each $j = 1, \ldots, d$, let $P_j = (P_j^{(1)}, P_j^{(2)}, \ldots, P_j^{(N)})$ be a random permutation of the integers $(1, 2, \ldots, N)$.[5] The P_j are independent of each other.

Step 2b. For $i = 1, \ldots, N$ and $j = 1, \ldots, d$, let $\xi_j^{(i)}$ be i.i.d. random variables with a uniform distribution on $[0, 1]$ and let

$$U_j^{(i)} = \frac{P_j^{(i)} - 1 + \xi_j^{(i)}}{N}.$$

Thus, given the $P_j^{(i)}$, each of the $U_j^{(i)}$ is uniformly distributed on an interval of length $1/N$, and, for each j, each of the intervals $[0, 1/N], [1/N, 2/N], \ldots,$ $[(N-1)/N, 1]$ contains exactly one of the $U_j^{(i)}$. So, for a given j, the sampling scheme is similar to stratified sampling with uniform randomness in each interval.

Basic Monte Carlo and Latin hypercube sampling are illustrated in Figure 10.14 for the bivariate uniform distribution on $[0, 1]^2$. The greater uniformity in the latter is evident for $N = 10$. This is still the case for $N = 1000$ although this is less obvious to the eye.

Step 2c. Let $\epsilon_j^{(i)} = \Phi^{-1}(U_j^{(i)})$.

A major drawback of stratified and Latin hypercube sampling is that if we wish to move even from N to $N + 1$ simulations we cannot just do one more simulation: instead we must start again and perform the full $N + 1$ simulations.

This drawback is avoided when we implement the quasi-Monte Carlo method.

10.5.3   Quasi-Monte Carlo or Low-Discrepancy Sequences

Quasi-Monte Carlo sequences are also often referred to as quasi-random sequences or low-discrepancy sequences. The use of the names *random* and *Monte Carlo* in the present context are quite misleading because the sequences generated by the various algorithms are entirely deterministic and totally lack any appearance of randomness.[6] The names *quasi-Monte Carlo* and *quasi-random* tend to be used quite widely, but experts generally prefer the term *low discrepancy*. The notion of discrepancy is most easily illustrated by looking at the one-dimensional variable on $[0, 1]$. Suppose that we have a sample of N points from this interval. Even if N is quite large, if we plot a histogram of a truly random set of values, then the plot will not be uniform, but will show a degree of unevenness. In rough terms, the discrepancy of the sequence of N values is a measure of how far its histogram is from being totally uniform.

---

[5]This random permutation can be achieved by simulating N uniform $[0, 1]$ random variables and then let $P_j^{(i)}$ be the rank of the jth random variable.

[6]In basic Monte Carlo we also, of course, use deterministically generated *pseudo-random* numbers. However, good sequences of pseudo-random numbers to all intents and purposes look random and tend to satisfy most tests of randomness and uniformity.

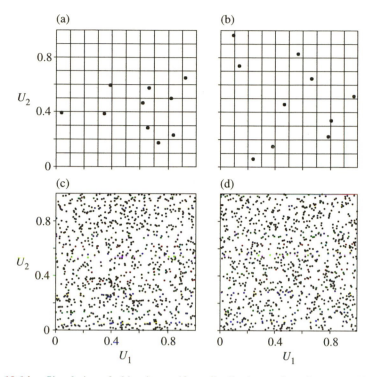

Figure 10.14. Simulation of a bivariate uniform distribution on the unit square. (a) and (c): Basic Monte Carlo. (b) and (d): Latin hypercube sampling. (a) and (b): 10 simulations. (c) and (d): 1000 simulations.

Applications of low-discrepancy sequences are documented in a variety of places, most notably Joy, Boyle and Tan (1996) and Boyle, Broadie and Glasserman (1997). They look at rates of convergence of derivative price estimates using low-discrepancy sequences compared with other numerical methods. Tan and Boyle (2000) introduce an element of true randomization into low-discrepancy sequences in order to improve convergence and also to permit calculation of standard errors. A helpful description of what low-discrepancy sequences are and of how to generate them can be found in Press et al. (1992).

As with the Latin hypercube technique, the only difference between quasi-Monte Carlo and basic Monte Carlo is in the generation of the $\epsilon_j^{(i)}$ in Step 2.

The sequences we will describe here are due to Halton (1960) and Faure (1982). Other sequences include those by Sobol' (1967) and Niederreiter (1987). Generation of the low-discrepancy sequences can be time-consuming. However, once the sequences have been generated they can be reused repeatedly, meaning that computing times for a given value of N should be no different from basic Monte Carlo.

Table 10.1.   Construction of entry 17 in Halton sequences with bases 2, 3 and 5.

j	p_j	i in base p_j notation	$U_j^{(17)}$ in base p_j notation	$U_j^{(17)}$ in fractional notation
1	2	10001 base 2	0.10001 base 2	17/32
2	3	122 base 3	0.221 base 3	25/27
3	5	32 base 5	0.23 base 5	13/25

Halton Sequences

Recall that we wish to simulate in discrete time from t_0 up to $t_n = t_0 + n\Delta t$ with N repetitions.

Step 2a.   Let $p_1, \ldots, p_n$ be the first n prime numbers.

Step 2b.   For each $j = 1, \ldots, n$, $i = 1, 2, \ldots, N$, write down the integer i in base p_j format. In particular, find the unique set of integers $a_j(i, l)$ with $0 \leqslant a_j(i, l) < p_j$ for $l = 0, 1, 2, \ldots$ such that

$$i = \sum_{l=0}^{\infty} a_j(i, l) p_j^l = \sum_{l=0}^{m(i, p_j)} a_j(i, l) p_j^l,$$

where $m(i, p_j)$ is the largest index of a non-zero digit in base p_j (that is, $a_j(i, l) = 0$ for all $l > m(i, p_j) \geqslant \log(i + 1)/\log(p_j)$).

As examples, consider $i = 17$ and $j = 1, 2, 3$ (see Table 10.1).

Step 2c.   Now reverse the order of the digits and precede by a decimal point: that is, define

$$U_j^{(i)} = \sum_{l=0}^{m(i, p_j)} a_j(i, l) p_j^{-l-1}.$$

Step 2d.   Let $\epsilon_j^{(i)} = \Phi^{-1}(U_j^{(i)})$.

In this final step we have used the inverse of the cumulative distribution function to generate a quasi-random sample from a standard Normal distribution. Joy, Boyle and Tan (1996) describe an algorithm for calculating $\Phi^{-1}(u)$. They also make it clear that the popular Box–Müller method for generating a bivariate standard Normal distribution should *not* be used as the non-linear transformation used destroys the low-discrepancy structure of the sequence.

As a further example, the first 10 entries in the Halton sequences with $p = 2, 3, 5, 7, 11$ are given in Table 10.2.

Faure Sequences

Step 2a.   Let p be the smallest prime number greater than or equal to n.

Table 10.2. First 10 entries for each of the Halton sequences with bases 2, 3, 5, 7 and 11.

i	$j = 1$ $p_j = 2$	$j = 2$ $p_j = 3$	$j = 3$ $p_j = 5$	$j = 4$ $p_j = 7$	$j = 5$ $p_j = 11$
1	1/2	1/3	1/5	1/7	1/11
2	1/4	2/3	2/5	2/7	2/11
3	3/4	1/9	3/5	3/7	3/11
4	1/8	4/9	4/5	4/7	4/11
5	5/8	7/9	1/25	5/7	5/11
6	3/8	2/9	6/25	6/7	6/11
7	7/8	5/9	11/25	1/49	7/11
8	1/16	8/9	16/25	8/49	8/11
9	9/16	1/27	21/25	15/49	9/11
10	5/16	10/27	2/25	22/49	10/11

Step 2b. Let $m = \min\{k \in \mathbb{Z} : k > \log(N)/\log(p)\} - 1$. This implies that $m + 1$ is the largest number of digits required to express any number from 1 to N in base p.

Step 2c. Define the $(m + 1) \times (m + 1)$ matrix $C = [c_{kl}]_{k,l=0}^{m}$, where $c_{kl} = 0$ if $k > l$ and, for $k \leqslant l$, $c_{kl} = l!/(k!(l - k)!)$. For numerical accuracy we calculate the entries of C as follows:

$$c_{kl} = 0 \quad \text{for } k > l,$$
$$c_{0l} = 1 \quad \text{for } l = 0, \ldots, m,$$
$$c_{kk} = 1 \quad \text{for } k = 1, \ldots, m,$$
$$c_{kl} = c_{k,l-1} + c_{k-1,l-1} \quad \text{for } k = 1, \ldots, m - 1 \text{ and } l = k + 1, \ldots, m.$$

Step 2d. For $j = 1$ and for each $i = 1, 2, \ldots, N$ write down the integer i in base p format as with the Halton sequence. Thus, we find $a_1(i, l)$ for $l = 0, \ldots, m$ with $0 \leqslant a_1(i, l) < p$ and

$$i = \sum_{l=0}^{m} a_1(i, l) p^l.$$

Define the column vector $a_1(i) = (a_1(i, 0), \ldots, a_1(i, m))'$.

Step 2e. For each $j = 2, \ldots, n$ and $k = 0, \ldots, m$ let

$$a_j(i, k) = \sum_{l=0}^{m} c_{kl} a_{j-1}(i, l) \bmod p$$

or, in vector notation,

$$a_j(i) = C a_{j-1}(i) \bmod p.$$

When we consider the sets $\{a_j(1), \ldots, a_j(p^k - 1)\}$ for each $k = 1, \ldots, m$ we can observe that the action of $C \bmod p$ is to permute the numbers that they represent.

Table 10.3.   First eight entries for each of the three Faure sequences when $n = p = 3$.

	$j = 1$	$j = 2$	$j = 3$
$i = 1$	1/3	1/3	1/3
2	2/3	2/3	2/3
3	1/9	4/9	7/9
4	4/9	7/9	1/9
5	7/9	1/9	4/9
6	2/9	8/9	5/9
7	5/9	2/9	8/9
8	8/9	5/9	2/9

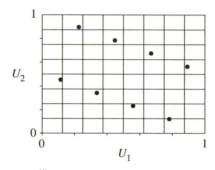

Figure 10.15. $U_1^{(i)}$ versus $U_2^{(i)}$ for $i = 1, \ldots, 8$ for the Faure sequences with $n = p = 3$.

We know that for $j = 1$, $\{a_j(1), \ldots, a_j(p^k - 1)\}$ represents in base p the integers $\{1, 2, \ldots, p^k - 1\}$. The same is true for $j = 2, 3, \ldots, n$; that is, the action of $C \bmod p$ never leads to duplication of numbers (see, for example, Table 10.3).

Step 2f.   Now reverse the order of the digits (as with the Halton sequence) and precede by a decimal point; that is, define

$$U_j^{(i)} = \sum_{l=0}^{m} a_j(i, l) p_j^{-l-1}.$$

Step 2g.   Let $\epsilon_j^{(i)} = \Phi^{-1}(U_j^{(i)})$.

The first two columns of Table 10.3 are illustrated in Figure 10.15. We can see here that (as with Latin hypercube sampling) for each positive integer k, if we plot the first $p^k - 1$ entries in the two sequences and overlay a regular lattice with $p^k - 1$ columns and $p^k - 1$ rows, then exactly one point lies in each column and exactly one point in each row.

In Figures 10.16 and 10.17 we take $n = p = 5$. In Figure 10.16 we take pairs of sequences and plot the first 124 ($= p^3 - 1$) points. We can see that each displays a

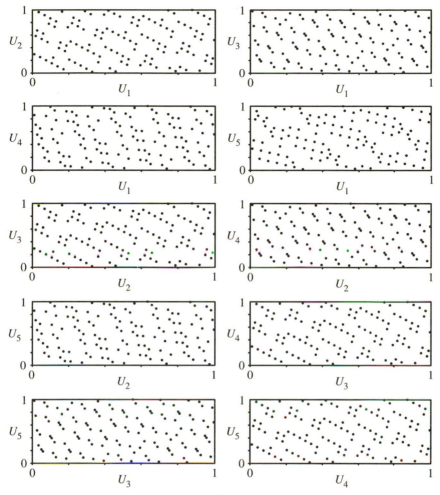

Figure 10.16. $U_j^{(i)}$ versus $U_k^{(i)}$ for $i = 1, \ldots, 124$ for pairs of Faure sequences with $n = p = 5$.

certain amount of regularity. As an example, we might speculate that the graph with pair $(1, 4)$ has some holes in amongst the points. However, when we add a further 500 points to this, as in Figure 10.17, then we see that these holes disappear.

As a final note, Joy, Boyle and Tan (1996) recommend that in any sequence the first $p^4 - 1$ points are discarded in order to improve accuracy.

10.5.4 Importance Sampling or Change of Measure

This technique provides a useful way of reducing variance when the usual distribution Q attaches only a low probability to those sample paths where X varies most.

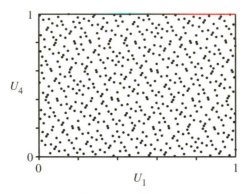

Figure 10.17. $U_1^{(i)}$ versus $U_4^{(i)}$ for $i = 1, \ldots, 624$ for
the Faure sequences with $n = p = 5$.

For example, with an option that is far out of the money most of the probability mass is attached to a payoff of zero, while the interesting, positive outcomes have low probability.

Importance sampling seeks to remedy this by altering the probability distribution. As a very simple example, consider a very simple binary option with no discounting:

$$X = \begin{cases} 1 & \text{if } Z > x, \\ 0 & \text{if } Z \leqslant x, \end{cases}$$

where $Z \sim N(0, 1)$ under Q. Let $p = \mathrm{Pr}_Q(X > x) = 1 - \Phi(x)$. Then $E_Q(X) = p$ and $\mathrm{SD}_Q(X) = \sqrt{p(1-p)}$. If p is small (that is, the option is far out of the money), then, clearly, the standard deviation of X is very high compared with its expectation. The density function for Z under Q is $f(z) = (2\pi)^{-0.5} e^{-z^2/2}$. Now consider an alternative (equivalent) distribution for Z, $\hat{P}$ with density $g(z) = (2\pi)^{-0.5} \exp[-(z - \mu)^2/2]$. Then note that

$$E_Q(X) = \int_{-\infty}^{\infty} I(z > x) f(z) \, dz$$
$$= \int_{-\infty}^{\infty} I(z > x) \frac{f(z)}{g(z)} g(z) \, dz$$
$$= E_{\hat{P}}\left[I(Z > x) \frac{f(Z)}{g(Z)} \right];$$

that is, we replace the original random variable $X = I(Z > x)$ with a new random variable $\hat{X} = I(Z > x) f(Z)/g(Z)$, where Z is now simulated under $\hat{P}$.[7] This has the same expected value as X. If we choose the new mean, μ, of Z carefully, then we can reduce substantially the variance of $\hat{X}$.

---

[7] Note that $f(z)/g(z)$ is nothing other than the Radon–Nikodým derivative for this change of measure.

Since $E_{\hat{P}}(\hat{X})$ is fixed we minimize its variance by minimizing

$$E_{\hat{P}}(\hat{X}^2) = \int_{-\infty}^{\infty} I(Z > x)^2 \frac{f(z)^2}{g(z)^2} g(z)\, dz = e^{\mu^2}(1 - \Phi(x + \mu)).$$

As an example, suppose that $x = 1.96$ so that $E_Q[X] = 0.025$. If we use importance sampling, the optimal value of μ is 2.18. Sampling under Q the variance of X is 0.0244. Sampling under $\hat{P}$ the variance of $\hat{X}$ is 0.0014. This means that the use of importance sampling would require only about 25% of the number of simulations to achieve the same degree of accuracy as basic Monte Carlo.

This, of course, is a very simple example to illustrate the technique. In the context of interest rate modelling, importance sampling is equivalent to choosing carefully a change of measure. Instead, then, of multiplying X by the ratio of two density functions, we need to multiply X by the Radon–Nikodým derivative. In the case of an out-of-the-money call option we would introduce a change of measure which results in a greater proportion of simulated paths ending up in the money.

10.5.5 Antithetic Variates

Consider the simple case where $X = f(Z)$ for some function $f(z)$ and $Z \sim N(0, 1)$. Suppose that under basic Monte Carlo we would carry out $2N$ simulations using i.i.d. $Z_1, \ldots, Z_{2N}$. It follows that $\mathrm{Var}(\bar{X}) = \mathrm{Var}(X_i)/2N$ as before.

Instead, now generate i.i.d. $Z_1, \ldots, Z_N$ and then define $\tilde{Z}_i = -Z_i$ for $i = 1, \ldots, N$. We then let $X_i = f(Z_i)$, $\tilde{X}_i = f(\tilde{Z}_i)$, $\hat{X} = N^{-1}\sum_{i=1}^{N} X_i$ and $\hat{\tilde{X}} = N^{-1}\sum_{i=1}^{N} \tilde{X}_i$. The $(Z_i, \tilde{Z}_i)$ are called *antithetic pairs* and the $(X_i, \tilde{X}_i)$ are *antithetic variates*. Finally, let $\bar{X} = (\hat{X} + \hat{\tilde{X}})/2$. It is straightforward to show that the variance of $\bar{X}$ is

$$\mathrm{Var}(\bar{X}) = \frac{1}{N}\,\mathrm{Var}\left(\frac{X_i + \tilde{X}_i}{2}\right) = \frac{1}{2N}\,\mathrm{Var}(X_i)(1 + \mathrm{cor}(X_i, \tilde{X}_i)).$$

The fact that $\tilde{Z}_i = -Z_i$ means that $\mathrm{cor}(X_i, \tilde{X}_i)$ will typically be negative, meaning that $\mathrm{Var}(\bar{X})$ will be lower than that using basic Monte Carlo. The biggest gains from the use of the antithetic variate technique come when the correlation, $\mathrm{cor}(X_i, \tilde{X}_i)$, is close to -1. This tends to happen when $f(z)$ is nearly linear over the main body of the distribution of Z.

Further discussion is given in Boyle, Broadie and Glasserman (1997) and references therein.

10.5.6 Control Variates

It is likely that the reason for simulating X is because the distribution of X is not known. However, in some cases there may be some other derivative with random discounted value Y (the *control variate*) where (a) Y is known to give a good approximation to X or where Y is known to be highly correlated with X, and (b) the expected value of Y is known.[8]

Let $E(Y)$ be the known (analytical) expected value of Y. Now carry out N basic Monte Carlo simulations resulting in simulated values of both the X_i and Y_i. Let $W_i = X_i - \beta(Y_i - E(Y))$ (for some constant β) and calculate its mean $\bar{W} = N^{-1} \sum_{i=1}^{N} W_i = \bar{X} - \beta(\bar{Y} - E(Y))$. Let our estimator for $E(X)$ be

$$\hat{X} = \bar{W}.$$

Then

$$E(\hat{X}) = E(\bar{W}) = E[\bar{X} - \beta(\bar{Y} - E(Y))] = E(X).$$

So $\hat{X}$ is an unbiased estimator for $E(X)$. Furthermore,

$$\text{Var}(\hat{X}) = \text{Var}(\bar{W}) = \frac{1}{N}(\text{Var}(X_i) + \beta^2 \text{Var}(Y_i) - 2\beta \text{Cov}(X_i, Y_i)).$$

This is minimized when

$$\beta^* = \frac{\text{Cov}(X_i, Y_i)}{\text{Var}(Y_i)},$$

at which

$$\text{Var}(\hat{X}) = \frac{\text{Var}(X_i)}{N}(1 - \text{cor}(X_i, Y_i)^2).$$

It follows that the biggest reduction in variance can be achieved when the correlation between X and Y is close to 1. In practice we may not know the covariance in order to calculate β^*. In this case we may be able to assume that $\beta = 1$ is reasonably good. Alternatively, we may apply linear regression to a test simulation to estimate the optimal β.

Again, further discussion can be found in Boyle, Broadie and Glasserman (1997, and references therein) and also in Clewlow and Strickland (1997).

10.6   Exercise

Example 10.1. Consider an out-of-the-money call option on the zero-coupon bond $P(t, U)$. We wish to value this using a combination of importance sampling and Monte Carlo simulation. How would the SDE for $r(t)$ be different under the new measure $\hat{P}$? In particular, is the mean-reversion level the same, higher or lower?

---

[8]A well-known example of this is Asian equity options, where the distribution of the arithmetic average is not known while the distribution of the geometric average is.

11

Credit Risk

11.1 Introduction

In this chapter we will give a brief introduction to the rapidly developing field of credit-risk modelling. In particular, we will introduce the reader to some of the techniques available for assessing the impact of credit risk on bonds and interest rate derivatives. For more detailed accounts of credit-risk modelling the interested reader is referred to the various texts cited in this chapter and also to the comprehensive books on credit-risk modelling by Bielecki and Rutkowski (2002) and Duffie and Singleton (2003).

Previously, we have assumed that bonds and other interest rate derivative contracts have been free of default. This is a reasonable assumption where we are dealing with many government bonds. However, in the simplest case of a fixed-interest bond, if the issuer is a corporate entity rather than a government there is the possibility that the company defaults at some time during the term of the contract. This means that the contracted payment stream might be

- totally wiped out;
- continued but at a reduced rate;
- cancelled by payment of an amount less than the default-free value of the contract;
- rescheduled.

Typically, default will be the result of cashflow difficulties within the company and should be regarded by the company as a last resort. In particular, the company must bear in mind that once it has defaulted it will be very difficult in the future to raise further capital through the issue of debt at competitive rates.

Credit risk is often an important factor to consider in the derivatives market. The traded-derivatives markets operate in a way which substantially reduces the adverse consequences of default through the use of margining[1], so that default risk is not

---

[1] With most traded derivatives both parties are required to put up a certain amount of cash or other assets (the *margin account*) as collateral to the agreement. These accounts act as substantial buffers against the adverse effects of default.

generally a substantial factor. On the other hand, over-the-counter derivatives do not have the same degree of protection, with the result that credit risk can have a significant impact on the price of a derivative contract.

The literature presents us with two distinct types of model.

Structural models. These models attempt to mimic the actual causes of default. As such they are explicit models for a firm with both equity and debt. Most models which exist in the literature are typically quite simple, however, so they cannot realistically be used for pricing credit risk. However, they give us considerable insight into the nature of default and the interaction between equityholders and bondholders.

Reduced-form models. These are essentially statistical models that make use only of observed market statistics rather than data specific to the company or corporate entity. Most often, the market statistics that are used are the credit ratings issued by various agencies such as Standard and Poor's (S&P) and Moody's. For example, S&P allocate each company to one of the grades AAA (least likely to default), AA, A, BBB, BB, B, and CCC (most likely to default) as well as D (already defaulted). It is taken as read that the credit-ratings agencies will themselves have used more-detailed, company-specific information before setting their ratings. Furthermore, it is assumed that this information is regularly reviewed to ensure that companies are rerated up or down whenever appropriate, and in a timely fashion. These statistics are used in conjunction with market data on the default-free bond market. The aim of a reduced-form model is then to mimic the behaviour of a company's credit rating over time, with the specific aim of establishing a distribution for the time to default.

Practitioners tend to use reduced-form models rather than structural models, partly because of the idealistic nature of the existing structural models in the academic literature. However, in time we may see models being developed which combine the best features of the two types of model.

11.1.1 Yield to Maturity

The impact of credit risk is easily seen when we look at quoted yields to maturity (see Section 1.2.8) on corporate bonds. As a simple example, consider two zero-coupon bonds with 10 years to maturity. The first bond is default free and has a price of 0.5, while the second has associated default risk and has a price of 0.4. The yields to maturity on these bonds (continuously compounding) are found by solving the equation $P_i = e^{-10\delta_i}$; that is, 6.93% and 9.16%—a *credit spread* of 2.23%. Although the yield on the latter is higher we do not *expect* to earn this rate. The key point is that yield to maturity is calculated on the artificial assumption that the bond is default free. In this example it is perhaps easier to compare the prices directly to gauge the effect of default. A direct comparison of prices is less easy with corporate bonds that pay coupons. In such cases, for bonds with a similar term to maturity, a comparison of yields to maturity is more enlightening.[2]

For high-quality corporate bonds (AAA, AA, A) with, say, 10 years to maturity, the yield to maturity is typically in the range 50–100 basis points (0.50–1.00%) *above* that for equivalent default-free government debt. Below the top three grades, credit spreads widen quite rapidly to 5% or more.

11.2   Structural Models

We will give here only a brief introduction to this particular class of model. The aim is to link default events explicitly to the fortunes of the underlying corporate entity.

11.2.1   The Merton Model

The earliest model of this type was developed by Merton (1974). Consider a company which has issued both debt (in the form of zero-coupon corporate bonds) and equity. The structure of the model is very simple.

The current total value of the firm (that is, attributable to both equityholders and bondholders) at time t is $F(t)$, which varies over time as a consequence of the company's operations. No dividends are payable to equityholders, or coupons to bondholders. At some fixed time T in the future, bondholders are entitled to a fixed payment of K. The remainder of the value of the company is distributed to equityholders and the company is wound up.

If the total value of the company at time T, $F(T)$, is insufficient to pay the bondholders (that is, $F(T) < K$) then there is a default. Bondholders will then receive $F(T)$ rather than K and equityholders will receive nothing at time T.

We now describe a simple extension of the Merton (1974) model (under which $F(t)$ is a geometric Brownian motion) by allowing the risk-free rate of interest $r(t)$ to be stochastic rather than constant. Let

$$dF(t) = F(t)(\mu(t)\,dt + \sigma_{FF}\,dW_F(t) + \sigma'_{Fr}\,dW_r(t)),$$

where $\mu(t)$ is some previsible process, σ_{FF} is a constant scalar, σ_{Fr} is a constant ($m \times 1$) vector, $W_r(t)$ is an m-dimensional Brownian motion under the real-world measure P and $W_F(t)$ is a one-dimensional Brownian motion under P that is independent of $W_r(t)$. Interest rate dynamics are driven by $W_r(t)$ according to the SDE

$$dr(t) = a(t)\,dt + b(t)'\,dW_r(t),$$

for previsible processes $a(t)$ and $b(t)$.

---

[2]Some authors refer to the *risk premium* instead of credit spread. Strictly this is incorrect, as risk premium is usually used to mean the expected extra return over cash in return for the risks being taken. As we have noted, the credit spread assumes no defaults will take place, so that this is an overestimate of the true risk premium.

Associated with $W_F(t)$ and $W_r(t)$ we have (previsible) market prices of risk $\gamma_F(t)$ and $\gamma_r(t)$ (an $(m \times 1)$ vector process) which transform us to the risk-neutral model

$$dr(t) = (a(t) - b(t)'\gamma_r(t))\, dt + b(t)'\, d\tilde{W}_r(t), \qquad (11.1)$$

$$dF(t) = F(t)(r(t)\, dt + \sigma_{FF}\, d\tilde{W}_F(t) + \sigma_{Fr}\, d\tilde{W}_r(t)), \qquad (11.2)$$

where $\tilde{W}_F(t)$ and $\tilde{W}_r(t)$ are Brownian motions under the risk-neutral measure Q. The interest rate model also gives us the usual SDEs for default-free bond prices

$$dP(t, T) = P(t, T)(r(t)\, dt + S(t, T)'\, d\tilde{W}_r(t)),$$

where $S(t, T)$ is the previsible vector of bond volatilities.

We return now to the company equityholders and bondholders. At time T the capital structure will result in cashflows of

$$\min\{K, F(T)\} = K - \max\{K - F(T), 0\} \qquad \text{to the bondholders,}$$
$$\max\{F(T) - K, 0\} \qquad \text{to the equityholders.}$$

The dynamics of $F(t)$ are such that equityholders have the equivalent of a European call option on $F(t)$ while bondholders have a combination of a long position in a risk-free bond and a short position in a put option on $F(t)$. The value of the corporate bond at time t is therefore $V(t, T) = K P(t, T) - p(t)$, where

$$p(t) = E_Q\left[\frac{B(t)}{B(T)}(K - F(T))_+ \,\middle|\, \mathcal{F}_t\right]$$

and $B(t) = \exp \int_0^t r(s)\, ds$ is the usual cash account value at time t.

Using the forward-measure technology of Chapter 7, we can write the value of the put option in the alternative form

$$p(t) = P(t, T)E_{P_T}[(K - F(T))_+ \mid \mathcal{F}_t],$$

where P_T is the forward measure under which the prices of all tradable assets discounted by $P(t, T)$ are martingales. In a similar fashion to Section 7.6, we can see that if the bond volatilities $S(t, T)$ are deterministic, then

$$p(t) = P(t, T)K\Phi(-d_2) - F(t)\Phi(-d_1),$$

where

$$d_1 = \frac{\log(F(t)/K P(t, T)) + \tfrac{1}{2}b^2}{b}, \qquad d_2 = d_1 - b,$$

$$b^2 = \int_t^T |\sigma_{Fr} - S(v, T)|^2\, dv + \sigma_{FF}^2(T - t).$$

Thus,

$$V(t, T) = K P(t, T) - (K P(t, T)\Phi(-d_2) - F(t)\Phi(-d_1))$$
$$= K P(t, T)\Phi(d_2) + F(t)\Phi(-d_1).$$

As a check on the reasonableness of this formula for $V(t, T)$ we can note that, for fixed t:

$$F(t) \to 0 \quad \Rightarrow \quad d_1, d_2 \to -\infty \quad \Rightarrow \quad V(t, T) \to F(t);$$
$$F(t) \to +\infty \quad \Rightarrow \quad d_1, d_2 \to +\infty \quad \Rightarrow \quad V(t, T) \to K P(t, T).$$

Intuitively, these are correct. In the first case, if $F(t)$ is very small, then default is almost certain and so bondholders will get $F(T)$ rather than K. In the second case, the probability of default is almost zero so the bondholders will be very likely to get the full K at time T.

11.2.2 Other Models

Clearly, the Merton model is very simple. However, it does open the way for a variety of more sophisticated models. Specific features of these extended models are that they

- allow the capital structure of the company to include a mixture of bonds with different maturity dates, with coupons and possibly perpetual;

- allow the equityholders to receive dividends;

- incorporate rules for default to be triggered in advance of some terminal date t or before the total value of the company hits zero;

- allow for rescheduling of debt repayments at the time of default and possible subsequent defaults;

- give some control to equityholders over the timing of defaults in order to maximize value to their group.

These models go beyond the scope of this book, but further details can be found in the papers by, for example, Black and Cox (1976), Leland (1994), Longstaff and Schwartz (1995) and Anderson and Sundaresan (1996). A brief review of these models can also be found in the paper by Lando (1997).

11.3 A Discrete-Time Model

We will consider here the simplest model for default-risk that extends beyond the binomial, default-free interest rate model discussed earlier in the book (Section 3.2). We have the following assets available:

- default-free zero-coupon bonds maturing at times $T = 1, 2, \ldots$ with associated prices $P(t, T)$;

- zero-coupon corporate bonds all issued by the same company maturing at times $T = 1, 2, \ldots$ with associated prices $V(t, T)$.

If the company defaults at time τ, then all corporate bonds maturing at time τ and later become worthless.

The default-free term structure of interest rates is the same as in Section 3.2. In particular, we have

$$
P(t+1, T) = \begin{cases} \dfrac{P(t, T)}{P(t, t+1)} u(t, T-t) & \text{if prices go up,} \\[3mm] \dfrac{P(t, T)}{P(t, t+1)} d(t, T-t) & \text{if prices go down,} \end{cases}
$$

where the $u(t, s)$ and $d(t, s)$ are possibly stochastic but known at time t. Under the risk-neutral measure Q we have

$$
q_u(t) = \Pr_Q (\text{up between } t \text{ and } t+1) = \frac{1 - d(t, s)}{u(t, s) - d(t, s)},
$$

$$
q_d(t) = \Pr_Q (\text{down between } t \text{ and } t+1) = \frac{u(t, s) - 1}{u(t, s) - d(t, s)}.
$$

(Recall that, in fact, $q_u(t)$ and $q_d(t)$ do not depend upon s.)

In addition, over each time period the company might or might not default. So over each time period (given the company has not yet defaulted) there are four outcomes. For a derivative contract linked to the credit status of the company or one of the $V(t, T)$ we therefore require four assets for hedging.

We will now consider a one-period hedging problem (which will demonstrate that the issue is already rather more complex than the default-free binomial model).

Example 11.1. You are given the following market information:

T	$P(0, T)$	$V(0, T)$
1	$\pi_1(0) = 0.95$	$\pi_3(0) = 0.9$
2	$\pi_2(0) = 0.9025$	$\pi_4(0) = 0.81$

In addition it is assumed that, for default-free bonds, $u(0, 2) = 1.04$ and $d(0, 2) = 0.92$, implying that $q_u = q_u(0) = 2/3$ and $q_d = q_d(0) = 1/3$.

The four outcomes at time 1 are as follows:

Outcome, ω	Description
$\omega_1 = (u, D)$	Default-free prices rise, company defaults
$\omega_2 = (d, D)$	Default-free prices fall, company defaults
$\omega_3 = (u, N)$	Default-free prices rise, company does not default
$\omega_4 = (d, N)$	Default-free prices fall, company does not default

This model still gives us some flexibility over the value of $V(t, 2)$ at $t = 1$. For convenience, in what follows we will use simpler notation for the prices at time 1, and also refer to the price process $\pi_i(t)$ for asset i (as defined in the table). Thus, at time 1,

Outcome	$\pi_1(1) =$ $P(1, 1)$	$\pi_2(1) =$ $P(1, 2)$	$\pi_3(1) =$ $V(1, 1)$	$\pi_4(1) =$ $V(1, 2)$
$\omega_1 = (u, D)$	1	p_u	0	0
$\omega_2 = (d, D)$	1	p_d	0	0
$\omega_3 = (u, N)$	1	p_u	1	v_u
$\omega_4 = (d, N)$	1	p_d	1	v_d

The risk-free rate of interest over the first time period is $r(0) = -\log \pi_1(0)$. The prices given at times 0 and 1 (with suitable constraints on v_u and v_d that we will discuss shortly) allow us to derive a unique risk-neutral measure Q under which the prices of assets 1, 2 and 3 when discounted by $\pi_1(t)$ are martingales. That is, there exists a set of real numbers $(q_{uD}, q_{dD}, q_{uN}, q_{dN})$ for which

$$q_{uD}\pi_i(1)(uD) + q_{dD}\pi_i(1)(dD) + q_{uN}\pi_i(1)(uN) + q_{dN}\pi_i(1)(dN)$$
$$= e^{r(0)}\pi_i(0) = \pi_i(0)/\pi_1(0),$$

for $i = 1, 2, 3, 4$.

The solution to this is

$$q_{uD} = \frac{1}{\pi_1(0)\Delta p \Delta v}(-p_d \Delta v \pi_1(0) + \Delta v \pi_2(0) + v_d \Delta p \pi_3(0) - \Delta p \pi_4(0)),$$

$$q_{dD} = \frac{1}{\pi_1(0)\Delta p \Delta v}(p_u \Delta v \pi_1(0) - \Delta v \pi_2(0) - v_u \Delta p \pi_3(0) + \Delta p \pi_4(0)),$$

$$q_{uN} = \frac{1}{\pi_1(0)\Delta p \Delta v}(-v_d \Delta p \pi_3(0) + \Delta p \pi_4(0)),$$

$$q_{dN} = \frac{1}{\pi_1(0)\Delta p \Delta v}(v_u \Delta p \pi_3(0) - \Delta p \pi_4(0)),$$

where $\Delta p = p_u - p_d$ and $\Delta v = v_u - v_d$.

The model will be arbitrage-free if each of these quantities is strictly positive. We will assume that $v_u > v_d$.[3] The requirements that each of the four quantities is strictly positive then means that (respectively)

$$v_u(-p_d \pi_1(0) + \pi_2(0)) + v_d(p_d \pi_1(0) - \pi_2(0) + \Delta p \pi_3(0)) > \Delta p \pi_4(0), \quad (11.3)$$

$$v_u(p_u \pi_1(0) - \pi_2(0) - \Delta p \pi_3(0)) + v_d(-p_u \pi_1(0) + \pi_2(0)) > -\Delta p \pi_4(0), \quad (11.4)$$

---

[3] This assumption follows from the fact that v_u corresponds to a high price for the default-free bond. However, under certain circumstances, when the default probabilities are highly dependent on the interest rate regime it is possible that $v_u < v_d$.

$$v_d < \pi_4(0)/\pi_3(0), \tag{11.5}$$

$$v_u > \pi_4(0)/\pi_3(0). \tag{11.6}$$

Since we know also that $(q_u p_u + q_d p_d)\pi_1(0) = \pi_2(0)$, inequalities (11.3) and (11.4) become

$$v_u q_u \pi_1(0) + v_d(-q_u \pi_1(0) + \pi_3(0)) > \pi_4(0), \tag{11.7}$$

$$v_u((1 - q_u)\pi_1(0) - \pi_3(0)) - v_d(1 - q_u)\pi_1(0) > -\pi_4(0). \tag{11.8}$$

Now recall the numerical information given at the outset of this example. The above inequalities then become (with the implied values $p_u = 0.988$, $p_d = 0.874$ and $\Delta p = 0.114$)

$$\tfrac{38}{60}v_u + \tfrac{16}{60}v_d > 0.81,$$

$$-\tfrac{35}{60}v_u - \tfrac{19}{60}v_d > -0.81,$$

$$v_d < 0.9,$$

$$v_u > 0.9.$$

These inequalities, in fact, result in an infinite wedge shape radiating from $(v_u, v_d) = (0.9, 0.9)$ (see Figure 11.1). To complete the picture we need to recognize that $V(1, 2)$ is a zero-coupon corporate bond, so we must also have $0 < v_u < p_u$ and $0 < v_d < p_d$.

Inequalities (11.7) and (11.8) imply that the two boundaries cross at $v_u = v_d = \pi_4(0)/\pi_3(0)$. In general, the sets of points for which inequalities (11.7) and (11.8) are satisfied also satisfy the earlier assumption that $v_u > v_d$.

The feasible region for (v_u, v_d) is plotted in Figure 11.1. In the diagram the solid line PQ represents those points where v_u and v_d are in the same proportion as p_u and p_d; that is, when default risk in the second time period is uncorrelated with the interest rate over the same period.

The basic risk-neutral default probability for the first time period under this model is $q_D = 1 - q_N = 1 - V(0, 1)/P(0, 1) = 0.9/0.95$. If the default event in the first time period is to be independent of the up or down movement of default-free bond prices, then we require $q_{uN} = q_u q_N$ (which then implies that $q_{uD} = q_u q_D$, $q_{dN} = q_d q_N$ and $q_{dD} = q_d q_D$). This imposes the constraint that

$$v_u(-p_d \pi_1(0) + \pi_2(0)) + v_d(p_u \pi_1(0) - \pi_2(0)) = \Delta p \pi_1(0)\pi_4(0)/\pi_3(0).$$

This line is represented by the dashed line RS in Figure 11.1. The point of intersection of the lines PQ and RS (that is, $(v_u, v_d) = (0.936, 0.828)$) is where the default event over both periods is independent under the risk-neutral measure Q of the default-free interest rate dynamics. However, it should be borne in mind that such independence may not be justified.

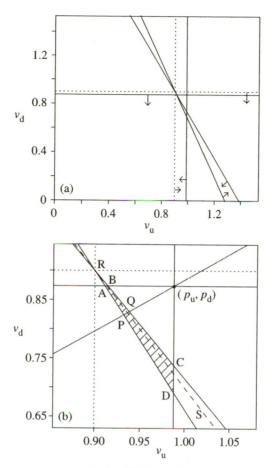

Figure 11.1. (a) Construction of the feasible (no-arbitrage) region for the prices $V(1, 2)$ under the outcomes (u, N) and (d, N). Lines indicate the boundaries for the various inequalities, while the arrows indicate in which direction feasible points must lie. (b) Enlargement of (a) with the feasible region hatched. Points on the boundary ABCD are not feasible.

Suppose then that we do, indeed, have $v_u = 0.936$ and $v_d = 0.828$, along with $p_u = 0.988$ and $p_d = 0.874$. This gives us the risk-neutral probabilities $q_{uD} = 2/57$, $q_{dD} = 1/57$, $q_{uN} = 36/57$ and $q_{dN} = 18/57$. This completes the rather arduous first step required in the pricing of a derivative contract.

Consider then a European put option on $V(t, 2)$ with a strike price of $K = 0.9$. Suppose that we hold ϕ_i units of asset i. Then, in order to replicate the payoff on the derivative, we require

$$\sum_{i=1}^{4} \phi_i \pi_i(1)(\omega_j) = \max\{K - \pi_4(1)(\omega_j), 0\},$$

for $j = 1, 2, 3, 4$, where, recall, $\omega_1 = (u, D)$, etc. This requires us to solve the linear problem

$$
\begin{pmatrix}
1 & p_u & 0 & 0 \\
1 & p_d & 0 & 0 \\
1 & p_u & 1 & v_u \\
1 & p_d & 1 & v_d
\end{pmatrix}
\begin{pmatrix}
\phi_1 \\
\phi_2 \\
\phi_3 \\
\phi_4
\end{pmatrix}
=
\begin{pmatrix}
K \\
K \\
(K - v_u)_+ \\
(K - v_d)_+
\end{pmatrix}.
$$

In the present example, this means that $\phi_1 = 0.9$, $\phi_2 = 0$, $\phi_3 = -0.276$ and $\phi_4 = -0.666\,667$. The value of this portfolio at time 0 is $\sum_i \phi_i \pi_i(0) = 0.0666$. By the *law of one price* this must equal the price for the derivative.

The alternative to the hedging approach if we wish only to establish a price for the derivative is to use the risk-neutral pricing formula

$$
\begin{aligned}
\text{Price} &= e^{-r(0)} E_Q[(K - V(1, 2))_+] \\
&= \pi_1(0)(q_{uD}K + q_{dD}K + q_{uN}(K - v_u)_+ + q_{dN}(K - v_d)_+) \\
&= 0.0666,
\end{aligned}
$$

as we expect.

11.4   Reduced-Form Models

We will now consider continuous-time models.

In this section we will concentrate on intensity-based models. In particular, much of both research and practice has concentrated on the use of multiple-state models in which the jump processes between states are governed by these intensities. Such models have been used widely for many years in both applied probability and statistics (see, for example, Brémaud 1981; Anderson et al. 1993; Fleming and Harrington 1991) and actuarial mathematics (see, for example, Sverdrup 1965; Hoem 1969, 1988; Hoem and Aalen 1978; Ramlau-Hansen 1988; Norberg 1991, 1992, 1999). This has resulted in a well-established body of both theoretical and applied knowledge.

11.4.1   Single Credit Class

In order to introduce some of the basic ideas we will look at the simplest model in continuous time. There are two states N (not previously defaulted) and D (previously defaulted). Clearly, D is an absorbing state. For further simplicity we will assume that the default-free term structure is deterministic with $r(t) = r$ for all t.

The transition intensity, under the real-world measure P, from N to D at time t is denoted by $\lambda(t)$ (see Figure 11.2).

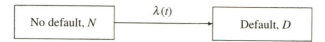

Figure 11.2. Two-state model for default risk.

Let $X(t)$ be the state at time t. The transition intensity is interpreted as follows:

$$\Pr_P(X(t + dt) = N \mid X(t) = N) = 1 - \lambda(t)\, dt + o(dt) \qquad \text{as } dt \to 0,$$
$$\Pr_P(X(t + dt) = D \mid X(t) = N) = \lambda(t)\, dt + o(dt) \qquad \text{as } dt \to 0.$$

Let us also introduce the stopping time

$$\tau = \inf\{t : X(t) = D\}$$

(with $\inf \varnothing = \infty$) and the counting process

$$N(t) = \begin{cases} 0 & \text{if } \tau > t, \\ 1 & \text{if } \tau \leqslant t. \end{cases}$$

These quantities have simple interpretations: τ is the time of default and $N(t)$ is the number of defaults up to and including time t.

We will assume that, once a company has defaulted, all contracted corporate-bond payments will be reduced by a known, deterministic factor $(1 - \delta)$. Thus, for a zero-coupon bond which is contracted to pay 1 at time T, the actual payment at time T will be 1 if $\tau > T$ and δ if $\tau \leqslant T$.

The factor δ is called the *recovery rate*. Note that if we look back to the basic Merton model (Section 11.2.1) the recovery rate is, in fact, random, because the default event can only occur at T. Other structural models can be constructed which ensure that default is triggered when the value of the firm is equal to some constant δ times the default-free value of the outstanding debt. In credit-risk modelling it is quite common to assume that the recovery rate is constant, although this is quite a strong assumption.

Besides the transition intensity $\lambda(t)$, under P there is an alternative probability measure Q with a different intensity $\tilde{\lambda}(t)$. For P and Q to be equivalent we require, for each $t > 0$, $\lambda(t) > 0$ if and only if $\tilde{\lambda}(t) > 0$.

As before, let $V(t, T)$ be the price at time t of a zero-coupon, corporate bond that matures at time T. Then (we will state this without proof) there exists a risk-neutral measure Q equivalent to P under which

$$V(t, T) = \mathrm{e}^{-r(T-t)} E_Q[\text{payoff at } T \mid \mathcal{F}_t] = \mathrm{e}^{-r(T-t)} E_Q[1 - (1 - \delta)N(T) \mid \mathcal{F}_t].$$

Now

$$E_Q[N(T) \mid N(t) = 0] = E_Q\left[1 - \exp\left(-\int_t^T \tilde{\lambda}(s)\, ds\right)\right].$$

If we assume that $\tilde{\lambda}(s)$ is deterministic, then this implies that

$$V(t, T) = e^{-r(T-t)}\left[1 - (1 - \delta)\left(1 - \exp\left(-\int_t^T \tilde{\lambda}(s)\,ds\right)\right)\right]$$

$$\Rightarrow \quad \tilde{\lambda}(s) = -\frac{\partial}{\partial s}\log[e^{r(s-t)}V(t, s) - \delta].$$

In more general circumstances, a full knowledge of both the default-free and corporate-bond term structures, coupled with an assumption about the recovery rate, allows us to back out the implied risk-neutral transition intensities.

It is relevant at this point to state that the Radon–Nikodým derivative for this problem is (see Brémaud 1981, Chapter VI, Section 2)

$$\frac{dQ}{dP}(0, T) = \begin{cases} \mu(\tau)\exp\displaystyle\int_0^\tau (1 - \mu(s))\lambda(s)\,ds & \text{if } \tau \leqslant T, \\[2ex] \exp\displaystyle\int_0^T (1 - \mu(s))\lambda(s)\,ds & \text{if } \tau > T, \end{cases}$$

where

$$\mu(s) = \begin{cases} \tilde{\lambda}(s)/\lambda(s) & \text{if } s < \tau, \\ 1 & \text{if } s \geqslant \tau. \end{cases}$$

An example of the effect of credit risk for this simple model is illustrated in Figure 11.3. Here we used $r = 0.06$, $\tilde{\lambda}(t) = \tilde{\lambda} = 0.1$ and $\delta = 0.75$. If, in addition, we have estimated from historical data that $\lambda = 0.05$, then we can infer that $\mu(t) = \mu = 2$. In this figure, the initial credit spread of 2.5%, or 0.025, is equal to the risk-neutral default intensity times the loss on default; that is, $0.025 = \tilde{\lambda} \times (1 - \delta)$.

We will now think about why it is appropriate to consider the possibility that P and Q might be different; that is, that $\lambda \neq \tilde{\lambda}$. Consider the case where $\lambda(t) = \lambda$ and $\tilde{\lambda}(t) = \tilde{\lambda}$ are constant, with $\mu = \tilde{\lambda}/\lambda$. Under Q the discounted price processes $P(t, T)/B(t)$ and $V(t, T)/B(t)$ are martingales, where $B(t) = e^{rt}$ is the usual cash account. For the present model let us introduce the compensated processes

$$M(t) = N(t) - \int_0^t \lambda(s)I(\tau > s)\,ds,$$

$$\tilde{M}(t) = N(t) - \int_0^t \tilde{\lambda}(s)I(\tau > s)\,ds,$$

where $I(\tau > s) = 1 - N(s)$ is the usual indicator random variable. Under P, $M(t)$ is a martingale. Under Q, $\tilde{M}(t)$ is a martingale. The respective stochastic differential equations are

$$\left.\begin{aligned} dM(t) &= dN(t) - \lambda(t)(1 - N(t))\,dt, \\ d\tilde{M}(t) &= dN(t) - \tilde{\lambda}(t)(1 - N(t))\,dt. \end{aligned}\right\} \tag{11.9}$$

Assume now that λ and $\tilde{\lambda}$ are constant and define

$$Z(t, T) = V(t, T)/B(t) = e^{-rT}[\delta + (1 - \delta)e^{-\tilde{\lambda}(T-t)}(1 - N(t))].$$

Then

$$dZ(t, T) = e^{-rT}(1 - \delta)\tilde{\lambda}e^{-\tilde{\lambda}(T-t)}(1 - N(t))\,dt - e^{-rT}(1 - \delta)e^{-\tilde{\lambda}(T-t)}\,dN(t)$$

$$= -e^{-rT}(1 - \delta)e^{-\tilde{\lambda}(T-t)}\,d\tilde{M}(t) \tag{11.10}$$

$$= e^{-rT}(1 - \delta)e^{-\tilde{\lambda}(T-t)}((\mu - 1)\lambda\,dt - dM(t)), \tag{11.11}$$

since the two equalities in (11.9) imply that $d\tilde{M}(t) = dM(t) + (\lambda(t) - \tilde{\lambda}(t))\,dt$ for $t < \tau$.

Since $\tilde{M}(t)$ is a martingale under Q, equation (11.10) reiterates the fact that $Z(t, T)$ is a martingale under Q. On the other hand, we can use equation (11.11) to determine the risk premium on a credit-risky bond when considering dynamics under Q. In particular, we can rewrite $dZ(t, T)$ as

$$dZ(t, T) = Z(t, T)\left[(\mu - 1)\lambda\left(1 - \frac{\delta e^{-rT}}{Z(t, T)}\right)dt - \left(1 - \frac{\delta e^{-rT}}{Z(t, T)}\right)dM(t)\right].$$

It follows that the risk premium on $V(t, T)$ (that is, the excess expected return over the risk-free rate r) is

$$(\mu - 1)\lambda\left(1 - \frac{\delta e^{-rT}}{Z(t, T)}\right).$$

(Note that Jarrow, Lando and Turnbull (1997) refer to μ as the risk premium. With the present interpretation of the term 'risk premium' it seems inappropriate to apply the term to μ.)

We refer to $\mu - 1$ as the *market price of risk* for the counting process $N(t)$. Note, though, that the exact meaning of the market price of risk here is less clear than when the term was applied to diffusion models (see Section 1.6.4 and Section 4.2). With a diffusion model there is a single aspect to the risk: volatility. Here there are two: the jump rate, λ, and the relative magnitude of the change in price at the time of a jump, $1 - \delta e^{-rT}/Z(t, T)$.

11.4.2 Multiple Credit Classes: the JLT Model

The single-credit-class model allowed us to introduce many of the main themes involved in intensity based models. Let us now move on to the more general and more realistic case popularized by Jarrow, Lando and Turnbull (1997) (JLT), where there are $n - 1$ credit ratings or classes plus default. We will, additionally, introduce a stochastic, default-free term structure. The basic structure of the model is represented in Figure 11.4, where $\lambda_{ij}(t)$ is the transition intensity under P from state i to state j at time t. We will assume that the $\lambda_{ij}(t)$ are deterministic.

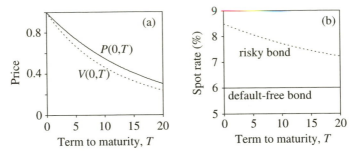

Figure 11.3. Default-free versus credit-risky term structures.
(a) Zero-coupon bond prices and (b) spot rates.

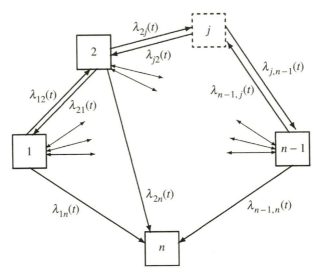

Figure 11.4. n-state model for default risk. All states potentially
communicate both ways except for the default state n, which is absorbing.

Let $X(t) \in \{1, 2, \ldots, n\}$ be the state or rating at time t. We then have, for
$i = 1, \ldots, n - 1$,

$$\Pr_P(X(t + dt) = j \mid X(t) = i)$$

$$= \begin{cases} \lambda_{ij}(t)\, dt + o(dt) & \text{for } j \neq i, \\ 1 - \sum_{j \neq i} \lambda_{ij}(t)\, dt + o(dt) = 1 + \lambda_{ii}(t)\, dt + o(dt) & \text{for } j = i, \end{cases}$$

as dt tends to 0, where $\lambda_{ii}(t) = -\sum_{j \neq i} \lambda_{ij}(t)$. Since state n (default) is absorbing,
we have $\lambda_{nj}(t) \equiv 0$ for all j and for all t.

Now let $\Lambda(t)$ be the $n \times n$ intensity matrix

$$\Lambda(t) = (\lambda_{ij}(t))_{i,j=1}^{n}.$$

Furthermore, for $s > t$, let

$$p_{ij}(t, s) = \Pr_P[X(s) = j \mid X(t) = i] \quad \text{and} \quad \Pi(t, s) = (p_{ij}(t, s))_{i,j=1}^{n}$$

be the matrix of transition probabilities. It is well known that

$$\Pi(t, s) = \exp\left[\int_t^s \Lambda(u)\, du\right],$$

where, for an $n \times n$ matrix B,

$$e^B \equiv I + \sum_{k=1}^{\infty} \frac{1}{k!} B^k.$$

In the description so far we have considered transition probabilities and intensities under the real-world measure P. There will also exist a set of deterministic transition intensities under a risk-neutral measure Q: $\tilde{\lambda}_{ij}(t)$ for $i, j = 1, \ldots, n$ with

$$\tilde{\lambda}_{ij}(t) > 0 \quad \Leftrightarrow \quad \lambda_{ij}(t) > 0 \quad \text{for all } j \neq i, \tag{11.12}$$

$$\tilde{\lambda}_{ii}(t) = \sum_{j \neq i} \tilde{\lambda}_{ij}(t),$$

$$\tilde{\lambda}_{nj}(t) = 0 \quad \text{for all } j \text{ and for all } t. \tag{11.13}$$

We also define the associated market prices of risk for $i, j = 1, \ldots, n$

$$\mu_{ij}(t) = \begin{cases} \tilde{\lambda}_{ij}(t)/\lambda_{ij}(t) & \text{if } \lambda_{ij}(t) > 0, \\ 1 & \text{otherwise.} \end{cases}$$

The condition (11.12) is necessary for P and Q to be equivalent measures. These equivalent intensities give us the equivalent matrix of transition probabilities

$$\tilde{\Pi}(t, s) = \exp\left[\int_t^s \tilde{\Lambda}(u)\, du\right].$$

Associated with this model we have the counting processes for $i, j = 1, \ldots, n$:

$$N_{ij}(t) = \text{number of transitions from state } i \text{ to state } j \text{ between } 0 \text{ and } t,$$

with

$$dN_{ij}(t) = \begin{cases} 1 & \text{if } \lim_{u \to t^-} X(u) = i \text{ and } X(t) = j, \\ 0 & \text{otherwise.} \end{cases}$$

This gives rise to a series of stopping times: for $i, j = 1, \ldots, n$ and for $k = 1, 2, \ldots$

$$\tau_{ij}(k) = \inf\{t : N_{ij}(t) = k\},$$

with $\inf \emptyset = \infty$. We also get the compensated processes $M_{ij}(t)$ and $\tilde{M}_{ij}(t)$, where

$$dM_{ij}(t) = dN_{ij}(t) - \lambda_{ij}(t)I(X(t) = i)\, dt,$$
$$d\tilde{M}_{ij}(t) = dN_{ij}(t) - \tilde{\lambda}_{ij}(t)I(X(t) = i)\, dt,$$

$$I(X(t) = i) = \begin{cases} 1 & \text{if } X(t) = i, \\ 0 & \text{otherwise.} \end{cases}$$

The processes $M_{ij}(t)$ and $\tilde{M}_{ij}(t)$ are martingales under P and Q respectively. The Radon–Nikodým derivative for the change of measure is

$$\frac{dQ}{dP}(0, T) = \left(\prod_{i,j=1}^{n} \prod_{k=1}^{N_{ij}(T)} \mu_{ij}(\tau_{ij}(k)) \right)$$
$$\times \exp\left[\int_0^T \sum_{i,j=1}^{n} (1 - \mu_{ij}(s))\lambda_{ij}(s)I(X(s) = i)\, ds \right],$$

where $\prod_{k=1}^{0} \equiv 1$.

Now assume that the default-free term structure gives us

$$dP(t, T) = P(t, T)[r(t)\, dt + S(t, T)'\, d\tilde{W}(t)]$$
$$= P(t, T)[(r(t) + S(t, T)'\gamma(t))\, dt + S(t, T)'\, dW(t)],$$

where $W(t)$ and $\tilde{W}(t)$ are Brownian motions under P and Q respectively and $\gamma(t)$ (the market price of risk) and $S(t, T)$ are previsible vectors. In addition let $B(t) = B(0)\exp[\int_0^t r(u)\, du]$ be the usual cash account.

Consider a zero-coupon corporate-bond structure where bonds maturing at time T pay 1 if default has not yet occurred or δ if default has occurred. The price for a zero-coupon bond where the credit rating of the underlying corporate entity is i is

$$V(t, T, X(t)) = E_Q\left[\frac{B(t)}{B(T)}(1 - (1 - \delta)I(X(T) = n)) \,\Big|\, \mathcal{F}_t \right]$$
$$= P(t, T)[1 - (1 - \delta)\mathrm{Pr}_Q(X(T) = n \mid \mathcal{F}_t)],$$

assuming that $X(T)$ is independent of $B(T)$.

The Market Risk Premium

We can now write down the SDE for the value $V(t, T, X(t))$ of a zero-coupon bond. Thus, given $X(t) = i$,

$$dV(t, T, X(t))$$
$$= V(t, T, i)(r(t)\, dt + S(t, T)'\, d\tilde{W}(t)) + \sum_{j \neq i}(V(t, T, j) - V(t, T, i))\, d\tilde{M}_{ij}(t)$$

$$= V(t, T, i)[(r(t) + S(t, T)'\gamma(t)) \, dt + S(t, T)' \, dW(t)]$$

$$+ \sum_{j \neq i} (V(t, T, j) - V(t, T, i))((1 - \mu_{ij}(t))\lambda_{ij}(t) \, dt + dM_{ij}(t))$$

$$= V(t, T, i)\left[(r(t) + \pi_i(t)) \, dt + S(t, T)' \, dW(t) \right.$$

$$\left. - \sum_{j \neq i} \left(1 - \frac{V(t, T, j)}{V(t, T, i)} \right) dM_{ij}(t) \right],$$

where the risk premium under P is

$$\pi_i(t) = S(t, T)'\gamma(t) + \sum_{j \neq i} (\mu_{ij}(t) - 1)\left(1 - \frac{V(t, T, j)}{V(t, T, i)} \right) \lambda_{ij}(t) \qquad (11.14)$$

$$= S(t, T)'\gamma(t) + \sum_{j \neq i} (\mu_{ij}(t) - 1)\left(1 - \frac{1 - (1 - \delta)\tilde{p}_{jn}(t, T)}{1 - (1 - \delta)\tilde{p}_{in}(t, T)} \right) \lambda_{ij}(t).$$

$$(11.15)$$

It follows that $\mu_{ij}(t) - 1$ can be regarded as the market price of credit risk for the jump risk from state i to state j, where the magnitude of the risk is the jump intensity, $\lambda_{ij}(t)$, multiplied by the relative jump in price $(V(t, T, i) - V(t, T, j))/V(t, T, i)$.

It is useful to discuss briefly the likely size and sign of the market price of credit risk. We make the following points.

- Bonds subject to default risk are inherently more risky than the equivalent default-free bonds. Therefore, it is natural for such assets to trade with a higher risk premium to compensate for the extra risk.

- Consider two bonds with identical characteristics issued by two companies with the same credit rating and identical probability laws under P for the development of their credit ratings. It is tempting to suggest that the two bonds should trade at the same price. However, they may command different market prices of risk and risk premiums for a variety of reasons, including the amount in issue of each bond and correlation with other assets.

- Countering the argument that default risk means positive risk premiums, we can argue that the impact of default risk can be reduced substantially by holding a well-diversified portfolio of credit-risky bonds. In the extreme, the additional variability in a bond portfolio caused by defaults might be eliminated by diversification (diversifiable risk). Under such circumstances equilibrium arguments suggest that default risk should be priced under the real-world measure P with no risk premiums. In reality there will be certain systematic risk factors which cause dependence between the credit ratings of different companies. Such systematic risk cannot be diversified completely and so the extent to which it remains gives rise to positive risk premiums.

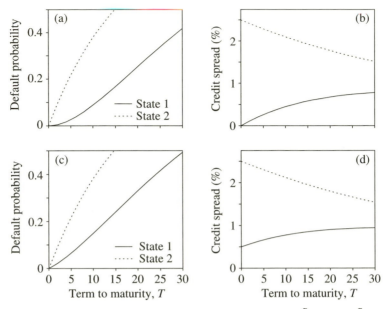

Figure 11.5. Default probabilities for the intensity matrices (a) $\tilde{\Lambda}_1$ and (c) $\tilde{\Lambda}_2$. Credit spreads when the recovery rate $\delta = 0.5$ for (b) $\tilde{\Lambda}_1$ and (d) $\tilde{\Lambda}_2$.

11.4.3 Examples

Example 11.2. We compare two sets of transition matrices (constant for all t):

$$\tilde{\Lambda}_1 = \tilde{\Lambda}_1(t) = \begin{pmatrix} -0.05 & 0.05 & 0 \\ 0.01 & -0.06 & 0.05 \\ 0 & 0 & 0 \end{pmatrix},$$

$$\tilde{\Lambda}_2 = \tilde{\Lambda}_2(t) = \begin{pmatrix} -0.05 & 0.04 & 0.01 \\ 0.01 & -0.06 & 0.05 \\ 0 & 0 & 0 \end{pmatrix}.$$

Default probabilities based upon $\tilde{\Lambda}_1$ and $\tilde{\Lambda}_2$ from each of states 1 and 2 are plotted in Figure 11.5(a),(c). We can note that some transitions out of state 1 are shifted towards default and this has the effect of raising default probabilities.

In Figure 11.5(b),(d) we plot the credit spreads against term to maturity for states 1 and 2 when the recovery rate, δ, is 0.5. The credit spread is defined as the difference between the yields to maturity (spot rates) on credit-risky and default-free zero-coupon bonds; that is,

$$\frac{1}{T} \log P(0, T) - \frac{1}{T} \log V(0, T, i) = -\frac{1}{T} \log(1 - (1 - \delta)\tilde{p}_{in}(0, T)).$$

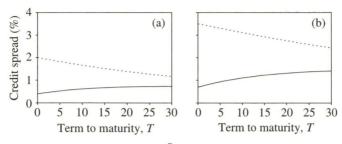

Figure 11.6. Credit spreads for $\tilde{\Lambda}_2$ when (a) $\delta = 0.6$ and (b) $\delta = 0.3$.

The principle feature of the comparison is that $\tilde{\lambda}_{13} = 0$ results in a zero credit spread for state 1 as the term to maturity tends to zero. In contrast, the possibility of an immediate default means that the credit spread is strictly above zero at all terms to maturity. Overall, the impact of the change from $\tilde{\Lambda}_1$ to $\tilde{\Lambda}_2$ is most marked for state 1.

In Figure 11.6 we concentrate on $\tilde{\Lambda}_2$ and demonstrate the effect of changing the recovery rate, δ, from 0.6 to 0.3. As might be expected, the recovery rate has a very significant impact on credit spreads.

Example 11.3 (JLT). We will now look at the more realistic model investigated by Jarrow, Lando and Turnbull (1997) with eight states. Using the Standard and Poor's notation, these are

State:	1	2	3	4	5	6	7	8
S&P Credit Rating:	AAA	AA	A	BBB	BB	B	CCC	D

Using publicly available historical data from Standard and Poor's, JLT estimated the constant transition matrix Λ to be

$$
\begin{pmatrix}
-0.1153 & 0.1019 & 0.0083 & 0.0020 & 0.0031 & 0.0000 & 0.0000 & 0.0000 \\
0.0091 & -0.1043 & 0.0787 & 0.0105 & 0.0030 & 0.0030 & 0.0000 & 0.0000 \\
0.0010 & 0.0309 & -0.1172 & 0.0688 & 0.0107 & 0.0048 & 0.0000 & 0.0010 \\
0.0007 & 0.0047 & 0.0713 & -0.1711 & 0.0701 & 0.0174 & 0.0020 & 0.0049 \\
0.0005 & 0.0025 & 0.0089 & 0.0813 & -0.2530 & 0.1181 & 0.0144 & 0.0273 \\
0.0000 & 0.0021 & 0.0034 & 0.0073 & 0.0568 & -0.1928 & 0.0479 & 0.0753 \\
0.0000 & 0.0000 & 0.0142 & 0.0142 & 0.0250 & 0.0928 & -0.4318 & 0.2856 \\
0 & 0 & 0 & 0 & 0 & 0 & 0 & 0
\end{pmatrix}.
$$

JLT separately considered how to estimate the market prices of risk, $\mu_{ij}(t)$, for all $j \neq i$, from market price data. In order for the problem to be manageable, JLT propose that, for each $i = 1, \ldots, n-1$ and for all t, $\mu_{ij}(t) = \mu_i(t) > 0$ for all j, and that the $\mu_i(t)$ are deterministic functions of time. If $\mu_i(t) > 1$, then this has the effect of speeding up the time to first exit from state i, while not affecting the distribution of the state jumped into at that time. (If, in addition, the $\mu_i(t)$ are all

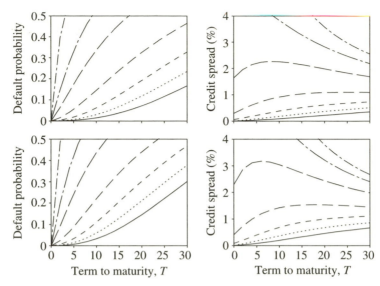

Figure 11.7. Default probabilities (left) and credit spreads (right) against term to maturity for the JLT model. The top two plots give results when the market prices of risk $\mu_i(t) - 1$ are zero for all i and t. The lower plots show the effect of introducing a market price of risk of $\mu_i(t) - 1 = 0.4$ for all i and t. The recovery rate, δ, is assumed to be 0.4. The seven curves correspond to the seven credit ratings AAA (lowest curve), AA (second lowest), A, BBB, BB, B, and CCC (highest).

equal, then the effect of a change from P to Q is merely to introduce a deterministic transformation of time with no effect on the jumps. In particular, this will reduce the time to default, thereby lowering bond prices.) We will not describe the procedure for estimating either Λ or the $\mu_i(t)$ (see, instead, Jarrow, Lando and Turnbull 1997). However, we will now investigate the numerical aspects of this model.

For simplicity, we will assume that the market price of risk does not depend either on the credit class or on time: that is, $\mu_i(t) = \mu$ for all i and t.

Results are plotted in Figures 11.7 and 11.8.

- In Figure 11.7 (top left) we see the real-world probabilities of default $p_{in}(0, T)$. The link between the credit rating and probability of default is clearly seen to be significant. For the worst-rated companies (e.g. CCC), we can see from the structure of Λ that if default occurs, then this will happen quickly, or the company will improve its credit rating and subsequently be subject to lower default intensities.

- In Figure 11.7 (bottom left) we give risk-neutral probabilities of default $\tilde{p}_{in}(0, T)$ when the market price of risk $\mu_i(t) - 1 = 0.4$ for all i and for all t. The effect of this change is most significant for AAA-rated companies.

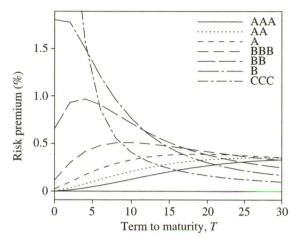

Figure 11.8. Risk premiums for the JLT model: $\delta = 0.4$; $\mu_i(t) - 1 = 0.4$ for all i and t.

Clearly, the effect is more or less marked depending on whether or not the market price of risk is higher or lower than 0.4.

- In Figure 11.7 (top right and bottom right) we have plotted credit spread as a function of term to maturity for each of the seven credit classes. The upper plot assumes that the market price of risk is zero while in the lower plot the market price of risk is 0.4. In both cases the recovery rate is 0.4 (similar to that suggested by JLT). Again, the impact of the introduction of a market price of risk is most significant for AAA-rated companies. For all credit classes, though, we can also note that the impact is more significant for bonds with a short term to maturity. This is a result of the application of the $\mu_i(t)$ speeding up all of the default processes.

- In Figure 11.8 we have plotted risk premiums (see equation (11.15) with the interest rate market price of risk $\gamma(t) = 0$) associated with default risk for $\mu_i(t) = 1.4$ and $\delta = 0.4$.

At first sight this graph is surprising, as for long-dated bonds the risk premium is lowest for what we perceive to be the riskiest bonds (CCC). However, this effect is easily explained. Over a term of, say, 30 years CCC bonds are almost certain to have defaulted. This means that their price will be approximately δ times the price of the equivalent default-free bond. If the bond then defaults in the near future, the effect on the price of the CCC bond will be quite small. So, in fact, the investment is rather less risky than a higher-rated bond with 30 years to maturity since a rerating will have a rather larger effect on its price.

11.4.4   Other Intensity-Based Models

The JLT approach assumes that there are n states with deterministic or constant transition intensities between states. An alternative approach assumes that there are just two states N and D, with the transition intensity from N to D, $\lambda(t)$, being stochastic and dependent on a separate state variable process, $X(t)$. The time of default, τ, is the first jump time in a Cox process with intensity $\lambda(t)$. In contrast to JLT, this allows $\lambda(t)$ to vary continuously in line with company fortunes and other economic factors. For example, default may be more likely when interest rates rise, so $X(t)$ should include relevant interest rates. Relevant papers taking this approach are Lando (1998) and Duffie and Singleton (1997, 1999) (see, also, Duffie and Singleton 2003). By their very nature, such a framework should allow the construction of models which combine the intensity-based approach with the structural modelling approach. In particular, the incorporation into the state-variable process, $X(t)$, of company-specific information and general economic information should give us further insight into the assessment of credit risk.

11.5   Derivative Contracts with Credit Risk

In this section we will only give a flavour of the extensive range of derivative contracts that involve credit risk. The majority of derivative contracts involving credit risk are over the counter. The terms of individual contracts vary considerably, therefore, with regard to how the contracted cashflows alter following a credit event. We will limit ourselves, therefore, to illustration of the issues with a few examples.[4]

Consider an interest rate derivative contract between parties A and B. The contract can involve two types of credit risk.

Counterparty risk, where one or both of A and B may default on all their obligations. In particular, this may result in the defaulter being unable to meet his obligations in the derivative contract being valued. In such an event the contract might be immediately cancelled with no further payments; there might be an immediate payment between A and B to close out the contract; or payments between A and B might continue but at renegotiated rates. Which of these outcomes prevails depends upon the terms of the derivative contract.

Reference risk, where A and B are considered to be free from default and the credit event is generated by a third party C. Such contracts tend to be simpler than those with counterparty risk. In particular, payments will continue after the credit event but the amounts of the individual payments will depend upon whether or not a credit event has taken place.

---

[4]More extensive coverage of the range of contracts can be found in Bielecki and Rutkowski (2002) and Duffie and Singleton (2003) and the references cited therein.

Clearly, there is scope also for contracts which involve both counterparty and reference risk.

It is straightforward to understand why contracts involving reference risk are entered into. As an example, suppose A and B are considered to be default free while C is subject to default risk. B has already entered into a derivative contract with C but the contract involves counterparty credit risk to which B does not wish to be exposed. B can therefore enter into an additional derivative contract with A where the reference event (default by C) triggers a payment or payments between A and B.

The evaluation of reference risk in the present context is very similar to the pricing of securities involving, for example, extreme weather risk or catastrophe derivatives. The simplest of such cases involves A paying B a premium at time 0 and in return B pays A an amount at the end of the year depending on whether or not a specified event has occurred during the year (see, for example, Schmock 1999; Lane 2000).

Let us focus now on contracts involving counterparty risk.

Some derivative contracts always have positive value to party A (for example, a swaption contract up to the exercise date). It follows that the terms of a contract dealing with a default event can be relatively simple. In particular, if A defaults, the contract can remain intact with no change of value. If B defaults, then the contract may be revalued to equal the recovery rate δ times the default-free value of the original contract.

Other contracts involving contracted cashflows in both directions may have a value which can fluctuate from positive to negative. The terms of the contract will, therefore, have to deal at the time of a default separately with each of A and B and with each of positive and negative contract values.

The example we will consider is a swap contract. The contract is initiated at time T_0 with regular payment dates $T_1, T_2, \ldots, T_m$, where $T_{i+1} = T_i + \tau$. Suppose that party A is the *payer* to B of the fixed rate K in return for which B (the *receiver* of the fixed rates) pays A the floating LIBOR rate. In a default-free environment this means that at time T_i (for $i = 1, \ldots, m$) A pays B $K\tau$ while B pays A $\tau L(T_{i-1}, T_{i-1}, T_i)$, where $L(t, T, S) = (S - T)^{-1}[P(t, T)/P(t, S) - 1]$ is the forward LIBOR rate (see Chapter 9).

Let $V_S^\delta(t)$ be the value of such a contract to the payer, A, at time t conditional on no default before time t and let T_A and T_B be the times of default of A and B respectively. Also let $V_S(t)$ be the value of the equivalent default-free contract. Then, for $T_i \leqslant t < T_{i+1}$,

$$V_S(t) = \frac{P(t, T_{i+1})}{P(T_i, T_{i+1})} - P(t, T_m) - K\tau \sum_{j=i+1}^{m} P(t, T_j). \qquad (11.16)$$

Bielecki and Rutkowski (2002, Chapter 14) list various contract terms for reacting to a counterparty default.

(a) If the payer, A, defaults, then payments continue between the two parties. After the default event A pays B the reduced amounts $\delta K \tau$ while B continues to pay A the full LIBOR rates originally specified in the contract. If B defaults, then A pays B in full while LIBOR payments from B to A are reduced by a factor of δ.

This type of contract can be easier to deal with, but it is perhaps unrealistic to the extent that if A defaults then B might be reluctant to continue making full payments while receiving much reduced payments in return from A.

(b) If either party defaults, then the contract is *marked to market* at the time of the default on terms defined in the original contract, resulting in a payment between A and B that closes out the contract.

We will consider approach (b) now in more detail and assume that both parties are subject to the possibility of default. We will assume that the recovery rates on default for A and B are δ_A and δ_B respectively. Recall that $V_S(t)$ is the value at time t to A of the default-free version of the contract.

Suppose that the first default time $T_D = \min\{T_A, T_B\}$ is less than T_m. One form of marking to market uses $V_S(T_D)$ and payments are made immediately at time T_D to close out the contract. (Variations on this delay the payment until the first payment date T_j after T_D with reference to either $V_S(T_D)$ or $V_S(T_j)$.)

- If A has defaulted first (that is, $T_D = T_A \leqslant T_m$), then

    if $V_S(T_D) > 0$, B pays A the full amount $V_S(T_D)$;

    if $V_S(T_D) < 0$, A pays B the reduced amount $-\delta_A V_S(T_D)$.

- If B has defaulted first (that is, $T_D = T_B \leqslant T_m$), then

    if $V_S(T_D) > 0$, B pays A the reduced amount $\delta_B V_S(T_D)$;

    if $V_S(T_D) < 0$, A pays B the full amount $-V_S(T_D)$.

Now note that each of the four cases listed above equates in value to a default-free continuation of the original interest rate swap contract scaled by the factors 1, δ_A, δ_B and 1 respectively.

For notational purposes we will write $j(t)$ to indicate the index that satisfies the inequality $T_{j(t)} < t \leqslant T_{j(t)+1}$.

The value to the payer, A, of the defaultable swap at time T_0 is then

$$V_S^\delta(T_0) = E_Q\left[\sum_{l=1}^{\min\{m, j(T_D)\}} \frac{B(T_0)}{B(T_l)}(L(T_{l-1}, T_{l-1}, T_l) - K)\tau \;\middle|\; \mathcal{F}_{T_0}\right]$$

$$+ E_Q\left[I(\{T_A = T_D \leqslant T_m\})\frac{B(T_0)}{B(T_A)}(V_S(T_A)^+ - \delta_A V_S(T_A)^-) \;\middle|\; \mathcal{F}_{T_0}\right]$$

$$+ E_Q\left[I(\{T_B = T_D \leqslant T_m\})\frac{B(T_0)}{B(T_B)}(\delta_B V_S(T_B)^+ - V_S(T_B)^-) \;\middle|\; \mathcal{F}_{T_0}\right],$$

$$(11.17)$$

where $x^+ \equiv \max\{x, 0\}$, $x^- = \max\{-x, 0\}$ and $\sum_{l=1}^0(\cdot) \equiv 0$.

Now note that

$$V_S(T_A)^+ - \delta_A V_S(T_A)^- = V_S(T_A) + (1 - \delta_A)V_S(T_A)^-$$

and

$$\delta_B V_S(T_B)^+ - V_S(T_B)^- = V_S(T_B) - (1 - \delta_B)V_S(T_B)^+.$$

We can then rearrange equation (11.17) to get

$$V_S^\delta(T_0) = E_Q\left[\sum_{l=1}^m \frac{B(T_0)}{B(T_l)}(L(T_{l-1}, T_{l-1}, T_l) - K)\tau \;\middle|\; \mathcal{F}_{T_0}\right]$$

$$+ (1 - \delta_A)E_Q\left[I(\{T_A = T_D \leqslant T_m\})\frac{B(T_0)}{B(T_A)}V_S(T_A)^- \;\middle|\; \mathcal{F}_{T_0}\right]$$

$$- (1 - \delta_B)E_Q\left[I(\{T_B = T_D \leqslant T_m\})\frac{B(T_0)}{B(T_B)}V_S(T_B)^+ \;\middle|\; \mathcal{F}_{T_0}\right].$$

$$(11.18)$$

We see from this equation that the first component is the value at time 0 of the equivalent default-free swap; that is, $V_S(0)$.

In the second component of equation (11.18),

$$E_Q\left[I(\{T_A = T_D \leqslant T_m\})\frac{B(T_0)}{B(T_A)}V_S(T_A)^- \;\middle|\; \mathcal{F}_{T_0}\right]$$

is the value of a *receiver swaption* with random exercise date T_A, provided $T_A < \min\{T_B, T_m\}$. If we assume that T_A and T_B are independent of each other (and that $\Pr_Q[T_A = T_B] = 0$) and of the stochastic, default-free term-structure model, then

$$E_Q\left[I(\{T_A = T_D \leqslant T_m\})\frac{B(T_0)}{B(T_A)}V_S(T_A)^- \;\middle|\; \mathcal{F}_{T_0}\right]$$

$$= \int_{T_0}^{T_m} E_Q\left[\frac{B(T_0)}{B(t)}V_S(t)^- \;\middle|\; \mathcal{F}_{T_0}\right] dF_A(t), \qquad (11.19)$$

where $F_A(t)$ is the defective, cumulative distribution function for $T_A < T_B$; that is, $F_A(t) = \Pr_Q[\{T_A \leqslant t\} \cap \{T_A < T_B\}]$. Finally, we note that within this integral the term $E_Q[(B(T_0)/B(t))V_S(t)^- \mid \mathcal{F}_{T_0}]$ is the value at T_0 of a default-free receiver swaption with exercise date t and payment dates $j(t)+1, \ldots, T_m$. The technology for the valuation of such a component has been established earlier, so we can insert relevant pricing formulae (see, for example, Theorem 9.7 and Exercise 11.5) into equation (11.19).

In the third component of equation (11.18),

$$E_Q\left[I(\{T_B = T_D \leqslant T_m\})\frac{B(T_0)}{B(T_B)}V_S(T_B)^+ \,\middle|\, \mathcal{F}_{T_0}\right]$$

is the value of a *payer swaption* with random exercise date T_B, provided $T_B < \min\{T_A, T_m\}$. With the same assumptions as before,

$$E_Q\left[I(\{T_B = T_D \leqslant T_m\})\frac{B(T_0)}{B(T_B)}V_S(T_B)^+ \,\middle|\, \mathcal{F}_{T_0}\right]$$
$$= \int_{T_0}^{T_m} E_Q\left[\frac{B(T_0)}{B(t)}V_S(t)^+ \,\middle|\, \mathcal{F}_{T_0}\right] dF_B(t), \qquad (11.20)$$

where $F_B(t)$ is the defective, cumulative distribution function for $T_B < T_A$; that is, $F_B(t) = \Pr_Q[\{T_B \leqslant t\} \cap \{T_B < T_A\}]$. Finally, we note that within this integral the term $E_Q[(B(T_0)/B(t))V_S(t)^+ \mid \mathcal{F}_{T_0}]$ is the value at T_0 of a default-free payer swaption with exercise date t and payment dates $j(t)+1, \ldots, T_m$.

11.6 Exercises

Exercise 11.1. Describe briefly why structural models for credit risk are not commonly used by market practitioners.

Exercise 11.2. Discuss why it is that credit ratings produced by reputable ratings agencies can be regarded as a suitable substitute for a good structural model.

Exercise 11.3. Consider the discrete-time binomial model in Section 11.3. You are given the following information about the default-free term structure and the credit-risky bonds: $r(0) = -\log P(0, 1) = 0.04$, and $r(1) = -\log P(1, 2) = 0.02$ if prices go up and 0.06 if prices go down relative to the risk-free return. Probabilities under Q are as follows.

- $\Pr_Q(\text{default-free prices rise at time } 1) = \Pr_Q(\{\omega_1, \omega_3\}) = 0.5$.
- $\Pr_Q(\text{default between times } 0 \text{ and } 1) = \Pr_Q(\{\omega_1, \omega_2\}) = 0.04$.
- The events that prices go up at time 1 and that the credit-risky bond defaults between times 0 and 1 are independent.

- If the credit-risky bond does not default before time 1, then the probability of default between times 1 and 2 is

    (i) 0.02 if default-free prices went up at time 1 (ω_3);

    (ii) 0.1 if default-free prices went down at time 1 (ω_4).

(a) Give a reason why the probability of default between times 1 and 2 depends on the change at time 1 in default-free bond prices.

(b) Find the values of $P(0, 1)$ and $P(0, 2)$.

(c) Show that $V(0, 1) = 0.912\,750$, $v_u = 0.960\,595$ and $v_d = 0.847\,588$.

(d) Find the value of $V(0, 2)$.

(e) For each of the four derivatives $i = 1, 2, 3, 4$, under which derivative i pays 1 at time 1 on outcome ω_i, calculate its price at time 0.

(f) Calculate the price of a European call option with underlying $V(t, 2)$, exercise date 1 and strike price 0.8.

Exercise 11.4. In the JLT model, what restrictions on the $\mu_{ij}(t)$ are sufficient to ensure that P and Q are equivalent.

Exercise 11.5. Consider the swap valuation formulae (11.18)–(11.20). In Theorem 9.7 we presented a formula for the pricing of a swaption when the exercise date is exactly time τ before a payment date. Equations (11.19) and (11.20) require a swaption valuation formula for when the exercise date falls between two payment dates. We will do this in the present exercise.

Suppose that $T_j < t < T_{j+1}$. Recall, now, equation (11.16) and define the swap rate $K_\tau(t, T_j, T_m)$ to be that at which $V_S(t) = 0$; that is,

$$K(t) = K_\tau(t, T_j, T_m) = \left(\frac{P(t, T_{j+1})}{P(T_j, T_{j+1})} - P(t, T_m) \right) \bigg/ \tau \sum_{l=j+1}^{m} P(t, T_l).$$

(a) Show that this implies

$$V_S(t) = (K(t) - K)\tau \sum_{l=j+1}^{m} P(t, T_l).$$

(b) Now return to Section 9.4. Consider a forward swap contract with fixed rate K which commences at time T_j and runs to time T_m. Consider next a swaption contract with exercise rate K and exercise date t such that $T_j < t < T_{j+1}$. In Chapter 9 we introduced the probability measure P_X under which the prices of all tradable assets divided by $X(u) = \tau \sum_{l=j+1}^{m} P(u, T_i)$ are martingales.

(i) Show that this martingale property applies not just to times $u < T_j$ but also to times $T_j \leqslant u < T_{j+1}$.

(ii) Show that $K(u)$ is a martingale under P_X for all $u < T_{j+1}$.

(c) Assume that $dK(u) = K(u)\gamma(u)\,dW_X(u)$, where $\gamma(u)$ is some deterministic function and $dW_X(u)$ is a standard Brownian motion under the measure P_X.

(i) For $T_j \leqslant t < T_{j+1}$ establish the distribution under P_X of $K(t)$ given $\mathcal{F}_{T_0}$.

(ii) Hence derive a formula for the price at time T_0 of the swaption contract with exercise date t.

Exercise 11.6. Estimation with complete information.

(a) Consider the two-state model in Figure 11.2 with a constant default intensity $\lambda(t) = \lambda$ per annum. We start at time 0 with n companies in state N. Over the next τ years m companies default at times $T_1, \ldots, T_m$; the remainder do not default. Show that the log-likelihood function for λ is

$$l(\lambda) = m \log \lambda - \lambda \left(\sum_{i=1}^{m} T_i + (n-m)\tau \right)$$

and that the maximum likelihood estimator (MLE) for λ is

$$\hat{\lambda} = \frac{m}{\sum_{i=1}^{m} T_i + (n-m)\tau}.$$

(b) Now consider the n-state JLT model with constant transition intensities λ_{ij}. There are two types of event.

(1) Type (i, j) events, where a company jumps from state i to state j. There are n_{ij} occurrences of this event and the only relevant information is that, for the company involved in a specific jump k (for $k = 1, \ldots, n_{ij}$), the duration spent in state i during the last visit to that state before the jump was T_{ijk}.

(2) Type (i) events, where a company is still in state i at the end of the period of observation. There are m_i occurrences of this event. For company k still in state i (for $k = 1, \ldots, m_i$), the only relevant information is that S_{ik}, the time spent in state i since it last arrived in this state.

(i) Show that the log-likelihood function for the transition intensity matrix Λ is

$$l(\Lambda) = \sum_{i=1}^{n-1} \sum_{j=1, j\neq i}^{n} n_{ij} \log \lambda_{ij} + \sum_{i=1}^{n-1} \lambda_{ii}(T_i + S_i),$$

where

$$T_i = \sum_{j=1, j\neq i}^{n} \sum_{k=1}^{n_{ij}} T_{ijk} \quad \text{and} \quad S_i = \sum_{k=1}^{m_i} S_{ik}.$$

(ii) Find the MLEs for the λ_{ij}, $i = 1, \ldots, n-1$, $j = 1, \ldots, n$, $j \neq i$.

(iii) Transitions for 1000 companies were simulated over a 10-year period. There are four credit classes: A (initially 412 companies), B (366), C (222) and D. The complete simulation is summarized as follows:

$n_{12} = 281,$	$n_{13} = 120,$	$n_{14} = 25,$
$n_{21} = 190,$	$n_{23} = 182,$	$n_{24} = 99,$
$n_{31} = 19,$	$n_{32} = 96,$	$n_{34} = 215,$
$T_1 = 1424.5,$	$T_2 = 1716.5,$	$T_3 = 1081.3,$
$S_1 = 1328.1,$	$S_2 = 1647.1,$	$S_3 = 1046.6.$

Find the MLE for Λ given these data.

(c) Use the estimated transition matrix in part (b) to calculate and plot credit spreads and risk premiums associated with credit risk for each of classes A, B and C, for range of terms to maturity.

Assume that the recovery rate is fixed and that initially it is $\delta = 0.6$. Also assume initially that the $\mu_{ij}(t) = \mu_i$ are constant and all equal to 1 (so the credit-risk premiums will be zero initially).

Now investigate the sensitivity of these results to variation in the values of δ and the μ_i.

12
Model Calibration

In earlier chapters we have described models that require input of the entire term structure of interest rates or an entire set of zero-coupon bond prices (for example, the no-arbitrage models in Chapter 5 including the HJM framework, and the market models in Chapter 9). We can also take this input one step further by deriving interest rate volatility from selected traded derivatives. As described in earlier chapters, this ensures that (a) market makers price derivatives in a way which is consistent with the market and (b) as interest rates evolve through time the quoted prices of derivatives will change in line with other market makers.

In this chapter we will concentrate more on the first of these inputs; that is, one of zero-coupon prices, the spot-rate curve or the forward-rate curve. We will conclude with a discussion of the issues surrounding calibration of the volatility structure.

12.1 Descriptive Models for the Yield Curve

A descriptive model takes a snapshot of the bond market as it is today. Generally, there is no reference to price data from other dates. The sole aim is to get a good description of today's prices; that is, of the rates of interest that are implicit in today's prices.

A descriptive model, on its own, gives us no indication of how the term structure might change in the future. We know that there is randomness in the future but this sort of model does not describe this feature. The description of the dynamics of the term structure falls into either the domain of arbitrage-free models or more general actuarial and econometric models that are not necessarily arbitrage free but which pay more attention to past history (see, for example, Wilkie 1995; Mills 1993).

Descriptive models have a number of uses.

- They can be used to assess which bonds are over- or under-priced (so called *cheap/dear* analysis).

- They give us a broad picture of market rates of interest which are implied by market prices (and, in particular, if the market contains only coupon bonds).

For example, see Nelson and Siegel (1987), Svensson (1994), Dalquist and Svensson (1996) and Cairns (1998).

- They can be used to price forward bond contracts.

- They can assist in the analysis of monetary policy (Dalquist and Svensson 1996).

- A forward-rate curve can be used as part of the input to a dynamic stochastic model based on the Heath, Jarrow and Morton (1992) framework or the more recent positive interest framework of Flesaker and Hughston (1996). Of course, here, it is also necessary to specify a volatility structure (for example, see Jarrow 1996). Once this has been added to the fitted curve using the descriptive model, we have a full model that describes the dynamics of the term structure. The input forward-rate curve is often recalibrated each day.

- They can be used in the construction of yield indices (Feldman et al. 1998).

- Finally, descriptive models provide sufficient information for us to get a precise market value of a non-profit insurance portfolio or to price, for example, annuity contracts.

Typically, it is much more important that the fitted curves are accurate when they are being used as part of the input for derivative pricing than for analysis of monetary policy. A curve which results in errors, say, of 10% in the prices of traded derivatives is not going to be of great help in pricing an over-the-counter derivative.

12.2 A General Parametric Model

With the majority of parametric models the aim is to achieve a parsimonious model of the yield curve. Thus, we aim to capture as much as possible of the structure of market interest rates with as few parameters as possible. Clearly, these are conflicting objectives and the models we will now describe attempt to achieve what the authors believe to be a balance between goodness of fit on the one hand and parsimony on the other. Later, in Section 12.4, we will describe the use of splines (rather than parametric curves), where greater emphasis is placed on goodness of fit at the expense of the requirement that we have to estimate more parameters.

The exponential-polynomial class for forward rates

This is a very general class of forward-rate curves (see Björk and Christensen 1999). They are made up of a combination of polynomials and exponentials with different rates of decay. Let $T = t + s$, where $s > 0$. Then we have

$$f(t, t + s) = L_0(t, s) + \sum_{i=1}^{n} L_i(t, s) \exp(-c_i(t)s),$$

where

$$L_i(t, s) = \sum_{j=0}^{k_i} b_{ij}(t)s^j.$$

Well-documented special cases are

$$f(t, t + s) = b_0(t) + (b_{10}(t) + b_{11}(t)s)\exp(-c_1(t)s)$$
$$\text{(Nelson and Siegel 1987)},$$

$$f(t, t + s) = b_0(t) + (b_{10}(t) + b_{11}(t)s)\exp(-c_1(t)s) + b_{21}(t)s\exp(-c_2(t)s)$$
$$\text{(Svensson 1994)},$$

$$f(t, t + s) = b_0(t) + \sum_{i=1}^{n} b_i(t)\exp(-c_i(t)s)$$
$$\text{(Wiseman 1994)},$$

$$f(t, t + s) = b_0(t) + \sum_{i=1}^{n} b_i(t)\exp(-c_i s)$$
$$\text{(Cairns (1998), restricted-exponential model)}.$$

In these models, any parameters that are functions of t are estimated as frequently as is required for the application in hand. This gives us the key distinction between the Wiseman (1994) model (where the $c_i(t)$ are estimated at each point in time) and the Cairns (1998) model (where the c_i are not to be re-estimated). In the latter model, only the linear b_i coefficients are estimated, while the c_i are fixed at the outset. This makes estimation of parameters much more reliable (and much quicker in the case of a zero-coupon bond market). With the Svensson (1994) model (which is popular with economists and central bankers), if it is felt that $c_1(t)$ and $c_2(t)$ might be very close to one another, then it is convenient to parametrize the model in the following form:

$$f(t, t + s) = b_0(t) + (b_{10}(t) + \bar{b}_{11}(t)s)e^{-c_1(t)s} + \frac{\bar{b}_{21}(t)s(e^{-c_1(t)s} - e^{-c_2(t)s})}{c_2(t) - c_1(t)}.$$

This prevents $\bar{b}_{21}(t)$ from becoming very large if $c_2(t)$ is very close to $c_1(t)$.

12.3 Estimation

Suppose that we have N zero-coupon bonds maturing at times $T_1, \ldots, T_N$. Let $R_i = R(t, T_i)$ be the observed spot rates. For a given form for $f(t, t + s)$ we have

$$R(t, t + s) = \sum_{i,j} b_{ij}(t)d_{ij}(c(t), s),$$

where

$$d_{ij}(c(t), s) = \frac{1}{s} \int_0^s u^j \exp(-c_i(t)u) \, du,$$

$$c(t) = (0, c_1(t), \ldots, c_n(t))',$$

or

$$R(t, t+s) = R(t, t+s)(b, c)$$
$$= b'd(c, s).$$

In the final step above we assume that the arrays for b and d are arranged into vectors. This simplifies computational procedures without affecting the numerical results.

For example, with the restricted-exponential model:

$$b = (b_0, b_1, \ldots, b_4)',$$

$$d(c, s) = \left(1, \frac{(1 - \exp(-c_1 s))}{c_1 s}, \ldots, \frac{(1 - \exp(-c_4 s))}{c_4 s}\right)'.$$

We aim to estimate the $R(t, t+s)$ which fits best the observed R_i. A common approach is the use of weighted least squares. Thus we minimize

$$S(b, c) = \sum_{i=1}^{N} w_i (R_i - R(t, T_i)(b, c))^2 = \sum_{i=1}^{N} w_i (R_i - b'd(c, T_i - t))^2,$$

where the w_i are the weights. In some applications the weights are all equal. In others (see, for example, Cairns 1998; Cairns and Pritchard 2001) the weights diminish to zero as the term to maturity of the security approaches zero.

As alternatives to the least-squares approach, Cairns (1998) and Cairns and Pritchard (2001) investigate the use of both maximum likelihood and full Bayesian methods. These approaches are shown to produce similar results in general to least squares, but have the useful characteristic of injecting a degree of intellectual rigour into the estimation process, notably absent previously. An important, additional feature of the maximum likelihood and Bayesian approaches is that they allow the construction of confidence intervals for both parameters and the various yield curves (see, for example, Cairns 1998; Sack 2000).

For simplicity, in the present chapter we will restrict ourselves to the weighted least-squares approach. Since $R(t, T)$ is linear in b we can get a simple solution for b given c: denoted $\hat{b}(c)$. Thus we get

$$\hat{b}(c) = \left(\sum_{i=1}^{N} w_i d(c, T_i - t)d(c, T_i - t)'\right)^{-1} \left(\sum_{i=1}^{N} R_i d(c, T_i - t)\right).$$

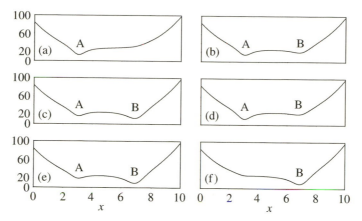

Figure 12.1. Example of a problem with the minimization algorithm; a catastrophic jump occurs between days 5 and 6. (a) Day 1, $S(1, x)$; (b) day 2, $S(2, x)$; (c) day 3, $S(3, x)$; (d) day 4, $S(4, x)$; (e) day 5, $S(5, x)$; (f) day 6, $S(6, x)$.

We are left with the problem of minimizing

$$S(\hat{b}(c), c) = \sum_{i=1}^{N} w_i (R_i - \hat{b}(c)'d(c, T_i - t))^2.$$

Under the restricted-exponential model we do not estimate the c_i so we are finished, and we have a unique solution $\hat{b}(c)$.

Where we also need to estimate the c_i, $S(\hat{b}(c), c)$ may have more than one maximum over the possible range of values for c. The result is that the estimates for b and c might jump about in a way which is inconsistent with price changes.

For example, in Figure 12.1, we have a stylized picture of what might happen (Cairns 1998). Suppose that the minimization algorithm starts from yesterday's minimum and finds the nearest local minimum. On day 1 the function to be minimized has a unique minimum at A and there is no problem. On day 2 a second minimum at B appears but A is still the global minimum. On day 3 B is lower but the algorithm stays at A. Days 4 and 5 tilt back and forward again with the located minimum remaining always at A regardless of whether or not it is the global minimum. Finally on day 6 the minimum at A disappears and the algorithm finds B. This causes a significant jump in the parameter estimates and also, for example, in the fitted forward-rate curve.

In contrast, the Cairns (1998) model is known to provide us with a unique minimum when we have zero-coupon bond data. Where we are provided with coupon-bond data Cairns and Pritchard (2001) demonstrate that although there may be a small risk of multiple minima, this risk is much smaller than that for the Nelson–Siegel and Svensson models.

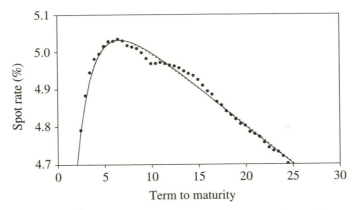

Figure 12.2. Zero-coupon bond data for 28 June 2002. Dots: observed spot rates for 7 June and 7 December maturity dates (source www.dmo.gov.uk). Svensson model with solid spot-rate curve: $c_1 = 0.510, c_2 = 0.123$. Dotted curve: $c_1 = 0.098, c_2 = 0.474$. Values for (c_1, c_2) correspond to the local minima in Figure 12.3. The solid and dotted curves are very similar despite the differences in the parameter values.

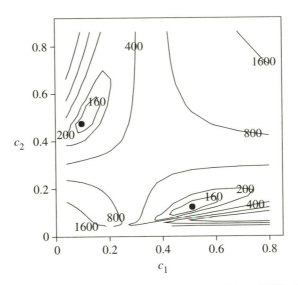

Figure 12.3. Svensson model, zero-coupon bond data for 28 June 2002. Contours for the least-squares function $S(\hat{b}(c), c)$ (with rates of interest used in $S(b, c)$ expressed in basis points) over a range of values for c_1 and c_2. The two black dots identify the locations of the two local minima.

A real example of this problem is given in Figures 12.2 and 12.3 where we use the Svensson (1994) model. Actual spot rates for 28 June 2002 (derived from STRIPS prices) are shown in Figure 12.2 as dots; the Svensson (1994) model was fitted to

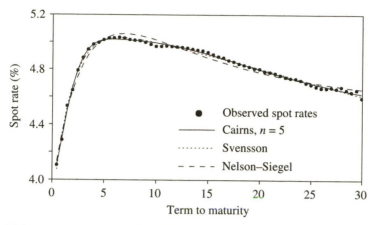

Figure 12.4. Zero-coupon bond data for 28 June 2002: Dots: observed spot rates for 7 June and 7 December maturity dates (source www.dmo.gov.uk). Solid spot-rate curve: Cairns model with $n = 5$, $c = (0.2, 0.4, 0.6, 0.8, 1.0)$. Dotted curve: Svensson model with $c = (0.098, 0.474)$. Dashed curve: Nelson–Siegel model with $c = 0.315$. Cairns and Svensson model results are very similar. Nelson–Siegel model with fewer parameters fits less well.

this data using least squares with equal weights. In Figure 12.3 we have plotted contours for the partially optimized, least-squares function $S(\hat{b}(c), c)$.

From this plot it can be seen that there are two local minima at $c_A = (0.510, 0.123)$ and $c_B = (0.098, 0.474)$. The two solutions fit the data almost equally well. Spot-rate curves corresponding to these two solutions are plotted in Figure 12.2. We can see that, although the two solutions c_A and c_B are far apart, they give very similar spot-rate curves, differing by, at most, 1 basis point. This difference may seem small but could lead to more substantial differences in derivative prices. We can note additionally that this difference is rather smaller than the errors between the observed and fitted values. On other occasions these differences between c_A and c_B can be more marked. When this happens we need to be careful to identify the risk that the optimal solution jumps from one local minimum to another.

In Figure 12.4 we plot the fitted curves for the models of Svensson (with $c_B = (0.098, 0.474)$), Cairns (with $n = 5$ and $c = (0.2, 0.4, 0.6, 0.8, 1.0)$) and Nelson–Siegel (with $c_1 = 0.315$). In general, we can see that none of the models is capable of fitting the 'kink' in the spot-rate curve at 10 years. However, the Cairns and Svensson models clearly manage to get a better fit at this duration than the Nelson–Siegel model. Since the Nelson–Siegel model (four parameters) is a special case of the Svensson model (six parameters), it fits less well over the full range (root mean squared error, RMSE = 2.8 versus RMSE = 1.5 basis points). The Cairns model, where we estimate six rather than four parameters, provides us with a significantly better fit again (RMSE = 1.1 basis points) than the Svensson model.

12.4 Splines

As an alternative to the parametric models described above, the method of splines is commonly employed. The basic idea of splines is that the fitted spot-rate curve (or some other curve of interest) is a linear combination of some base functions $B_1(s), \ldots, B_N(s)$; that is,

$$R(s) \equiv R(t, t + s) = \sum_{i=1}^{N} \beta_i(t) B_i(s). \qquad (12.1)$$

Splines can be easily applied to the forward-rate curve, the par-yield curve or the zero-coupon price curve.

A spline that is described as having *order k* means that $R(s)$ has continuous derivatives up to order $k - 1$. Throughout this section, we will consider splines of order 3; that is, the fitted curve will have continuous first and second derivatives.

12.4.1 Cubic Splines

For many applications the basic cubic splines are used. Suppose that we have a set of m bonds which mature at times $0 < T_1 < \cdots < T_m$. The first thing that we need to specify are the *knots*: $\xi_0 < \cdots < \xi_n$. We first require that $\xi_0 < T_1$ and $\xi_n > T_m$. Additionally, ξ_0 is often set to 0. The base functions are then

$$B_{-3}^c(s) = 1, \qquad B_{-2}^c(s) = s, \qquad B_{-1}^c(s) = s^2, \qquad B_0^c(s) = s^3,$$

and, for $k = 1, \ldots, n - 1$,

$$B_k^c(s) = (s - \xi_k)_+^3.$$

The spot-rate curve is then $R(s) = \sum_{i=-3}^{n-1} \beta_i B_i^c(s)$ for $\xi_0 < s < \xi_n$. At the knots $\xi_1, \ldots, \xi_{n-1}$, $R(s)$ has continuous first and second derivatives but jumps of size $6\beta_k$ in the third derivative.

Cubic splines are not, however, regarded as very suitable for fitting interest rates as the base functions become unbounded as $s \to \infty$.

12.4.2 B-Splines

A much-more-widely used form of spline is the B-spline. These require the additional specification of three knots below ξ_0 and three above ξ_n giving us $\xi_{-3} < \xi_{-2} < \cdots < \xi_{n+2} < \xi_{n+3}$.

The splines are defined as follows. For $p = -3, \ldots, n - 1$,

$$B_p(s) = \sum_{j=p}^{p+4} \left(\prod_{i=p, i \neq j}^{p+4} \frac{1}{\xi_i - \xi_j} \right) (s - \xi_j)_+^3.$$

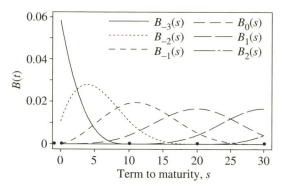

Figure 12.5. B-splines, $B_k(s)$, for knots at $-3, -2, -1$, 0, 10, 20, 30, 40, 50 and 60. The dots show the location of the knots.

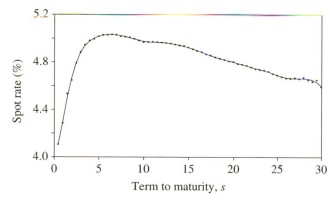

Figure 12.6. Spot rates on 28 June 2002. Fitted B-spline using evenly spaced knots at $-6, -4, -2, 0, 2, \ldots, 28, 30, 32, 34, 36$.

With this formulation $B_p(s)$ is zero outside the interval (ξ_p, ξ_{p+4}). An example showing the base functions when knots are placed at $-3, -2, -1, 0, 10, 20, 30, 40$, 50 and 60 is given in Figure 12.5. A key contrast with cubic splines is that the base functions are bounded.

12.4.3 Estimation

Two forms of estimation are employed.

The first of these is essentially the same as that used for the parametric model: that is, fit the spot-rate curve using weighted least squares. This approach is sometimes referred to as *regression splines* for the simple fact that, once the knots have been specified, the problem comes down to a straightforward multiple linear regression.

An example of a regression spline fitted to STRIPS data on 28 June 2002 is plotted in Figure 12.6. Compared with Figure 12.4 we can see that the substantially greater number of parameters in combination with the flexibility of splines means that we get a much better fit (RMSE = 0.7 basis points compared with the best of the parametric models). In particular, the spline curve fits the kink at around 10 years quite well.

One problem with the use of splines is that results can be quite sensitive to the number and placement of the knots. The subjectivity surrounding this decision can be reduced by ensuring that the number of knots is rather less than the number of observed spot rates and that the intervals between consecutive knots contain roughly equal numbers of observed spot rates. In Figure 12.6 the regularity of the STRIPS data makes this straightforward, with four points in each interval. If, on the other hand, we are using coupon-bond data, then the observations tend to be quite unevenly spaced and so greater care must be taken when choosing the location of the knots.

A second problem is that splines with a large number of knots can end up fitting the data 'too' well; that is, they begin to eliminate errors which are, in fact, genuine errors in observed spot rates. Such errors do exist in practice because of liquidity problems in the market and bid–offer spreads. The result of this is that the fitted spot-rate curve can, in practice, become rather more lumpy than we believe to be the case. This is difficult to see in the spot-rate curve in Figure 12.6, except above 25 years to maturity. The lumpiness is much more evident when we look at the implied forward-rate curve in Figure 12.7(b) (solid curve). The bulge at around 11 years is, perhaps, a genuine feature related to the kink at 10 years in the spot-rate curve. However, the implied forward-rate curve becomes unstable above $s = 20$ in a way which does not seem realistic or reasonable.

The solution to the latter problem is to use *smoothing splines* (also known as P-splines). This approach, popularized by Fisher, Nychka and Zervos (1995), combines the least-squares function with a component that penalizes curves which are too rough in some sense. The result is that the optimization procedure favours curves which are smooth. The penalty function proposed by Fisher, Nychka and Zervos (1995) measures the total curvature of the fitted spot-rate curve

$$C_S(\beta) = \int_{\xi_0}^{\xi_n} (R''(s))^2 \, ds. \tag{12.2}$$

We then minimize over the β_k the objective function

$$S(\beta) = \sum_{i=1}^{N} \left[R_i - \sum_{k=-3}^{n-1} \beta_k B_k (T_i - t) \right]^2 + \lambda C_S(\beta).$$

The parameter λ is first set in a way which avoids the minimization being dominated by either the first or the second component. Second, it is varied to attach more or less importance on the roughness.

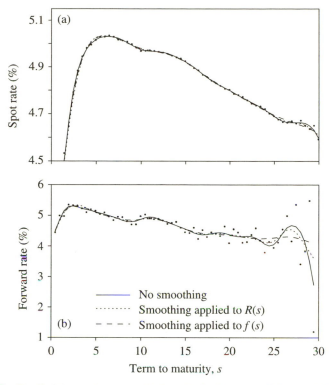

Figure 12.7. Implied forward rates on 28 June 2002. Dots: actual 6-month forward rates derived from zero-coupon bonds maturing at 6-monthly intervals. Solid line: implied forward-rate curve with no smoothing. Dotted curve: implied forward-rate curve with a roughness penalty applied to the spot-rate curve. Dashed line: implied forward-rate curve with a roughness penalty applied to the forward-rate curve.

Now, given the knots have been specified, $R''(s)$ is linear in the β_k. Hence, $S(\beta)$ is quadratic in the β_k meaning that it is straightforward to minimize. (In theory, this involves inversion of an $(n+3) \times (n+3)$ matrix. This matrix is often nearly singular, resulting in computational problems. In practice, therefore, numerical methods are employed to minimize $S(\beta)$.)

As an alternative, we can aim for a smooth implied forward-rate curve using the penalty function

$$C_F(\beta) = \int_{\xi_0}^{\xi_n} (f''(s))^2 \, ds, \tag{12.3}$$

where $f(s) = f(t, t+s)$.

Smoothing splines fitted to STRIPS data on 28 June 2002 are illustrated in Figure 12.7: spot-rates in the upper graph and implied forward rates in the lower graph. For comparison, the unsmoothed regression spline is also plotted. In both graphs the dotted curve shows the result if a significant penalty is applied when the fitted spot-rate curve is rough. It is difficult to see the effect of this on the spot rate itself. The implied forward-rate curve, however, is still lumpy at the long end. The dashed curve shows the results when the penalty function measures the roughness of the implied forward-rate curve. Here the effect of the penalty function is much more evident: the implied forward-rate curve is much more smooth.

12.5 Volatility Calibration

The topic of how to calibrate the volatility component of a model is an important one. However, the topic is also a very extensive one, with different methods being required for each model. Here, then, we will just mention a few issues and point to some further references.

The simplest type of example to consider is whether the model is intended for pricing and hedging a single derivative contract or for pricing and hedging in a consistent way a range of derivatives. In the former case, calibration of the volatility is relatively simple. Consider, for example, the Hull and White model (Section 5.2.2). There we priced, as a simple example, a European call option to be exercised at time T with strike price K and underlying $P(t, S)$. The formula for the price of this option (equation (5.2)) was

$$V(t) = P(t, S)\Phi(d_1) - KP(t, T)\Phi(d_2),$$

where

$$d_1 = \frac{1}{\sigma_p} \log \frac{P(t, S)}{KP(t, T)} + \frac{\sigma_p}{2}, \qquad d_2 = d_1 - \sigma_p,$$

$$\sigma_p = \frac{\sigma}{\alpha}(1 - e^{-\alpha(S-T)})\sqrt{\frac{1 - e^{-2\alpha(T-t)}}{2\alpha}}.$$

In this equation $P(t, T)$ and $P(t, S)$ can be obtained directly from STRIPS prices or by using a descriptive yield curve model as discussed earlier in this chapter. The remaining component is σ_p. In a similar fashion to equity derivatives, if we know the price, $V(t)$, of the derivative, then we can calculate the implied cumulative variance σ_p^2. This implied value is then used to derive the dynamic hedging strategy for the same derivative.

A more interesting situation occurs when we wish to match the prices of a range of traded derivative contracts (for example, caplets with a range of maturity dates). There are two aims. First, we wish to price and hedge these derivatives in a consistent fashion. Second, we wish to be able to price and hedge, in a consistent way, a range of over-the-counter derivatives with similar characteristics to the original traded contracts.

This problem is tackled directly by Black and Karasinski (1991). In their model (equation (5.3)) the volatility of $r(t)$ is $\sigma(t)r(t)$, where $\sigma(t)$ is a deterministic function. Black and Karasinski propose that $\sigma(t)$ can be calibrated directly from knowledge of current spot-rate volatilities and caplet prices using numerical methods. This is discussed further in Rebonato (1996).

The same problem can be tackled using the Heath–Jarrow–Morton framework. In the one-factor model we have

$$\mathrm{d}P(t, T) = P(t, T)[r(t)\,\mathrm{d}t + S(t, T)\,\mathrm{d}\tilde{W}(t)].$$

The function $S(t, T)$ gives us what is called the *volatility term structure*. In the Gaussian case $S(t, T)$ is a deterministic function of t and T. In other cases $S(t, T)$ might also depend upon the current term structure of interest rates. The Hull and White model is an example of a Gaussian model (Section 5.2.2). There we had $S(t, T) = -\sigma(1 - \mathrm{e}^{-\alpha(T-t)})/\alpha$. However, we can be much more flexible with $S(t, T)$ while still retaining the Gaussian nature of the term structure of interest rates. In particular, we can estimate the deterministic function $S(t, T)$ from a range of current derivative prices.

As a final remark, we can recall that the market-models framework (Chapter 9) is set up in a way which makes calibration very straightforward. First, we model directly observable LIBOR rates. Second, the cumulative volatility is derived from caplet or swaption prices. Market models are generally set up in a way which has as many factors as there are caplets. This ensures that all derivatives are priced consistently with one another.

12.6 Exercises

Exercise 12.1. Consider the estimation of parameters using zero-coupon-bond price data for the Wiseman (1994) exponential model for the forward-rate curve. Will the same problem as with the Svensson (1994) model arise, where the objective function to be minimized might have more than one local minimum?

Exercise 12.2. Explain why the Vasicek (1977) arbitrage-free model for the term structure of interest rates is a special case of the Cairns (1998) restricted-exponential model. What constraints are placed on the parameters $b_i(t)$ and on the c_i?

Exercise 12.3. Suppose that you are provided with price data (and full technical information) for coupon bonds with no option characteristics. Use one of the parametric models for the forward-rate curve to develop formulae for estimated prices. Then write down a suitable objective function (for example, based on weighted least squares) to be optimized.

Exercise 12.4. Consider least-squares estimation using STRIPS data. On the one hand, we can perform least-squares estimation using the observed spot rates, R_i, attaching equal weight to each observation. On the other hand, we can perform weighted least-squares estimation using the observed zero-coupon prices, P_i. Use a Taylor expansion to show that weights of $w_i = 1/P_i^2$ will give approximately the same answer as the spot-rate estimation process.

Exercise 12.5. Recall the simple relationship between forward rates and spot rates:

$$f(t, t + s) = \frac{\partial}{\partial s} s R(t, t + s).$$

Use this identity to derive a relationship between the two roughness penalty functions, $C_S(\beta)$ and $C_F(\beta)$ in equations (12.2) and (12.3). Discuss this relationship.

Exercise 12.6. Suggest some other ways of measuring roughness as alternatives to equations (12.2) and (12.3).

Exercise 12.7. Suppose that you are provided with spot-rate data that has evenly spaced maturity dates $t_{i+1} = t_i + \Delta t$, where, for example, $\Delta t = 6$ months. Write down a discrete-time approximation to the integral in equation (12.2).

Exercise 12.8. The use of a roughness penalty function is a way of incorporating prior beliefs about the general shape of the spot-rate or forward-rate curve. Another prior belief could be that the spot-rate curve is always strictly positive.

 (a) Describe a possible drawback of this constraint. Suggest an alternative with a similar aim.

 (b) What other prior beliefs might you wish to incorporate?

Appendix A
Summary of Key Probability and SDE Theory

In this appendix we will summarize the basic mathematical toolkit required for this book. For further details the reader is referred to, for example, Karatzas and Shreve (1991), Øksendal (1998) and Mikosch (1998).

A.1 The Multivariate Normal Distribution

Consider a set of random variables $X = (X_1, \ldots, X_n)'$, where $E[X_i] = \mu_i$ and $\mathrm{Cov}[X_i, X_j] = v_{ij}$ for $i, j = 1, \ldots, n$. Let $\mu = (\mu_1, \ldots, \mu_n)'$ and $V = (v_{ij})_{i,j=1}^n$ be the $n \times n$ covariance matrix. We will assume that V is non-singular.

X has a *multivariate Normal distribution* if it has the density function

$$f(x) = (2\pi)^{-n/2}|V|^{-1/2} \exp[-\tfrac{1}{2}(x - \mu)'V^{-1}(x - \mu)].$$

This is sometimes written as $X \sim \mathrm{MVN}(\mu, V)$.

The multivariate *Laplace transform* of X is

$$L(\theta_1, \ldots, \theta_n) = E\left[\exp\left(-\sum_{i=1}^n \theta_i X_i\right)\right] = E[e^{-\theta'X}] \qquad \text{(definition)}$$

$$= \exp[-\theta'\mu + \tfrac{1}{2}\theta'V\theta].$$

If V is non-singular, then there is a unique $n \times n$ matrix S which is lower triangular (that is, the (i, j) element $s_{ij} = 0$ for all $j > i$) with $V = SS'$. S is called the *Cholesky decomposition* of V. We can then write $X = \mu + SZ$, where $Z = (Z_1, \ldots, Z_n)'$ and the Z_i are independent and identically distributed (i.i.d.) standard normal random variables (that is, $Z_i \sim N(0, 1)$).

A.2 Brownian Motion

Definition A.1. A stochastic process $\{W(t) : 0 \leqslant t < \infty\}$ is called a *standard Brownian motion* if

1. it has continuous sample paths;

2. $W(0) = 0$;

3. it has *independent, Normally distributed increments* such that for any $0 < t_1 < t_2 < \cdots < t_n$ the vector of random variables $(X_1, \ldots, X_n)$, where $X_i = W(t_i) - W(t_{i-1})$, has a multivariate Normal distribution with $E[X_i] = 0$, $\mathrm{Var}[X_i] = (t_i - t_{i-1})$ and $\mathrm{Cov}[X_i, X_j] = 0$ for all $i \neq j$—in particular, $W(t)$ is Normally distributed with mean 0 and variance t.

Now let $(\Omega, \mathcal{F}, P)$ be a probability triple associated with $W(t)$. Let $\mathcal{F}_t = \sigma(\{W(u) : 0 \leqslant u \leqslant t\})$ be the sigma-algebra generated by $W(u)$ up to time t. We can interpret $\mathcal{F}_t$ as the history of the process up to time t.

We can note the following properties of standard Brownian motion.

- $W(t)$ is a Markov process; that is, the law of $\{W(u) : u \geqslant t\}$ given $\mathcal{F}_t$ is the same as the law of $\{W(u) : u \geqslant t\}$ given $W(t)$ only.
- $W(t)$ is a martingale; that is, for $s > t$, $E[W(s) \mid \mathcal{F}_t] = W(t)$.
- $X(t) = W(t)^2 - t$ is a martingale; that is, for $s > t$, $E[X(s) \mid \mathcal{F}_t] = W(t)^2 - t$.
- For any constant $\theta \in \mathbb{R}$, $X(t) = \exp[\theta W(t) - \frac{1}{2}\theta^2 t]$ is a martingale.
- For any $h > 0$, $X(t) = W(t + h) - W(h)$ is a standard Brownian motion.
- $W(t)$ is nowhere differentiable. In particular, given $W(t) = w$, for all $\epsilon > 0$, the set $A = \{s \in (t, t+\epsilon) : W(s) = w\}$ is always non-empty. In other words, $W(s)$ crosses w infinitely often, infinitely close to t.

Consider the n-dimensional process $W(t) = (W_1(t), \ldots, W_n(t))'$. If each of the $W_i(t)$ is a standard one-dimensional Brownian motion and if the $W_i(t)$ are all independent of one another, then $W(t)$ is called a *standard n-dimensional Brownian motion*.

A.3 Itô Integrals

We start with the probability triple $(\Omega, \mathcal{F}, P)$. Let $W(t, \omega)$ be a sample path of Brownian motion given $\omega \in \Omega$ under P and let $\mathcal{F}_t = \sigma(\{W(u, \omega) : 0 \leqslant u \leqslant t\})$ be the sigma-algebra generated by the sample paths $W(u, \omega)$ up to time t.

Let $X(t, \omega)$ be some stochastic process which is adapted to $\mathcal{F}_t$. How do we define the integral $\int_0^T X(t, \omega) \, dW(t, \omega)$? In this context $dW(t, \omega)$ can be interpreted as $W(t + dt, \omega) - W(t, \omega)$ for small dt.

First, define $L^2(0, T)$ to be the class of functions $g(t, \omega) : [0, \infty) \times \Omega \to \mathbb{R}$ such that $g(t, \omega)$ is $\mathcal{F}_t$-adapted, and $E[\int_0^T g(t, \omega)^2 \, dt] < \infty$.

Second, consider any *elementary function* $h(t, \omega)$ which is piecewise constant; that is, there exists a sequence of times $0 = t_0 < t_1 < t_2 < \cdots < t_N$, where $0 < N \leqslant \infty$, and a set of function values $e_j(\omega)$, where each $e_j(\omega)$ is $\mathcal{F}_{t_j}$-measurable, for which

$$h(t, \omega) = e_j(\omega) \quad \text{if } t_j \leqslant t < t_{j+1}$$

or

$$h(t, \omega) = \sum_{j=0}^{N} e_j(\omega) I_{[t_j, t_{j+1})}(t).$$

Definition A.2. The Itô integral of $h(t, \omega)$ is defined as

$$\int_0^T h(t, \omega) \, dW(t, \omega) = \sum_{j=0}^{N} e_j(\omega)(W(t_{j+1}, \omega) - W(t_j, \omega)).$$

Lemma A.3. *For each $g(t, \omega) \in L^2(0, T)$ there exists a sequence of approximating elementary functions $\{h_n(t, \omega)\} \subset L^2(0, T)$ such that*

$$E\left[\int_0^T (g(t, \omega) - h_n(t, \omega))^2 \, dt\right] \to 0 \quad as\, n \to \infty.$$

Definition A.4. The Itô integral of $g(t, \omega)$ is defined as

$$\int_0^T g(t, \omega) \, dW(t, \omega) = \lim_{n \to \infty} \int_0^T h_n(t, \omega) \, dW(t, \omega).$$

Theorem A.5 (Itô's isometry). *For any $g(t, \omega) \in L^2(0, T)$*

$$E\left[\left(\int_0^T g(t, \omega) \, dW(t, \omega)\right)^2\right] = E\left[\int_0^T g(t, \omega)^2 \, dt\right].$$

In particular, if $g(t, \omega) = g(t)$ is deterministic, then

$$E\left[\left(\int_0^T g(t) \, dW(t, \omega)\right)^2\right] = \int_0^T g(t)^2 \, dt.$$

A.4 One-Dimensional Itô and Diffusion Processes

A stochastic process, $X(t)$, is called an *Itô process* if

$$X(t) = X(0) + \int_0^t a(u) \, du + \int_0^t b(u) \, dW(u),$$

where $a(u)$ and $b(u)$ are $\mathcal{F}_t$-adapted processes and $\int_0^t b(u) \, dW(u)$ is the Itô integral. This can be expressed in the alternative form of a stochastic differential equation (SDE):

$$dX(t) = a(t) \, dt + b(t) \, dW(t).$$

If $X(t)$ is Markov, then we can write $a(t) \equiv a(t, X(t))$ and $b(t) \equiv b(t, X(t))$. If, in addition, $X(t)$ is a continuous process, then it is referred to as a *diffusion process*.

The *quadratic variation* of $X(t)$ is defined as

$$\langle X \rangle(t) = \int_0^t b(u)^2 \, du.$$

This can be written in the differential form

$$d\langle X\rangle(t) = b(t)^2 \, dt.$$

(Sometimes this is written in the form $(dX(t))^2 = b(t)^2 \, dt$. In particular, this means we equate $(dW(t))^2$ with dt, $dt \, dW(t)$ with 0, and dt^2 with 0.)

Theorem A.6 (Itô's formula for one-dimensional diffusions). *Suppose $X(t)$ is some diffusion process with SDE*

$$dX(t) = a(t) \, dt + b(t) \, dW(t).$$

Let $f(t, x) : [0, \infty) \times \mathbb{R} \to \mathbb{R}$ be a function which is once continuously differentiable in t and twice continuously differentiable in x (that is, a $C^{1,2}$ function).

Let $Y(t) = f(t, X(t))$. Then $Y(t)$ is a diffusion process with

$$
\begin{aligned}
dY(t) &= \left.\frac{\partial f}{\partial t}\right|_{(t, X(t))} dt + \left.\frac{\partial f}{\partial x}\right|_{(t, X(t))} dX(t) + \frac{1}{2}\left.\frac{\partial^2 f}{\partial x^2}\right|_{(t, X(t))} d\langle X\rangle(t) \\
&= \left[\frac{\partial f}{\partial t}(t, X(t)) + \frac{\partial f}{\partial x}(t, X(t))a(t) + \frac{1}{2}\frac{\partial^2 f}{\partial x^2}(t, X(t))b(t)^2\right] dt \\
&\quad + \frac{\partial f}{\partial x}(t, X(t))b(t) \, dW(t).
\end{aligned}
$$

A.5   Multi-Dimensional Diffusion Processes

An n-dimensional diffusion process $X(t)$ is a vector $(X_1(t), \ldots, X_n(t))'$ of diffusion processes with

$$dX_i(t) = a_i(t) \, dt + \sum_{j=1}^{n} b_{ij}(t) \, dW_j(t) = a_i(t) \, dt + b_i(t)' \, dW(t)$$

or

$$dX(t) = a(t) \, dt + B(t) \, dW(t),$$

where

$$
\begin{aligned}
a(t) &= (a_1(t), \ldots, a_n(t))', \\
b_i(t) &= (b_{i1}(t), \ldots, b_{in}(t))' \quad \text{(used below)}, \\
B(t) &= (b_{ij}(t))_{i,j=1}^{n},
\end{aligned}
$$

and $W(t) = (W_1(t), \ldots, W_n(t))'$ is a standard n-dimensional Brownian motion. The quadratic covariation process for $X_i(t)$ and $X_j(t)$ is

$$\langle X_i, X_j\rangle(t) = \int_0^t b_i(u)'b_j(u) \, du \quad \text{or} \quad d\langle X_i, X_j\rangle(t) = b_i(t)'b_j(t) \, dt.$$

Sometimes this is written in the form $dX_i(t) \, dX_j(t)$. (This means we equate $(dW_i(t))^2$ with dt, and $dW_i(t) \, dW_j(t)$ with 0 for all $i \neq j$.)

Theorem A.7 (Itô's formula for n-dimensional diffusions). *Let $X(t)$ be an n-dimensional diffusion as defined above. Let $f(t, x) : [0, \infty) \times \mathbb{R}^n \to \mathbb{R}$ be a $C^{1,2}$ function. Then $Y(t) = f(t, X(t))$ is a diffusion process with*

$$
\begin{aligned}
dY(t) &= \left.\frac{\partial f}{\partial t}\right|_{(t, X(t))} dt + \sum_{i=1}^{n} \left.\frac{\partial f}{\partial x_i}\right|_{(t, X(t))} dX_i(t) \\
&\qquad + \frac{1}{2} \sum_{i,j=1}^{n} \left.\frac{\partial^2 f}{\partial x_i \partial x_j}\right|_{(t, X(t))} d\langle X_i, X_j \rangle(t) \\
&= \left[\frac{\partial f}{\partial t}(t, X(t)) + \sum_{i=1}^{n} \frac{\partial f}{\partial x_i}(t, X(t))a_i(t) \right. \\
&\qquad \left. + \frac{1}{2} \sum_{i,j=1}^{n} \frac{\partial^2 f}{\partial x_i \partial x_j}(t, X(t))b_i(t)'b_j(t)\right] dt \\
&\qquad + \sum_{j=1}^{n} \left(\sum_{i=1}^{n} \frac{\partial f}{\partial x_i}(t, X(t))b_{ij}(t)\right) dW_j(t).
\end{aligned}
$$

Corollary A.8 (the product rule). *Suppose $X(t)$ and $Y(t)$ are one-dimensional diffusion processes. Let $R(t) = X(t)Y(t)$. Then*

$$
dR(t) = X(t)\,dY(t) + Y(t)\,dX(t) + d\langle X, Y \rangle(t).
$$

Note that if

$$
dX(t) = a_X(t)\,dt + b_X(t)'\,dW(t) \quad \text{and} \quad dY(t) = a_Y(t)\,dt + b_Y(t)'\,dW(t),
$$

then $d\langle X, Y \rangle(t) = b_X(t)'b_Y(t)\,dt$. $d\langle X, Y \rangle(t)$ is often written as $dX(t)\,dY(t)$, which we 'simplify' by equating $dW(t)\,dW(t)'$ with $I\,dt$, $dW(t)\,dt$ with 0, and dt^2 with 0.

A.6 The Feynman–Kac Formula

Theorem A.9. *Suppose that the function $P(t, x)$ satisfies the following partial differential equation*

$$
\frac{\partial P}{\partial t} + f(t, x)\frac{\partial P}{\partial x} + \tfrac{1}{2}\rho^2(t, x)\frac{\partial^2 P}{\partial x^2} - R(x)P + h(t, x) = 0
$$

subject to the boundary condition $P(T, x) = \psi(x)$.

Then there exists a process $\tilde{W}(t)$ and a measure Q under which $\tilde{W}(t)$ is a standard Brownian motion and where $P(t, x)$ has the solution

$$
P(t, x) = E_Q\left[\int_t^T V(t, u)h(u, X(u))\,du + V(t, T)\psi(X(T)) \,\middle|\, \mathcal{F}_t\right] \quad \text{for } t < T,
$$

where $dX(t) = f(t, X(t)) \, dt + \rho(t, X(t)) \, d\tilde{W}(t)$ *and*

$$V(t, u) = \exp\left(-\int_t^u R(X(s)) \, ds\right),$$

provided

$$\int_t^T E_Q\left[\left(\rho(s, X(s))\frac{\partial P}{\partial x}(s, X(s))\right)^2 \middle| \mathcal{F}_t\right] ds < \infty.$$

A.7 The Martingale Representation Theorem

Let $(\Omega, \mathcal{F}, P)$ be a probability triple. Let $W(t)$ be a standard n-dimensional Brownian motion under P and define $\mathcal{F}_t = \sigma(\{W(u) : 0 \leqslant u \leqslant t\})$ to be the sigma-algebra generated by $W(u)$ up to time t.

Theorem A.10 (martingale representation).

(a) *Suppose $M(t)$ is an $\mathcal{F}_t$-martingale under P and that $E_P[\int_0^T M(t)^2 \, dt] < \infty$. Then there exists a unique n-dimensional, previsible process $m(t)$ (that is, $m(t)$ is $\mathcal{F}_t$- measurable) such that*

$$M(t) = M(0) + \int_0^t m(u)' \, dW(t) \quad \text{or} \quad dM(t) = m(t)' \, dW(t).$$

(b) *Suppose that the n-dimensional diffusions $M^{(1)}(t)$ and $M^{(2)}(t)$ are $\mathcal{F}_t$-martingales under P with SDEs*

$$dM^{(i)}(t) = S_i(t) \, dW(t).$$

$S_1(t)$ and $S_2(t)$ are $n \times n$ volatility matrices that can also be dependent on $M^{(1)}(t)$ and $M^{(2)}(t)$ respectively. Suppose also that $S_1(t)$ is non-singular for all $0 \leqslant t \leqslant T$ almost surely. Then there exists a unique, previsible $n \times n$ matrix process $\phi(t) = (\phi_{ij}(t))_{i,j=1}^n$ such that

$$dM^{(2)}(t) = \phi(t) \, dM^{(1)}(t) \quad \text{or} \quad dM_i^{(2)}(t) = \sum_{j=1}^n \phi_{ij}(t) \, dM_j^{(1)}(t).$$

(Note the non-singularity of $S_1(t)$ allows us to derive the $dW(t)$. We can then apply part (a).)

A.8 Change of Probability Measure

When we consider the probability triple $(\Omega, \mathcal{F}, P)$, the sample space Ω defines the possible sample paths that can exist. Given Ω, there are many probability measures that are consistent with Ω. We will now describe how these probability measures can be related to one another.

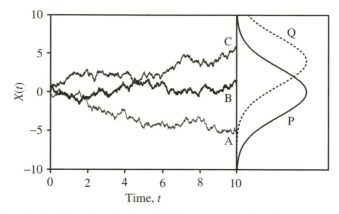

Figure A.1. Sample paths for a diffusion process $X(t)$. Under P, $X(t)$ is a standard Brownian motion. Under Q, $X(t) - 0.4t$ is a standard Brownian motion.

As an example, consider Figure A.1. We have plotted three sample paths for a process $X(t)$ taken from Ω. Under the measure P, $X(t)$ is a standard Brownian motion and so we can see that sample path B is more likely to happen (in some sense) than paths A and C (which are, approximately, equally likely under P). Under the alternative probability measure Q, $X(t) - 0.4t$ is a standard Brownian motion. This means that sample path A is more likely than path B, which in turn is more likely than path C. Under P and Q all of these paths are possible. It is just the relative likelihood of each path which changes.

Let $0 < T < \infty$. Suppose that $W(t, \omega) : [0, T] \times \Omega \to \mathbb{R}^n$ is an n-dimensional Brownian motion under P.[1] Let $\gamma(t, \omega) = (\gamma_1(t, \omega), \dots, \gamma_n(t, \omega))'$, where the $\gamma_i(t, \omega)$ are previsible diffusions satisfying the *Novikov condition*

$$E_P\left[\exp\left(\frac{1}{2}\int_0^T |\gamma(t, \omega)|^2 \, dt\right)\right] < \infty.$$

Define

$$Z(t, \omega) = \exp\left[-\int_0^t \gamma(u, \omega)' \, dW(u, \omega) - \frac{1}{2}\int_0^t |\gamma(u, \omega)|^2 \, du\right].$$

Given $\gamma(t, \omega)$ satisfies the Novikov condition, $Z(t, \omega)$ is a martingale under P for $0 \leqslant t \leqslant T$.

Let $\mathcal{F}_T$ be the sigma-algebra generated by $W(t, \omega)$ up to time T.

---

[1] Normally, we suppress the ω and write $W(t)$, etc. However, we write out $W(t, \omega)$ in full here to emphasize the functional form of $W(\cdot)$ and its independence from P; that is, given ω, the sample path $W(t, \omega)$ is not affected by the use of P or another measure Q.

Definition A.11 (equivalent measures). Consider two measures P and Q on $(\Omega, \mathcal{F}_T)$. If for any event $A \subset \Omega$

$$\text{Pr}_P(A) > 0 \quad \Longleftrightarrow \quad \text{Pr}_Q(A) > 0,$$

then P and Q are said to be *equivalent measures*.

Theorem A.12 (Cameron–Martin–Girsanov or Girsanov Theorem).

(a) *Suppose that $W(t, \omega)$ is a standard n-dimensional Brownian motion under P and $\gamma(t, \omega)$ is an n-dimensional diffusion that satisfies the Novikov condition. Then there exists a measure Q such that*

    (i) *Q is equivalent to P;*

    (ii) *$\tilde{W}(t, \omega) = W(t, \omega) + \int_0^t \gamma(u, \omega)\, du$ is a standard n-dimensional Brownian motion under Q.*

(b) *If Q is known to be equivalent to P on $[0, T] \times \Omega$, then there exists an n-dimensional previsible diffusion $\gamma(t, \omega)$ such that $\tilde{W}(t, \omega) = W(t, \omega) + \int_0^t \gamma(u, \omega)\, du$ is a Brownian motion under Q.*

Definition A.13 (Radon–Nikodým derivatives). For the framework described in Theorem A.12 we define the *Radon–Nikodým derivative* to be

$$\frac{dQ}{dP}(\omega) = Z(T, \omega) \quad \text{or} \quad \frac{dQ}{dP} = Z(T).$$

The Radon–Nikodým derivative measures the relative likelihood of a given sample point ω under the two measures P and Q over the interval $[0, T]$.

The Radon–Nikodým derivative is applied as follows. Suppose that X is any $\mathcal{F}_T$-measurable random variable. Then

$$E_Q[X] = E_P\left[\frac{dQ}{dP} X\right] = E_P[Z(T)X].$$

Furthermore, if X is $\mathcal{F}_t$-measurable for some $t < T$, then

$$E_Q[X] = E_P[Z(T)X] = E_P[E_P\{Z(T)X \mid \mathcal{F}_t\}]$$
$$= E_P[X E_P\{Z(T) \mid \mathcal{F}_t\}] = E_P[X Z(t)],$$

since $Z(s)$ is a martingale under P.

Finally, it is convenient to introduce a more general form for the Radon–Nikodým derivative. For $0 \leqslant t < s \leqslant T$,

$$\frac{dQ}{dP}(t, s) = \frac{Z(s)}{Z(t)}.$$

Then, for any $\mathcal{F}_s$-measurable random variable X we have

$$E_Q[X \mid \mathcal{F}_t] = E_P\left[\frac{dQ}{dP}(t, s)X \,\middle|\, \mathcal{F}_t\right] = Z(t)^{-1} E_P[Z(s)X \mid \mathcal{F}_t].$$

The Vasicek and CIR Models: Proofs

B.1   The Vasicek Model

We prove Theorem 4.7 making use of the following lemma.

Lemma B.1. (a) *The bivariate Laplace transform for $\int_t^T r(s)\,ds$ and $r(T)$ given $r(t)$ is*

$$
P_{\rm L}(t, T, r, v, \omega) = E_Q\left[\exp\left(-v\int_t^T r(s)\,ds - \omega r(T)\right)\,\bigg|\,r(t) = r\right]
$$
$$
= \exp[A(t, T, v, \omega) - B(t, T, v, \omega)r], \qquad (\rm B.1)
$$

where

$$
\tau = T - t, \qquad B(t, T, v, \omega) = v B_1(t, T) + \omega B_2(t, T),
$$
$$
B_1(t, T) = \frac{1 - e^{-\alpha\tau}}{\alpha}, \qquad B_2(t, T) = e^{-\alpha\tau},
$$
$$
A(t, T, v, \omega) = -v A_1(t, T) - \omega A_2(t, T)
$$
$$
+ \tfrac{1}{2}v^2 C_{11}(t, T) + v\omega C_{12}(t, T) + \tfrac{1}{2}\omega^2 C_{22}(t, T),
$$
$$
A_1(t, T) = \mu\left(\tau - \frac{1 - e^{-\alpha\tau}}{\alpha}\right), \qquad A_2(t, T) = \mu(1 - e^{-\alpha\tau}),
$$
$$
C_{11}(t, T) = \frac{\sigma^2}{2\alpha^3}[2\alpha\tau - 3 + 4e^{-\alpha\tau} - e^{-2\alpha\tau}],
$$
$$
C_{12}(t, T) = \frac{\sigma^2}{2\alpha^2}(1 - e^{-\alpha\tau})^2, \qquad C_{22}(t, T) = \frac{\sigma^2}{2\alpha}(1 - e^{-2\alpha\tau}).
$$

(b) *Hence $\int_t^T r(s)\,ds$ and $r(T)$ given $r(t)$ have a bivariate Normal distribution under Q with*

$$
E_Q[r(T)\mid r(t)] = B_2(t, T)r(t) + A_2(t, T) = \mu + (r(t) - \mu)e^{-\alpha\tau},
$$
$$
E_Q\left[\int_t^T r(s)\,ds\,\bigg|\,r(t)\right] = B_1(t, T)r(t) + A_1(t, T)
$$
$$
= \mu\tau + (r(t) - \mu)\frac{(1 - e^{-\alpha\tau})}{\alpha},
$$

$$\text{Var}_Q[r(T) \mid r(t)] = C_{22}(t, T) = \sigma^2 \frac{(1 - e^{-2\alpha\tau})}{2\alpha},$$

$$\text{Var}_Q\left[\int_t^T r(s)\, ds \;\middle|\; r(t)\right] = C_{11}(t, T) = \frac{\sigma^2}{2\alpha^3}[2\alpha\tau - 3 + 4e^{-\alpha\tau} - e^{-2\alpha\tau}],$$

$$\text{Cov}_Q\left[r(T), \int_t^T r(s)\, ds \;\middle|\; r(t)\right] = C_{12}(t, T) = \frac{\sigma^2}{2\alpha^2}(1 - e^{-\alpha\tau})^2.$$

Proof of Lemma B.1 (sketch). We can apply the converse of the Feynman–Kac formula (Theorem A.9) to see that $P_L(t, T, r, v, \omega)$ satisfies the partial differential equation

$$\frac{\partial P_L}{\partial t} + \alpha(\mu - r)\frac{\partial P_L}{\partial r} + \tfrac{1}{2}\sigma^2 \frac{\partial^2 P_L}{\partial r^2} - vr P_L = 0 \tag{B.2}$$

with boundary condition $P_L(T, T, r, v, \omega) = \exp(-\omega r)$. It is straightforward (but rather tedious) to show that the expression for $P_L(t, T, r, v, \omega)$ given in equation (B.1) does indeed satisfy this PDE.

The bivariate normality of $\int_t^T r(u)\, du$ and $r(T)$ follows when we observe that $\log P_L(t, T, r, v, \omega)$ is quadratic in v and ω. □

Proof of Theorem 4.7. (a) The price at time t given $r(t) = r$ of a zero-coupon bond maturing at time T is

$$P(t, T, r) = E_Q\left[\exp\left(-\int_t^T r(s)\, ds\right) \;\middle|\; r(t) = r\right] = P_L(t, T, r, 1, 0)$$

$$= \exp[A(t, T, 1, 0) - B(t, T, 1, 0)r]$$

$$= \exp[A(t, T) - B(t, T)r], \quad \text{say, for notational convenience,}$$

where

$$B(t, T) = B_1(t, T) = \frac{1 - e^{-\alpha(T-t)}}{\alpha},$$

$$A(t, T) = -A_1(t, T) + \tfrac{1}{2}C_{11}(t, T)$$

$$= \left(\mu - \frac{\sigma^2}{2\alpha^2}\right)(B(t, T) - (T - t)) - \frac{\sigma^2}{4\alpha}B(t, T)^2$$

(after some algebraic rearrangement) as required.

This result can, of course, be derived directly by showing that $P(t, T, r) = P_L(t, T, r, 1, 0)$ satisfied the PDE (B.2) with boundary condition $P(T, T, r) = 1$ for all r.

(b) The payoff on the call option at time T is $(P(T, S) - K)_+$. Define the indicator random variable

$$I = I(P(T, S, r(T)) > K) = \begin{cases} 1 & \text{if } P(T, S, r(T)) > K, \\ 0 & \text{otherwise.} \end{cases}$$

The value at time $t < T$ of the option is

$$V(t) = E_Q[e^{-\int_t^T r(s)\,ds}(P(T, S, r(T)) - K)_+ \mid r(t)]$$

$$= E_Q[Ie^{-\int_t^T r(s)\,ds}P(T, S, r(T)) \mid r(t)] - KE_Q[Ie^{-\int_t^T r(s)\,ds} \mid r(t)]$$

$$= E_Q[Ie^{-\int_t^S r(s)\,ds} \mid r(t)] - KE_Q[Ie^{-\int_t^T r(s)\,ds} \mid r(t)]. \quad (B.3)$$

Let P_1 and P_2 be two new measures, equivalent to Q with Radon–Nikodým derivatives:

$$\frac{dP_1}{dQ} = \frac{e^{-\int_t^S r(s)\,ds}}{E_Q[e^{-\int_t^S r(s)\,ds} \mid r(t) = r]}, \qquad \frac{dP_2}{dQ} = \frac{e^{-\int_t^T r(s)\,ds}}{E_Q[e^{-\int_t^T r(s)\,ds} \mid r(t) = r]}.$$

Then we have, from equation (B.3),

$$V(t) = E_Q[e^{-\int_t^S r(s)\,ds} \mid r(t) = r]E_Q\left[\frac{dP_1}{dQ}I \,\middle|\, r(t) = r\right]$$

$$- KE_Q[e^{-\int_t^T r(s)\,ds} \mid r(t) = r]E_Q\left[\frac{dP_2}{dQ}I \,\middle|\, r(t) = r\right]$$

$$= P(t, S, r)E_{P_1}[I \mid r(t) = r] - KP(t, T, r)E_{P_2}[I \mid r(t) = r]$$

$$= P(t, S, r)\mathrm{Pr}_{P_1}(P(T, S) > K \mid r(t) = r)$$

$$- KP(t, T, r)\mathrm{Pr}_{P_2}(P(T, S) > K \mid r(t) = r). \quad (B.4)$$

It remains for us to establish the distribution of $r(T)$ under P_1 and P_2. Let us look first at P_2:

$$P_L(t, T, r, 1, \omega) = \exp[A(t, T, 1, \omega) - B(t, T, 1, \omega)r]$$

$$= E_Q[e^{-\int_t^T r(s)\,ds - \omega r(T)} \mid r(t) = r]$$

$$= P(t, T, r)E_{P_2}[e^{-\omega r(T)} \mid r(t) = r].$$

Hence

$$E_{P_2}[e^{-\omega r(T)} \mid r(t) = r]$$

$$= \exp[A(t, T, 1, \omega) - A(t, T, 1, 0) - (B(t, T, 1, \omega) - B(t, T, 1, 0))r]$$

$$= \exp[-\omega A_2(t, T) + \omega C_{12}(t, T) + \tfrac{1}{2}\omega^2 C_{22}(t, T) - \omega B_2(t, T)r].$$

Hence $r(T)$ given $r(t) = r$ has a Normal distribution under P_2 with

$$E_{P_2}[r(T) \mid r(t) = r] = A_2(t, T) - C_{12}(t, T) + B_2(t, T)r$$

$$= \mu + (r - \mu)e^{-\alpha(T-t)} - \frac{\sigma^2}{\alpha^2}(1 - e^{-\alpha(T-t)})^2$$

$$= r_2, \quad \text{say,}$$

and

$$\text{Var}_{P_2}[r(T) \mid r(t) = r] = C_{22}(t, T) = \frac{\sigma^2}{2\alpha}(1 - e^{-2\alpha(T-t)}).$$

Hence

$$\text{Pr}_{P_2}(r(T) < r^* \mid r(t) = r) = \Phi(d_2),$$

where

$$d_2 = \frac{r^* - r_2}{\sqrt{C_{22}(t, T)}}.$$

Now in the present context (equation (B.4))

$$r^* = \frac{A(T, S) - \log K}{B(T, S)}.$$

Also,

$$
\begin{aligned}
P(t, S) &= E_Q[e^{-\int_t^T r(s)\,ds - \int_T^S r(s)\,ds} \mid r(t) = r] \\
&= P(t, T) E_{P_2}[P(T, S, r(T)) \mid r(t) = r] \\
&= P(t, T) E_{P_2}[e^{A(T,S) - B(T,S)r(T)} \mid r(t) = r] \\
&= P(t, T) \exp[A(T, S) - B(T, S)r_2 + \tfrac{1}{2} B(T, S)^2 C_{22}(t, T)]
\end{aligned}
$$

$$\Rightarrow \quad \log \frac{P(t, S)}{K P(t, T)} = A(T, S) - B(T, S)r_2 + \tfrac{1}{2} B(T, S)^2 C_{22}(t, T) - \log K.$$

Therefore,

$$
\begin{aligned}
d_2 &= \frac{r^* - r_2}{\sqrt{C_{22}(t, T)}} = \frac{A(T, S) - \log K - B(T, S)r_2}{B(T, S)\sqrt{C_{22}(t, T)}} \\
&= \frac{\log(P(t, S)/K P(t, T)) - \tfrac{1}{2} B(T, S)^2 C_{22}(t, T)}{B(T, S)\sqrt{C_{22}(t, T)}} \\
&= \frac{1}{\sigma_p} \log \frac{P(t, S)}{K P(t, T)} - \frac{\sigma_p}{2},
\end{aligned}
$$

where

$$\sigma_p = B(T, S)^2 C_{22}(t, T) = \sigma^2 \frac{(1 - e^{-\alpha(S-T)})^2}{\alpha^2} \frac{(1 - e^{-2\alpha(T-t)})}{2\alpha}.$$

Now consider the distribution of $r(T)$ under P_1:

$$E_Q[e^{-\int_t^S r(s)\,ds - \omega r(T)} \mid r(t) = r] = P(t, S) E_{P_1}[e^{-\omega r(T)} \mid r(t) = r],$$

but

$$
\begin{aligned}
E_Q[e^{-\int_t^S r(s)\,ds - \omega r(T)} &\mid r(t) = r] \\
&= E_Q[e^{-\int_t^T r(s)\,ds - \omega r(T)} E_Q[e^{-\int_T^S r(s)\,ds} \mid r(T)] \mid r(t) = r] \\
&= E_Q[e^{-\int_t^T r(s)\,ds - \omega r(T) + A(T,S,1,0) - B(T,S,1,0)r(T)} \mid r(t) = r]
\end{aligned}
$$

$$= e^{A(T,S,1,0)} E_Q[e^{-\int_t^T r(s)\,ds - \omega_2 r(T)} \mid r(t) = r]$$
$$= \exp[A(T, S, 1, 0) + A(t, T, 1, \omega_2) - B(t, T, 1, \omega_2)r],$$

where

$$\omega_2 = \omega + B(T, S, 1, 0) = \omega + \frac{1 - e^{-\alpha(S-T)}}{\alpha}.$$

Hence

$$E_{P_1}[e^{-\omega r(T)} \mid r(t) = r]$$
$$= \exp[A(T, S, 1, 0) + A(t, T, 1, \omega_2) - A(t, S, 1, 0)$$
$$\qquad - (B(t, T, 1, \omega_2) - B(t, S, 1, 0))r]$$
$$= \exp[-A_1(T, S) + \tfrac{1}{2}C_{11}(T, S) - A_1(t, T) - \omega_2 A_2(t, T)$$
$$\qquad + \tfrac{1}{2}C_{11}(t, T) + \omega_2 C_{12}(t, T) + \tfrac{1}{2}\omega_2^2 C_{22}(t, T) + A_1(t, S)$$
$$\qquad - \tfrac{1}{2}C_{11}(t, S) - \{B_1(t, T) + \omega_2 B_2(t, T) - B_1(t, S)\}r]$$
$$= \exp[-\omega A_2(t, T) + \omega C_{12}(t, T) + \omega B_1(T, S)C_{22}(t, T)$$
$$\qquad + \tfrac{1}{2}\omega^2 C_{22}(t, T) - \omega B_2(t, T)r]$$

(where proof of the final step requires a certain amount of tedious algebra).

Thus, $r(T)$ given $r(t) = r$ is Normally distributed under P_1 with

$$E_{P_1}[r(T) \mid r(t) = r] = A_2(t, T) - C_{12}(t, T) - B_1(T, S)C_{22}(t, T) + B_2(t, T)r$$
$$= r_2 - \sigma^2 \frac{(1 - e^{-\alpha(S-T)})}{\alpha} \frac{(1 - e^{-2\alpha(T-t)})}{2\alpha}$$
$$= r_1, \quad \text{say,}$$

and

$$\mathrm{Var}_{P_1}[r(T) \mid r(t) = r] = C_{22}(t, T) = \frac{\sigma^2}{2\alpha}(1 - e^{-2\alpha(T-t)}).$$

Hence

$$\mathrm{Pr}_{P_1}(r(T) < r^*) = \Phi(d_1),$$

where

$$d_1 = \frac{r^* - r_1}{\sqrt{C_{22}(t, T)}} = \frac{r^* - r_2}{\sqrt{C_{22}(t, T)}} + \frac{B(T, S)C_{22}(t, T)}{\sqrt{C_{22}(t, T)}}$$
$$= d_2 + \sigma_p.$$

$\square$

B.2 The Cox–Ingersoll–Ross Model

We give here the proof of Theorem 4.8.

B.2.1   Proof of Theorem 4.8

Proof of part (a).   Recall that

$$P_L(t, T, r, v, \omega) = E_Q[e^{-v \int_t^T r(s)\,ds - \omega r(T)} \mid r(t) = r],$$

where

$$dr(t) = \alpha(\mu - r(t))\,dt + \sigma\sqrt{r(t)}\,d\tilde{W}(t).$$

We can apply the converse of the Feynman–Kac formula (Theorem A.9) to see that $P_L(t, T, r, v, \omega)$ must satisfy the PDE

$$\frac{\partial P_L}{\partial t} + \alpha(\mu - r)\frac{\partial P_L}{\partial r} + \tfrac{1}{2}\sigma^2 r \frac{\partial^2 P_L}{\partial r^2} - vr P_L = 0,$$

with boundary condition $P_L(T, T, r, v, \omega) = e^{-\omega r}$. It remains for us to show that the expression for $P_L(t, T, r, v, \omega)$ given in equation (4.12) satisfies this PDE. The solution to the PDE can be verified quickly using a mathematical computing package such as Maple or slowly using pen and paper!

Proof of part (b).   For zero-coupon bond prices:

$$P(t, T, r) = E_Q[e^{-\int_t^T r(s)\,ds} \mid r(t) = r] = P_L(t, T, r, 1, 0).$$

Substituting $v = 1$ and $\omega = 0$ into equation (4.12) we get the expression for $P(t, T, r)$ given in equation (4.13).

Proof of part (c).   The Laplace transform for $r(T)$ given $r(t) = r$ is

$$E_Q[e^{-\omega r(T)} \mid r(t) = r] = P_L(t, T, r, 0, \omega)$$

$$= \left(\frac{2\alpha e^{\alpha(T-t)}}{(\sigma^2\omega + 2\alpha)(e^{\alpha(T-t)} - 1) + 2\alpha} \right)^{2\alpha\mu/\sigma^2}$$

$$\times \exp\left(-\frac{2\alpha\omega r}{(\sigma^2\omega + 2\alpha)(e^{\alpha(T-t)} - 1) + 2\alpha} \right)$$

(noting that $\gamma(v) = \alpha$ when $v = 0$).

Now

$$(\sigma^2\omega + 2\alpha)(e^{\alpha(T-t)} - 1) + 2\alpha = 2\alpha e^{\alpha(T-t)}\left(1 + 2\frac{\sigma^2(e^{\alpha(T-t)} - 1)}{4\alpha e^{\alpha(T-t)}}\omega \right)$$

$$= 2\alpha e^{\alpha(T-t)}(1 + 2k_Q\omega),$$

where

$$k_Q = \sigma^2(1 - e^{-\alpha(T-t)})/4\alpha.$$

Furthermore, it is straightforward to verify that

$$\exp\left(-\frac{2\alpha\omega r}{(\sigma^2\omega + 2\alpha)(e^{\alpha(T-t)} - 1) + 2\alpha}\right)$$

$$= \exp\left(-\omega r e^{-\alpha(T-t)}\frac{2\alpha e^{\alpha(T-t)}}{2\alpha e^{\alpha(T-t)}(1 + 2k_Q\omega)}\right) = \exp\left(-\frac{\lambda_Q}{2} + \frac{\lambda_Q}{2(1 + 2k_Q\omega)}\right),$$

where

$$\lambda_Q = \frac{4\alpha r}{\sigma^2(e^{\alpha(T-t)} - 1)}.$$

Hence we have

$$E_Q[e^{-\omega r(T)} \mid r(t) = r] = E_Q[e^{-(k_Q\omega)(r(T)/k_Q)} \mid r(t) = r]$$

$$= \left(\frac{1}{1 + 2k_Q\omega}\right)^{2\alpha\mu/\sigma^2} \exp\left(-\frac{\lambda_Q}{2}\right)\exp\left(\frac{\lambda_Q}{2(1 + 2k_Q\omega)}\right).$$

Now compare this with the Laplace transform for the non-central chi-squared distribution given in Section 4.6.3. From this we infer that $r(t)/k_Q$ has a non-central chi-squared distribution with $d = 4\alpha\mu/\sigma^2$ degrees of freedom and non-centrality parameter λ_Q.

Proof of part (d). Technically, this is the most difficult part of Theorem 4.8 (and can be skipped by the faint hearted!). We follow the outline suggested in Lamberton and Lapeyre (1996, Exercise 34). We will state the key steps in the proof first, followed by a detailed development filling out these initial statements.

(i) Define the function $s(r) = \int_1^r \exp(2\alpha v/\sigma^2)v^{-2\alpha\mu/\sigma^2}\,dv$ for $0 < r < \infty$. Then

$$\alpha(\mu - r)\frac{\partial s}{\partial r} + \tfrac{1}{2}\sigma^2 r\frac{\partial^2 s}{\partial r^2} = 0.$$

(ii) For each t, and given $r(0) = r$,

$$s(r(t)) = s(r(0)) + \int_0^t \frac{ds}{dr}(r(u))\sigma\sqrt{r(u)}\,d\tilde{W}(u).$$

In particular, $s(r(t))$ is a local martingale under Q.

(iii) Define $\tau_x = \inf\{t : r(t) = x\}$, and $p \wedge q = \inf\{p, q\}$. Let ϵ, M be such that $0 < \epsilon < r(0) < M < \infty$. Then we exploit the local-martingale properties of $s(r(t))$ and the boundedness of ds/dr, where $\epsilon < r < M$ to demonstrate that $\Pr_Q(\tau_\epsilon \wedge \tau_M < \infty) = 1$.

(iv) The martingale property then implies that

$$s(r(0)) = s(\epsilon)\Pr_Q(\tau_\epsilon < \tau_M) + s(M)\Pr_Q(\tau_M < \tau_\epsilon).$$

(v) If $2\alpha\mu \geqslant \sigma^2$, then $s(\epsilon) \to -\infty$ as $\epsilon \to 0$. This implies $\Pr(\tau_0 < \tau_M) = 0$ for all $0 < r(0) < M < \infty$. Hence $\Pr(\tau_0 < \infty) = 0$.

(vi) If $0 < 2\alpha\mu < \sigma^2$, with $\alpha, \mu, \sigma^2 > 0$, then $-\infty < \lim_{\epsilon\to 0} s(\epsilon) < 0$. Hence $s(r(0)) = s(0)\Pr_Q(\tau_0 < \tau_M) + s(M)\Pr_Q(\tau_M < \tau_0)$, where $s(0)$ is defined as $\lim_{\epsilon\to 0} s(\epsilon)$. Since, in addition, $s(M) \to +\infty$ as $M \to +\infty$, $\Pr(\tau_0 < \infty) = 1$.

Let us now work through these steps more rigorously.

Consider a general twice continuously differentiable function $s(r)$. By Itô's formula,

$$ds(r(t)) = \frac{\partial s}{\partial r}(\alpha(\mu - r(t))\,dt + \sigma\sqrt{r(t)}\,d\tilde{W}(t)) + \frac{1}{2}\frac{\partial^2 s}{\partial r^2}\sigma^2 r(t)\,dt.$$

Thus, $s(r(t))$ is a local martingale if and only if

$$0 = \alpha(\mu - r(t))\frac{\partial s}{\partial r} + \tfrac{1}{2}\sigma^2 r(t)\frac{\partial^2 s}{\partial r^2}$$

$$\Rightarrow \quad s(r) = k_1 \int_1^r e^{2\alpha v/\sigma^2} v^{-2\alpha\mu/\sigma^2}\,dv + k_0.$$

Without loss of generality take $k_1 = 1$ and $k_0 = 0$.

We then have

$$ds(r(t)) = s'(r(t))\sigma\sqrt{r(t)}\,d\tilde{W}(t)$$

$$\Rightarrow \quad s(r(t)) = s(r(0)) + \int_0^t s'(r(u))\sigma\sqrt{r(u)}\,d\tilde{W}(u),$$

where $s'(r) = e^{2\alpha r/\sigma^2} r^{-2\alpha\mu/\sigma^2}$. It follows that $s(r(t))$ is a local martingale.

For $0 < \epsilon < r(0) < M < \infty$ we can note that $s(\epsilon) \leqslant s(r) \leqslant s(M)$ for $\epsilon \leqslant r \leqslant M$ since the derivative $s'(r) = e^{2\alpha r/\sigma^2} r^{-2\alpha\mu/\sigma^2}$ is positive for all $r > 0$.

Similarly, we note that for $0 < \epsilon < r(0) < M < \infty$, $s'(r)$ is bounded below by $M^{-2\alpha\mu/\sigma^2} = \delta$, say, where $\delta > 0$.

Let us now consider

$$s(r(t \wedge \tau_\epsilon \wedge \tau_M)) = s(r(0)) + \int_0^t I(u)s'(r(u))\sigma\sqrt{r(u)}\,d\tilde{W}(u),$$

where $\tau_x = \inf\{t : r(t) = x\}$, $p \wedge q = \inf\{p, q\}$ and

$$I(u) = \begin{cases} 1 & \text{if } u < \tau_\epsilon \wedge \tau_M, \\ 0 & \text{if } u \geqslant \tau_\epsilon \wedge \tau_M. \end{cases}$$

This is a martingale because of the boundedness of $s'(r(u))\sigma\sqrt{r(u)}$ for $u \leqslant t \wedge \tau_\epsilon \wedge \tau_M$. This implies that

$$s(r(0)) = E_Q[s(r(t \wedge \tau_\epsilon \wedge \tau_M)) \mid r(0)].$$

Furthermore,

$$\mathrm{Var}_Q[s(r(t \wedge \tau_\epsilon \wedge \tau_M))]$$

$$= E_Q\left[\int_0^t I(u)^2 s'(r(u))^2 \sigma^2 r(u)\,du \,\middle|\, r(0)\right] \quad \text{(Itô isometry)}$$

$$= E_Q\left[\int_0^{t \wedge \tau_\epsilon \wedge \tau_M} s'(r(u))^2 \sigma^2 r(u)\,du \,\middle|\, r(0)\right]$$

$$\geqslant \delta^2 \sigma^2 \epsilon \, E_Q[t \wedge \tau_\epsilon \wedge \tau_M]. \tag{B.5}$$

But we also have $\mathrm{Var}_Q[s(r(t \wedge \tau_\epsilon \wedge \tau_M))] \leqslant (s(M) - s(\epsilon))^2 < \infty$. Hence $\delta^2 \sigma^2 \epsilon \, E_Q[t \wedge \tau_\epsilon \wedge \tau_M] \leqslant (s(M) - s(\epsilon))^2$ for all t, implying that $E_Q[\tau_\epsilon \wedge \tau_M] < \infty$ and, therefore, that $\mathrm{Pr}_Q[\tau_\epsilon \wedge \tau_M < \infty] = 1$.

Now

$$s(r(0)) = E_Q[s(r(t \wedge \tau_\epsilon \wedge \tau_M))]$$

$$= s(\epsilon)\mathrm{Pr}_Q(\tau_\epsilon \leqslant t \wedge \tau_M) + s(M)\mathrm{Pr}_Q(\tau_M \leqslant t \wedge \tau_\epsilon)$$

$$\quad + E_Q[s(r(t)) \mid t < \tau_\epsilon \wedge \tau_M]\mathrm{Pr}_Q(t < \tau_\epsilon \wedge \tau_M).$$

In the limit as $t \to \infty$, $E_Q[s(r(t)) \mid t < \tau_\epsilon \wedge \tau_M]$ is bounded below and above by $s(\epsilon)$ and $s(M)$ respectively, while $\mathrm{Pr}_Q(t < \tau_\epsilon \wedge \tau_M) \to 0$, $\mathrm{Pr}_Q(\tau_\epsilon < t \wedge \tau_M) \to \mathrm{Pr}_Q(\tau_\epsilon < \tau_M)$ and $\mathrm{Pr}_Q(\tau_M < t \wedge \tau_\epsilon) \to \mathrm{Pr}_Q(\tau_M < \tau_\epsilon)$. Hence

$$s(r(0)) = s(\epsilon)\mathrm{Pr}_Q(\tau_\epsilon < \tau_M) + s(M)\mathrm{Pr}_Q(\tau_M < \tau_\epsilon).$$

Suppose that $2\alpha\mu \geqslant \sigma^2$ and $0 < \epsilon < 1$. Then

$$-s(\epsilon) = \int_\epsilon^1 e^{2\alpha v/\sigma^2} v^{-2\alpha\mu/\sigma^2}\,dv \geqslant \int_\epsilon^1 e^{2\alpha v/\sigma^2} \frac{1}{v}\,dv \geqslant \int_\epsilon^1 \frac{1}{v}\,dv$$

$$\to \infty \quad \text{as } \epsilon \to 0.$$

So we have the two results $s(r(0)) = s(\epsilon)\mathrm{Pr}_Q(\tau_\epsilon < \tau_M) + s(M)\mathrm{Pr}_Q(\tau_M < \tau_\epsilon)$ and $s(\epsilon) \to -\infty$ as $\epsilon \to 0$. Hence for fixed M, as $\epsilon \to 0$ we must have $\mathrm{Pr}_Q(\tau_\epsilon < \tau_M) \to 0$. That is, $\mathrm{Pr}_Q(\tau_0 < \tau_M) = 0$ for all M such that $0 < r(0) < M < \infty$.

Now consider the event that $r(t)$ hits zero in finite time:

$$\Omega_0 = \left\{\omega : \tau_0(\omega) < \infty, \sup_{0 < t < \tau_0(\omega)} r(t)(\omega) < \infty\right\}.$$

Note that we have excluded from Ω_0 sample paths, $r(t)(\omega)$, which explode before $\tau_0(\omega)$; that is,

$$\Omega_e = \left\{\omega : \tau_0(\omega) < \infty, \sup_{0 < t < \tau_0(\omega)} r(t)(\omega) = \infty\right\}.$$

Since $dr(t) = \alpha(\mu - r(t))\,dt + \sigma\sqrt{r(t)}\,d\tilde{W}(t)$ it is straightforward to establish that $r(t)$ does not explode with probability 1 (for example, see Øksendal 1998, Section 5.2, Condition 5.2.1). Thus, $\mathrm{Pr}_Q(\Omega_e) = 0$.

Now let, for integers n,

$$\Omega_n = \left\{ \omega : \tau_0(\omega) < \tau_n(\omega), \sup_{0 < t < \tau_0(\omega)} r(t)(\omega) < \infty \right\}.$$

We can see that for all n, $\Omega_n \subseteq \Omega_{n+1}$.

For each $\omega \in \Omega_0$ there exists $n_0(\omega)$ such that $\tau_n(\omega) > \tau_0(\omega)$ for all $n \geqslant n_0(\omega)$. Thus,

$$\Omega_0 = \bigcup_{n=1}^{\infty} \Omega_n.$$

But

$$\mathrm{Pr}_Q(\Omega_n) \leqslant \mathrm{Pr}_Q(\{\omega : \tau_0(\omega) < \tau_n(\omega)\}) = 0$$

$$\Rightarrow \quad \mathrm{Pr}_Q(\Omega_0) \leqslant \sum_{n=1}^{\infty} \mathrm{Pr}_Q(\Omega_n) = 0;$$

that is, $\mathrm{Pr}_Q(\tau_0 < \infty) = 0$.

Now suppose $0 < 2\alpha\mu < \sigma^2$. We see that, for $0 < \epsilon < 1$,

$$0 > s(\epsilon) = -\int_\epsilon^1 e^{2\alpha v/\sigma^2} v^{-2\alpha\mu/\sigma^2}\,dv > -\int_\epsilon^1 e^{2\alpha/\sigma^2} v^{-2\alpha\mu/\sigma^2}\,dv.$$

But the limit as $\epsilon \to 0$ of $\int_\epsilon^1 v^{-2\alpha\mu/\sigma^2}\,dv$ is finite since $2\alpha\mu/\sigma^2 < 1$. Thus, the limit as ϵ tends to zero of $s(\epsilon)$ lies strictly between $-\infty$ and 0. Define

$$s(0) = \lim_{\epsilon \to 0} s(\epsilon).$$

We have noted earlier for $0 < \epsilon < r(0) < M < \infty$ that

$$s(r(0)) = s(\epsilon)\mathrm{Pr}_Q(\tau_\epsilon < \tau_M) + s(M)\mathrm{Pr}_Q(\tau_M < \tau_\epsilon).$$

We now modify the arguments above to show that this is also true for $\epsilon = 0$ when $\frac{1}{2} \leqslant 2\alpha\mu/\sigma^2 < 1$.

Recall that (equation (B.5))

$$\mathrm{Var}_Q(s(r(t \wedge \tau_0 \wedge \tau_M))) = E_Q\left[\int_0^{t \wedge \tau_0 \wedge \tau_M} \sigma^2 s'(r(u))^2 r(u)\,du \right]$$

$$= E_Q\left[\int_0^{t \wedge \tau_0 \wedge \tau_M} \sigma^2 e^{4\alpha r(u)/\sigma^2} r(u)^{-4\alpha\mu/\sigma^2+1}\,du \right].$$

Let

$$f(r) = s'(r)^2 r = e^{4\alpha r/\sigma^2} r^{1-d} = e^{dr/\mu} r^{1-d},$$

where $d = 4\alpha\mu/\sigma^2$. We have already specified that $1 \leqslant d < 2$. If $1 < d < 2$, then $f(r)$ is minimized in the range $0 \leqslant r < \infty$ at $\hat{r} = (d-1)\mu/d$ with

$$f(\hat{r}) = e^{d-1}\left(\frac{(d-1)\mu}{d}\right)^{1-d} > 0.$$

If $d = 1$, then $f(r)$ is minimized in the range $0 \leqslant r < \infty$ at $\hat{r} = 0$ with $f(0) = 1$. Let δ be the minimum value $f(\hat{r})$ in either case. Hence

$$\sigma^2 s'(r(u))^2 r(u) \geqslant \sigma^2\delta \quad \text{for all } 0 < u < t \wedge \tau_0 \wedge \tau_M,$$

$$\Rightarrow \quad \mathrm{Var}_Q(s(r(t \wedge \tau_0 \wedge \tau_M))) \geqslant \sigma^2\delta E_Q[t \wedge \tau_0 \wedge \tau_M].$$

But $\mathrm{Var}_Q(s(r(t \wedge \tau_0 \wedge \tau_M))) \leqslant (s(M) - s(0))^2$, so

$$E_Q[t \wedge \tau_0 \wedge \tau_M] \leqslant \frac{(s(M) - s(0))^2}{\sigma^2\delta} \quad \text{for all } t,$$

$$\Rightarrow \quad E_Q[\tau_0 \wedge \tau_M] < \infty$$

$$\Rightarrow \quad \mathrm{Pr}_Q(\tau_0 \wedge \tau_M < \infty) = 1.$$

Hence

$$s(r(0)) = s(0)\mathrm{Pr}_Q(\tau_0 < \tau_M) + s(M)\mathrm{Pr}_Q(\tau_M < \tau_0).$$

Also $e^{2\alpha v/\sigma^2} v^{-2\alpha\mu/\sigma^2} \to \infty$ as $v \to \infty$, so $s(M) \to \infty$ as $M \to \infty$. Hence $\mathrm{Pr}_Q(\tau_M < \tau_0) \to 0$ and $\mathrm{Pr}_Q(\tau_0 < \tau_M) \to 1$ as $M \to \infty$.

Finally, suppose that $0 < 2\alpha\mu/\sigma^2 < \frac{1}{2}$ (that is, $0 < d < 1$). Let $X(t) = \sqrt{r(t)}$. By Itô's formula we then have

$$dX(t) = \left[\frac{\sigma^2}{4X(t)}\left(\frac{2\alpha\mu}{\sigma^2} - \frac{1}{2}\right) - \frac{\alpha X(t)}{2}\right]dt + \frac{\sigma}{2}\,d\tilde{W}(t).$$

Now let $Y(t)$ be the Ornstein–Uhlenbeck process governed by the SDE:

$$dY(t) = -\tfrac{1}{2}\alpha Y(t)\,dt + \tfrac{1}{2}\sigma\,d\tilde{W}(t), \quad Y(0) = X(0).$$

Define

$$\tau_0^X = \inf\{t : X(t) = 0\} \quad \text{and} \quad \tau_0^Y = \inf\{t : Y(t) = 0\}.$$

For each outcome, ω, for all $0 < t < \tau_0^X(\omega)$,

$$\frac{\sigma^2}{8X(t)(\omega)}(d-1) - \frac{\alpha X(t)(\omega)}{2} < -\frac{\alpha X(t)(\omega)}{2}.$$

It follows that $X(t)(\omega) < Y(t)(\omega)$ for all $0 < t < \tau_0^X(\omega)$. This implies that if $\tau_0^Y(\omega) < \infty$, then $\tau_0^X(\omega) < \infty$. However, we know from the basic properties of the Ornstein–Uhlenbeck process that $\mathrm{Pr}_Q(\tau_0^Y < \infty) = 1$. It follows that $\mathrm{Pr}_Q(\tau_0^X < \infty) = 1$: that is, $r(t)$ will hit zero with probability 1, under Q.

Proof of part (e).   For notational convenience and without loss of generality we will assume that $t = 0$.

The price at time 0 of a European call option with exercise time T and strike price K with the zero-coupon bond maturing at time $U = T + \tau > T$ as the underlying is

$$C = E_Q[e^{-\int_0^T r(s)\,ds}(P(T, U, r(T)) - K)_+ \mid r(0) = r].$$

Let us consider under what circumstances will the call option be exercised; that is, if and only if

$$P(T, U, r(T)) > K$$

$$\Leftrightarrow \quad e^{\bar{A}(U-T)-\bar{B}(U-T)r(T)} > K$$

$$\Leftrightarrow \quad r(T) < \frac{\bar{A}(U - T) - \log K}{\bar{B}(U - T)} = r^*, \quad \text{say.}$$

Thus,

$$C = E_Q[e^{-\int_0^T r(s)\,ds} P(T, U, r(T)) I(r(T) < r^*) \mid r(0) = r]$$

$$\quad - E_Q[e^{-\int_0^T r(s)\,ds} K I(r(T) < r^*) \mid r(0) = r]$$

$$= E_Q[e^{-\int_0^U r(s)\,ds} I(r(T) < r^*) \mid r(0) = r]$$

$$\quad - E_Q[e^{-\int_0^T r(s)\,ds} K I(r(T) < r^*) \mid r(0) = r]. \tag{B.6}$$

Let P_1 and P_2 be two new measures, equivalent to Q with Radon–Nikodým derivatives:

$$\frac{dP_1}{dQ} = \frac{e^{-\int_0^U r(s)\,ds}}{E_Q[e^{-\int_0^U r(s)\,ds} \mid r(0) = r]}, \qquad \frac{dP_2}{dQ} = \frac{e^{-\int_0^T r(s)\,ds}}{E_Q[e^{-\int_0^T r(s)\,ds} \mid r(0) = r]}.$$

Then we have from equation (B.6)

$$C = E_Q[e^{-\int_0^U r(s)\,ds} \mid r(0) = r]E_Q\left[\frac{dP_1}{dQ} I(r(T) < r^*) \,\middle|\, r(0) = r\right]$$

$$\quad - E_Q[e^{-\int_0^T r(s)\,ds} \mid r(0) = r]E_Q\left[\frac{dP_2}{dQ} K I(r(T) < r^*) \,\middle|\, r(0) = r\right]$$

$$= P(0, U, r)E_{P_1}[I(r(T) < r^*) \mid r(0) = r]$$

$$\quad - K P(0, T, r)E_{P_2}[I(r(T) < r^*) \mid r(0) = r]$$

$$= P(0, U, r)\mathrm{Pr}_{P_1}(r(T) < r^* \mid r(0) = r)$$

$$\quad - K P(0, T, r)\mathrm{Pr}_{P_2}(r(T) < r^* \mid r(0) = r). \tag{B.7}$$

It remains for us to establish the distribution of $r(T)$ under P_1 and P_2. Let us look at P_2 first. We have

$$E_Q[e^{-\int_0^T r(s)\,ds-\omega r(T)} \mid r(0) = r] = P(0, T, r)E_{P_2}[e^{-\omega r(T)} \mid r(0) = r].$$

Thus (with $\gamma = \sqrt{\alpha^2 + 2\sigma^2}$),

$$E_{P_2}[e^{-\omega r(T)} \mid r(0) = r] = P_L(0, T, r, 1, \omega)/P_L(0, T, r, 1, 0)$$
$$= \exp[A(0, T, 1, \omega) - A(0, T, 1, 0) - (B(0, T, 1, \omega) - B(0, T, 1, 0))r]$$
$$= \left(\frac{(\gamma + \alpha)(e^{\gamma T} - 1) + 2\gamma}{(\sigma^2\omega + \gamma + \alpha)(e^{\gamma T} - 1) + 2\gamma}\right)^{2\alpha\mu/\sigma^2}$$
$$\times \exp\left[-\left(\frac{\omega(2\gamma + (\gamma - \alpha)(e^{\gamma T} - 1)) + 2(e^{\gamma T} - 1)}{(\sigma^2\omega + \gamma + \alpha)(e^{\gamma T} - 1) + 2\gamma}\right.\right.$$
$$\left.\left. - \frac{2(e^{\gamma T} - 1)}{(\gamma + \alpha)(e^{\gamma T} - 1) + 2\gamma}\right)r\right].$$

Now concentrate on the terms involving ω to establish the form of the Laplace transform. Within the exponential term we have

$$X = \frac{\omega(2\gamma + (\gamma - \alpha)(e^{\gamma T} - 1)) + 2(e^{\gamma T} - 1)}{(\sigma^2\omega + \gamma + \alpha)(e^{\gamma T} - 1) + 2\gamma}$$
$$= \frac{\omega(2\gamma + (\gamma - \alpha)(e^{\gamma T} - 1)) + 2(e^{\gamma T} - 1)}{(2\gamma + (\gamma + \alpha)(e^{\gamma T} - 1))(1 + 2k_2\omega)},$$

where

$$k_2 = \frac{\sigma^2(e^{\gamma T} - 1)}{2(2\gamma + (\gamma + \alpha)(e^{\gamma T} - 1))}$$
$$\Rightarrow \quad X = \frac{\theta(1 + 2k_2\omega) + \phi}{(2\gamma + (\gamma + \alpha)(e^{\gamma T} - 1))(1 + 2k_2\omega)},$$

where

$$\theta = \frac{r}{\sigma^2(e^{\gamma T} - 1)}(4\gamma^2 e^{\gamma T} + 2\sigma^2(e^{\gamma T} - 1)^2),$$

$$\phi = 2(e^{\gamma T} - 1)r - \theta = -\frac{4\gamma^2 e^{\gamma T} r}{\sigma^2(e^{\gamma T} - 1)}.$$

Similarly,

$$Y = \left(\frac{(\gamma + \alpha)(e^{\gamma T} - 1) + 2\gamma}{(\sigma^2\omega + \gamma + \alpha)(e^{\gamma T} - 1) + 2\gamma}\right)^{2\alpha\mu/\sigma^2}$$
$$= \left(\frac{1}{1 + 2k_2\omega}\right)^{d/2} \times \text{const.},$$

where

$$d = 4\alpha\mu/\sigma^2.$$

Hence

$$E_{P_2}[e^{-(k_2\omega)(r(T)/k_2)} \mid r(0) = r] = \left(\frac{1}{1 + 2k_2\omega}\right)^{d/2} \exp\left(\frac{\lambda_2}{2(1 + 2k_2\omega)}\right) \times \text{const.},$$

where
$$\lambda_2 = \frac{8\gamma^2 e^{\gamma T} r}{\sigma^2 (e^{\gamma T} - 1)(2\gamma + (\gamma + \alpha)(e^{\gamma T} - 1))}.$$

Compare this Laplace transform with that in equation (4.19). We can then see that, under P_2, $r(T)/k_2$ has a non-central chi-squared distribution with d degrees of freedom and non-centrality parameter λ_2. It follows that

$$\Pr_{P_2}(r(T) < r^* \mid r(0) = r) = \Pr_{P_2}(r(T)/k_2 < r^*/k_2 \mid r(0) = r)$$
$$= \chi^2(d, \lambda_2; y_2),$$

where $y_2 = r^*/k_2$ and $\chi^2(d, \lambda; y)$ is the cumulative distribution function of the non-central chi-squared distribution with d degrees of freedom and non-centrality parameter λ.

Next, consider the distribution of $r(T)$ under P_1:

$$E_Q[e^{-\int_0^U r(s)\,ds - \omega r(T)} \mid r(0) = r] = P(0, U, r) E_{P_1}[e^{-\omega r(T)} \mid r(0) = r].$$

Now

$$E_Q[e^{-\int_0^U r(s)\,ds - \omega r(T)} \mid r(0) = r]$$
$$= E_Q[e^{-\int_0^T r(s)\,ds - \omega r(T)} E_Q[e^{-\int_T^U r(s)\,ds} \mid r(T)] \mid r(0) = r]$$
$$= E_Q[e^{-\int_0^T r(s)\,ds - \omega r(T) + A(T,U,1,0) - B(T,U,1,0)r(T)} \mid r(0) = r]$$
$$= e^{A(T,U,1,0)} E_Q[e^{-\int_0^T r(s)\,ds - \omega_1 r(T)} \mid r(0) = r]$$
$$= \exp(A(T, U, 1, 0) + A(0, T, 1, \omega_1) - B(0, T, 1, \omega_1)r),$$

where
$$\omega_1 = \omega + B(T, U, 1, 0) = \omega + \bar{B}(U - T).$$

Hence
$$E_{P_1}[e^{-\omega r(T)} \mid r(0) = r] = E_{P_1}[e^{-\int_0^U r(s)\,ds - \omega r(T)} \mid r(0) = r]/P(0, U, r)$$
$$= \exp[A(T, U, 1, 0) + A(0, T, 1, \omega_1) - A(0, U, 1, 0)$$
$$- (B(0, T, 1, \omega_1) - B(0, U, 1, 0))r].$$

Now concentrate on the terms involving ω (that is, ω_1) to establish the form of the Laplace transform. First, we have

$$\exp(A(0, T, 1, \omega_1)) = \left(\frac{2\gamma e^{(\gamma+\alpha)T/2}}{(\sigma^2\omega_1 + \gamma + \alpha)(e^{\gamma T} - 1) + 2\gamma} \right)^{2\alpha\mu/\sigma^2}$$
$$= \left(\frac{2\gamma e^{(\gamma+\alpha)T/2}}{(\sigma^2\omega + \gamma + \alpha + \sigma^2\bar{B}(U - T))(e^{\gamma T} - 1) + 2\gamma} \right)^{2\alpha\mu/\sigma^2}$$
$$= \text{const.} \times \left(\frac{1}{1 + 2k_1\omega} \right)^{2\alpha\mu/\sigma^2},$$

where

$$k_1 = \frac{\sigma^2(e^{\gamma T} - 1)}{2(2\gamma + (\gamma + \alpha + \sigma^2 \bar{B}(U - T))(e^{\gamma T} - 1))}.$$

Second, consider

$$B(0, T, 1, \omega_1)r = \frac{((\omega + \bar{B}(U - T))(2\gamma + (\gamma - \alpha)(e^{\gamma T} - 1)) + 2(e^{\gamma T} - 1))r}{(\sigma^2\omega + \gamma + \alpha + \sigma^2\bar{B}(U - T))(e^{\gamma T} - 1) + 2\gamma}$$

$$= \frac{\theta(1 + 2k_1\omega) + \phi}{(2\gamma + (\gamma + \alpha + \sigma^2\bar{B}(U - T))(e^{\gamma T} - 1))(1 + 2k_1\omega)},$$

where

$$\theta = \frac{(2\gamma + (\gamma - \alpha)(e^{\gamma T} - 1))(2\gamma + (\gamma + \alpha + \sigma^2\bar{B}(U - T))(e^{\gamma T} - 1))r}{\sigma^2(e^{\gamma T} - 1)}$$

$$= \frac{r(4\gamma^2 e^{\gamma T} + 2\sigma^2(e^{\gamma T} - 1)^2}{\sigma^2(e^{\gamma T} - 1)}$$
$$\frac{+\sigma^2\bar{B}(U - T)(e^{\gamma T} - 1)(2\gamma + (\gamma - \alpha)(e^{\gamma T} - 1)))}{\sigma^2(e^{\gamma T} - 1)}$$

and

$$\phi = 2(e^{\gamma T} - 1)r + \bar{B}(U - T)(2\gamma + (\gamma - \alpha)(e^{\gamma T} - 1))r - \theta$$

$$= -\frac{4\gamma^2 e^{\gamma T} r}{\sigma^2(e^{\gamma T} - 1)}.$$

Hence

$$E_{P_1}[e^{-(k_1\omega)(r(T)/k_1)} \mid r(0) = r] = \left(\frac{1}{1 + 2k_1\omega}\right)^{d/2} \exp\left(\frac{\lambda_1}{2(1 + 2k_1\omega)}\right) \times \text{const.},$$

where

$$\lambda_1 = \frac{8\gamma^2 e^{\gamma T} r}{\sigma^2(e^{\gamma T} - 1)(2\gamma + (\gamma + \alpha + \sigma^2\bar{B}(U - T))(e^{\gamma T} - 1))}.$$

(Note that if $U = T$, $k_1 = k_2$ and $\lambda_1 = \lambda_2$.)

So, under P_1, $r(T)/k_1$ has a non-central chi-squared distribution with d degrees of freedom and non-centrality parameter λ_1. Thus,

$$\Pr_{P_1}(r(T) < r^*) = \chi^2(d, \lambda_1; y_1),$$

where

$$y_1 = \frac{r^*}{k_1}.$$

References

Ahn, A.-H., Dittmar, R. F. and Gallant, A. R. (2002). Quadratic term structure models: theory and evidence. *Review of Financial Studies* **15**, 243–288.

Andersen, P. K., Borgan, Ø., Gill, R. D. and Keiding, N. (1993). *Statistical Models Based on Counting Processes*. Springer.

Anderson, R. W. and Sundaresan, S. (1996). Design and valuation of debt contracts. *Review of Financial Studies* **9**, 37–68.

Babbs, S. H. and Nowman, K. B. (1999). Kalman filtering of generalised Vasicek term-structure models. *Journal of Financial and Quantitative Analysis* **34**, 115–130.

Baxter, M. and Rennie, A. (1996). *Financial Calculus*. Cambridge University Press.

Beaglehole, D. R. and Tenney, M. S. (1991). General solutions of some interest-rate-contingent claim-pricing equations. *Journal of Fixed Income* **1** (September), 69–83.

Beaglehole, D. R., and Tenney, M. S. (1992). A non-linear equilibrium model of the term structure of interest rates: corrections and additions. *Journal of Financial Economics* **32**, 345–353.

Bielecki, T. R. and Rutkowski, M. (2002). *Credit Risk: Modeling, Valuation and Hedging*. Springer.

Björk, T. and Christensen, B. J. (1999). Interest rate dynamics and consistent forward rate curves. *Mathematical Finance* **9**, 323–348.

Black, F. and Cox, J. (1976). Valuing corporate securities: some effects of bond indenture provisions. *Journal of Finance* **5**, 187–195.

Black, F., Derman, E. and Toy, W. (1990). A one-factor model of interest rates and its application to treasury bond options. *Financial Analysts Journal* **46** (January–February), pp. 33–39.

Black, F. and Karasinski, P. (1991). Bond and option pricing when short rates are log-normal. *Financial Analysts Journal* **47** (July–August), pp. 52–29.

Black, F. and Scholes, M. (1973). The pricing of options and corporate liabilities. *Journal of Political Economy* **81**, 637–659.

Bolder, D. J. (2001). Affine term-structure models: theory and implementation. Bank of Canada, Working Paper 2001-15.

Boyle, P. P. (1977). Options: a Monte Carlo approach. *Journal of Financial Economics* **4**, 323–338.

Boyle, P. P., Broadie, M. and Glasserman, P. (1997). Monte Carlo methods for security pricing. *Journal of Economic Dynamics and Control* **21**, 1267–1321.

Brace, A. and Musiela, M. (1994). A multifactor Gauss Markov implementation of Heath, Jarrow and Morton. *Mathematical Finance* **4**, 259–283.

Brace, A., Gatarek, D. and Musiela, M. (1997). The market model of interest rate dynamics. *Mathematical Finance* **7**, 127–155.

Brémaud, P. (1981). *Point Processes and Queues—Martingale Dynamics*. Springer.

Brennan, M. and Schwartz, E. (1979). A continuous-time approach to the pricing of bonds. *Journal of Banking and Finance* **3**, 133–155.

Brigo, D. and Mercurio, F. (2001). *Interest Rate Models: Theory and Practice*. Springer.

Brown, P. J. (1998). *Bond Markets: Structures and Yield Calculations*. Gilmour Drummond Publishing, Edinburgh.

Brown, R. H. and Schaefer, S. (2000). Why long-term forward interest rates (almost) always slope downwards. London Business School, Working Paper.

Cairns, A. J. G. (1998). Descriptive bond-yield and forward-rate models for the British government securities' market. *British Actuarial Journal* **4**, 265–321.

Cairns, A. J. G. (1999). A multifactor model for the term structure and inflation. In *Proceedings of the 9th AFIR Colloquium, Tokyo*, Vol. 3, pp. 93–113.

Cairns, A. J. G. (2004). A family of term-structure models for long-term risk management and derivative pricing. (To appear in *Mathematical Finance*.)

Cairns, A. J. G. and Pritchard, D. J. (2001). Stability of descriptive models for the term-structure of interest rates with application to German market data. *British Actuarial Journal* **7**, 467–507.

Chan, K. C., Karolyi, G. A., Longstaff, F. A. and Sanders, A. B. (1992). An empirical comparison of alternative models of the short-term interest rate. *Journal of Finance* **47**, 1209–1227.

Clewlow, L. and Strickland, C. (1997). Monte Carlo valuation of interest-rate derivatives under stochastic volatility. *Journal of Fixed Income* **7**, 35–45.

Constantinides, G. M. (1992). A theory of the nominal term structure of interest rates. *Review of Financial Studies* **5**, 531–552.

Cox, J., Ingersoll, J. and Ross, S. (1985). A theory of the term-structure of interest rates. *Econometrica* **53**, 385–408.

Crank, J. and Nicolson, P. (1947). A practical method for numerical evaluation of solutions of partial differential equations of the heat-conduction type. *Proceedings of the Cambridge Philosophical Society* **43**, 50–67.

Dai, Q. and Singleton, K. J. (2000). Specification analysis of affine term structure models. *Journal of Finance* **55**, 1943–1978.

Dalquist, M. and Svensson, L. E. O. (1996). Estimating the term structure of interest rates for monetary policy analysis. *Scandinavian Journal of Economics* **98**, 163–183.

Deacon, M. and Derry, A. (1994). Estimating the term structure of interest rates. Bank of England Working Paper Series, Number 24.

Dobbie, G. M. and Wilkie, A. D. (1978). The FT-actuaries fixed interest indices. *Journal of the Institute of Actuaries* **105**, 15–27, and *Transactions of the Faculty of Actuaries* (1979) **36**, 203–213.

Dothan, M. U. (1978). On the term structure of interest rates. *Journal of Financial Economics* **6**, 59–69.

Duffie, D. (1996). *Dynamic Asset Pricing Theory* (2nd edn). Princeton University Press.

Duffie, D. and Kan, R. (1996). A yield-factor model of interest rates. *Mathematical Finance* **6**, 379–406.

Duffie, D. and Singleton, K. (1997). An econometric model of the term structure of interest-rate swap yields. *Journal of Finance* **52**, 1287–1321.

Duffie, D. and Singleton, K. (1999). Modeling term structures of defaultable bonds. *Review of Financial Studies* **12**, 687–720.

Duffie, D. and Singleton, K. (2003). *Credit Risk: Pricing Measurement and Management*. Princeton University Press.

Duffie, D., Ma, J. and Yong, J. (1995). Black's consol rate conjecture. *Annals of Applied Probability* **5**, 356–382.

Dybvig, P. H., Ingersoll, J. E. and Ross, S. A. (1996). Long forward and zero-coupon rates can never fall. *Journal of Business* **69**, 1–25.

Faure, H. (1982). Discrépence de suites associées à un système de numération (en dimension *s*). *Acta Arithmetica* **41**, 337–351.

Feldman, K. S., Bergman, B., Cairns, A. J. G., Chaplin, G. B., Gwilt, G. D., Lockyer, P. R. and Turley, F. B. (1998). Report of the fixed interest working group. *British Actuarial Journal* **4**, 213–263.

Fisher, M., Nychka, D. and Zervos, D. (1995). Fitting the term structure of interest rates with smoothing splines. The Federal Reserve Board, Finance and Economics Discussion Series, Working Paper 1995-1.

Fleming, T. R. and Harrington, D. P. (1991). *Counting Processes and Survival Analysis*. Wiley.

Flesaker, B. and Hughston, L. P. (1996). Positive interest. *Risk* **9**(1), 46–49.

Halton, J. H. (1960). On the efficiency of certain quasi-random sequences of points in evaluating multi-dimensional integrals. *Numerische Mathematik* **2**, 84–90.

Harrison, J. M. and Kreps, D. (1979). Martingales and arbitrage in multiperiod securities markets. *Journal of Economic Theory* **20**, 381–408.

Harrison, J. M. and Pliska, S. R. (1981). Martingales and stochastic integrals in the theory of continuous trading. *Stochastic Processes and Their Applications* **11**, 215–260.

Harrison, J. M. and Pliska, S. R. (1983). A stochastic calculus model of continuous trading: complete markets. *Stochastic Processes and Their Applications* **15**, 313–316.

Heath, D., Jarrow, R. and Morton, A. (1992). Bond pricing and the term structure of interest rates: a new methodology for contingent claims valuation. *Econometrica* **60**, 77–105.

Ho, S. Y. and Lee, S.-B. (1986). Term structure movements and pricing interest rate contingent claims. *Journal of Finance* **41**, 1011–1029.

Hoem, J. M. (1969). Markov chain models in life insurance. *Blätter der Deutschen Gesellschaft für Versicherungsmathematik* **9**, 91–107.

Hoem, J. M. (1988). The versatility of the Markov chain as a tool in the mathematics of life insurance. *Transactions of the 23rd International Congress of Actuaries, Helsinki* **S**, 171–202.

Hoem, J. M. and Aalen, O. O. (1978). Actuarial values of payment streams. *Scandinavian Actuarial Journal* **1978**, 38–47.

Hogan, M. (1993). Problems in certain two-factor term-structure models. *Annals of Applied Probability* **3**, 576–581.

Hogan, M. and Weintraub, K. (1993). The log-normal interest-rate model and Eurodollar futures. Working Paper, Citibank, New York.

Hubalek, F., Klein, I. and Teichmann, J. (2002). A general proof of the Dybvig–Ingersoll–Ross Theorem: long forward rates can never fall. *Mathematical Finance* **12**, 447–451.

Hull, J. C. (2000). *Options, Futures, and Other Derivatives* (4th edn). Prentice Hall.

Hull, J. C. and White, A. D. (1990). Pricing interest rate derivative securities. *Review of Financial Studies* **3**, 573–592.

Hull, J. C. and White, A. D. (1994a). Numerical procedures for implementing term structure models I: Single-factor models. *Journal of Derivatives* **2** (Fall), pp. 7–16.

Hull, J. C. and White, A. D. (1994b). Numerical procedures for implementing term structure models II: Two-factor models. *Journal of Derivatives* **2** (Winter), pp. 37–48.

Hunt, P. J. and Kennedy, J. E. (2000). *Financial Derivatives in Theory and Practice*. Wiley.

James, J. and Webber, N. (2000). *Interest Rate Modelling*. Wiley.

Jamshidian, F. (1989). An exact bond option formula. *Journal of Finance* **44**, 205–209.

Jamshidian, F. (1997). LIBOR and swap market models and measures. *Finance and Stochastics* **1**, 293–330.

Jarrow, R. (1996). *Modelling Fixed Income Securities and Interest Rate Options*. McGraw-Hill.

Jarrow, R. A., Lando, D. and Turnbull, S. M. (1997). A Markov model for the term structure of credit risk spreads. *Review of Financial Studies* **10**, 481–523.

Johnson, N. L., Kotz, S. and Balakrishnan, N. (1995). *Continuous Univariate Distributions*, Vol. 2. Wiley.

Joy, C., Boyle, P. P. and Tan, K. S. (1996). Quasi-Monte Carlo methods in numerical finance. *Management Science* **42**, 926–938.

Karatzas, I. and Shreve, S. (1991). *Brownian Motion and Stochastic Calculus*. Springer.

Lamberton, D. and Lapeyre, B. (1996). *Introduction to Stochastic Calculus Applied to Finance*. Chapman and Hall.

Lando, D. (1997). Modelling bonds and derivatives with default risk. In *Mathematics of Derivative Securities* (ed. M. Dempster and S. Pliska), pp. 369–393. Cambridge University Press.

Lando, D. (1998). On Cox processes and credit-risky securities. *Review of Derivatives Research* **2**, 99–120.

Lane, M. (2000). Pricing risk transfer functions. *ASTIN Bulletin* **30**, 259–293.

Langetieg, T. (1980). A multivariate model of the term structure. *Journal of Finance* **35**, 71–91.

Leland, H. E. (1994). Corporate debt value, bond covenants, and optimal capital structure. *Journal of Finance* **49**, 1213–1252.

Longstaff, F. A. (1989). A nonlinear general equilibrium-model of the term structure of interest rates. *Journal of Financial Economics* **23**, 195-224.

Longstaff, F. A. and Schwartz, E. S. (1992). Interest-rate volatility and the term structure—a 2-factor general equilibrium-model. *Journal of Finance* **47**, 1259–1282.

Longstaff, F. A. and Schwartz, E. S. (1995). A simple approach to valuing risky fixed and floating-rate debt. *Journal of Finance* **50**, 789–819.

Luenberger, D. G. (1998). *Investment Science*. Oxford University Press.

McCulloch, J. H. (1971). Measuring the term structure of interest rates. *Journal of Business* **44**, 19–31.

McCulloch, J. H. (1975). The tax-adjusted yield curve. *Journal of Finance* **30**, 811–830.

McKay, M. D., Conover, W. J. and Beckman, R. J. (1979). A comparison of three methods for selecting values of input variables in the analysis of output from a computer code. *Technometrics* **21**, 239–245.

Merton, R. C. (1973). Theory of rational option pricing. *Bell Journal of Economics and Management Science* **4**, 141–183.

Merton, R. C. (1974). On the pricing of corporate debt: the risk structure of interest rates. *Journal of Finance* **2**, 449–470.

Mikosch, T. (1998). *Elementary Stochastic Calculus, with Finance in View*. World Scientific.

Mills, T. C. (1993). *The Econometric Modelling of Financial Time Series*. Cambridge University Press.

Miltersen, K., Sandmann, K. and Sondermann, D. (1997). Closed-form solutions for term-structure derivatives with log-normal interest rates. *Journal of Finance* **52**, 409–430.

Musiela, M. and Rutkowski, M. (1997). *Martingale Methods in Financial Modelling*. Springer.

Nelson, C. R. and Siegel, A. F. (1987). Parsimonious modeling of yield curves. *Journal of Business* **60**, 473–489.

Niederreiter, H. (1987). Point sets and sequences with small discrepancy. *Monatshefte für Mathematik* **104**, 273–337.

Norberg, R. (1991). Reserves in life and pension insurance. *Scandinavian Actuarial Journal* **1991**, 3–24.

Norberg, R. (1992). Hattendorf's theorem and Thiele's differential equation generalized. *Scandinavian Actuarial Journal* **1992**, 2–14.

Norberg, R. (1999). A theory of bonus in life insurance. *Finance and Stochastics* **3**, 373–390.

Øksendal, B. (1998). *Stochastic Differential Equations* (5th edn). Springer.

Owen, A. B. (1994). Controlling correlations in Latin hypercube samples. *Journal of the American Statistical Association* **89**, 1517–1522.

Park, J.-S. (1994). Optimal Latin hypercube designs for computer experiments. *Journal of Statistical Planning and Inference* **39**, 95–111.

Pearson, N. and Sun, T.-S. (1994). An empirical examination of the Cox–Ingersoll–Ross model of the term structure of interest rates using the method of maximum likelihood. *Journal of Finance* **54**, 929–959.

Pelsser, A. (2000). *Efficient Methods for Valuing Interest Rate Derivatives*. Springer.

Pelsser, A. (2003). Pricing and hedging guaranteed annuity options via static option replication. (To appear in *Insurance: Mathematics and Economics*.)

Press, W., Teukolsky, S., William, T. V. and Brian, P. F. (1992). *Numerical Recipes in C* (2nd edn). Cambridge University Press.

Ramlau-Hansen, H. (1988). Hattendorff's Theorem: a Markov chain and counting process approach. *Scandinavian Actuarial Journal* **1988**, 143–156.

Rebonato, R. (1998). *Interest Rate Option Models* (2nd edn). Wiley.

Rhee, J. (1999). Interest rate models. PhD Thesis, Warwick University.

Rogers, L. C. G. (1995). Which model for the term-structure of interest rates should one use? In *Proceedings of the IMA Workshop on Mathematical Finance* (ed. D. Duffie and S. Shreve). Springer.

Rogers, L. C. G. (1997). The potential approach to the term-structure of interest rates and foreign exchange rates. *Mathematical Finance* **7**, 157–164.

Rutkowski, M. (1997). A note on the Flesaker and Hughston model of the term structure of interest rates. *Applied Mathematical Finance* **4**, 151–163.

Sack, B. (2000). Using Treasury STRIPS to measure the yield curve. The Federal Reserve Board, Finance and Economics Discussion Series, Working Paper 2000-42.

Sandmann, K. and Sondermann, D. (1997). A note on the stability of lognormal interest rate models and the pricing of Eurodollar futures. *Mathematical Finance* **7**, 119–128.

Schaefer, M. S. and Schwartz, E. S. (1984). A two-factor model of the term structure—an approximate analytical solution. *Journal of Financial and Quantitative Analysis* **19**, 413–424.

Schmock, U. (1999). Estimating the value of Wincat coupons of the Witherthur Insurance Convertible Bond. *ASTIN Bulletin* **29**, 101–163.

Smith, G. (1985). *Numerical Solution of Partial Differential Equations: Finite Difference Methods* (3rd edn). Oxford University Press.

Sobol', I. M. (1967). The distribution of points in a cube and the approximate evaluation of integrals. *USSR Computational Mathematics and Mathematical Physics* **7**, 86–112.

Stein, M. (1987). Large sample properties of simulations using Latin hypercube sampling. *Technometrics* **28**, 143–151.

Svensson, L. E. O. (1994). Estimating and interpreting forward interest rates: Sweden 1992–1994. Working Paper of the International Monetary Fund 94.114, 33 pp.

Sverdrup, E. (1965). Estimates and test procedures in connection with stochastic models for deaths, recoveries and transfers between states of health. *Skandinavisk Aktuaritidskrift* **48**, 184–211.

Tan, K. S. and Boyle, P. P. (2000). Applications of randomized low-discrepancy sequences to the valuation of complex securities. *Journal of Economic Dynamics and Control* **24**, 1747–1782.

Vasicek, O. (1977). An equilibrium characterisation of the term structure. *Journal of Financial Economics* **5**, 177–188.

Vasicek, O. E. and Fong, H. G. (1982). Term structure modelling using exponential splines. *Journal of Finance* **37**, 339–348.

Vetzal, K. (1998). An improved finite difference approach to fitting the initial term structure. *Journal of Fixed Income* **7** (March), pp. 62–81.

Wilkie, A. D. (1995). More on a stochastic asset model for actuarial use (with discussion). *British Actuarial Journal* **1**, 777–964.

Wiseman, J. (1994). The exponential yield curve model. J. P. Morgan, 16 December.

Index